Repairing the Climate
A Plea for Carbon Capture

L.J. Reinders

Leiden University
Leiden, Netherlands

CRC Press
Taylor & Francis Group
Boca Raton London New York

CRC Press is an imprint of the
Taylor & Francis Group, an **informa** business

A SCIENCE PUBLISHERS BOOK

First edition published 2025
by CRC Press
2385 NW Executive Center Drive, Suite 320, Boca Raton FL 33431

and by CRC Press
4 Park Square, Milton Park, Abingdon, Oxon, OX14 4RN

Library of Congress Cataloging-in-Publication Data (applied for)

ISBN: 978-1-032-68988-3 (hbk)
ISBN: 978-1-032-68990-6 (pbk)
ISBN: 978-1-032-68994-4 (ebk)

DOI: 10.1201/9781032689944

Typeset in Times New Roman
by Prime Publishing Services

Preface

The Earth is warming. This is due to human action, with the continuing emissions of CO_2 as the main culprit. The more warming, the greater the risks. The most logical way to proceed is to remove the culprit and bring this warming to an end.

This book is meant for those who agree (1) that we are in a climate emergency. If you think we are not, send this book back or keep it for later and buy a more elementary one that tells you more about climate change and global warming; (2) that this emergency is human made, caused by the constant emission of greenhouse gases into the atmosphere. If you think that is not so, send this book back and buy a more elementary one. Those who persist in thinking that the climate, and consequently the weather, is not changing as a result of human behaviour are delusional and make a mockery of the word 'thinking'; they just have lost it; and (3) that we can do something about it. If you think that is not the case and the whole situation is hopeless, please read on and I hope that at the end of the book you will be more optimistic. More in particular that we can indeed stop (most of) these emissions, and repair (part of) the damage done over the last 150 years without turning everything upside down.

When a problem shows up and you want to solve it, it is in general a good idea to try to do away with its cause. As regards climate change, the main cause is the high concentration of greenhouse gases in the air, in particular carbon dioxide, CO_2, which reinforces the natural greenhouse effect. This is primarily due to the burning of fossil fuels and the best solution at first sight seems to be to stop burning the stuff. That is indeed the case. It is easy to agree with this, and keep shouting that fossil fuels must be kept in the ground but unfortunately not realistic in the short term, i.e., in this century. And, what is more, it only partly helps. If your house burns down, it makes sense to extinguish the fire. This indeed stops the burning but does not repair the damage. The same is true for global warming. It is the current concentration of greenhouse gases in the atmosphere that is already too high and causes the climate to change. Since we will not be able to stop burning fossil fuels in the short term, this concentration will become higher still in the next few decades and make the problem worse. There is no escape from this. Even if we were able to stop burning fossil fuels today or tomorrow, it will not lower the concentration of carbon dioxide in the air. That is where its removal comes in to play a vital role. Scrubbing the surplus carbon dioxide out of the air, by Direct Air Capture (DAC) or other forms of carbon removal, which I will group together in this book under the term Direct Carbon Removal, is at present the only method we have that can actually lower this concentration, apart from patiently waiting for a few centuries after having stopped burning fossil fuels and even that may not help. Direct Carbon Removal makes only sense though if at the same time we also lower emission levels as much as possible, for instance by capturing CO_2 from the smokestacks of power stations and other point-source emitters, normally called CCS (carbon capture and storage). Not to do the latter would be akin to "mopping up while the tap is still running", as the Dutch saying goes.

Not only do I believe that Direct Carbon Removal, combined with reducing emissions as mentioned above, is a means to reverse global warming, I am also convinced that we hardly have any other choice. As always, in theory there are quite a number of possibilities to address a problem. The one touted most for the one at hand is switching, preferably immediately, to renewable sources of energy, like wind and solar. Although I am all in favour of this, I am also quite certain that it is not going to happen any time soon and certainly not immediately, as the data on global energy usage clearly shows. In 2021 close to 80% of global primary energy was still generated from fossil fuels, with real renewables, i.e., solar and wind, providing only 4% of the energy, hydropower and biomass each providing 6%, and nuclear 4%. The figures can be massaged a bit, so that they look a little better for renewables (especially the concept of primary energy should not be used here) but whatever you do and no matter how you look at it, there is no escape from the conclusion that the contribution from renewables is still pitifully small, almost negligible actually, and will not be able to clinch the matter in the short term; they are just too far behind. Perhaps mankind will shut the last coal-fired power station in 2075 or so, but certainly not much long before. That implies that the concentration of greenhouse gases in the atmosphere, already too high now, will for sure continue to rise, and that global warming will worsen with it.

In this book I will set out the case for removing carbon dioxide as set out above. Stop asking the impossible but do something useful. Start the mopping up and try to turn off the tap at the same time. After the introduction, I will first set out the current global picture as regards carbon, emphasizing where the real problems are, followed by a discussion of the promise of 100% renewable options. In subsequent chapters the need for carbon capture, in the form of both Direct Carbon Removal and carbon capture and storage (CCS) for emissions from point sources, and the closely connected urgent need for *global* carbon pricing are discussed. Carbon pricing is a must as carbon capture, being a cost factor, will never be able to come of age if pollution can continue without paying for the costs. All this will be covered in the first six chapters of the book, after which the various forms of carbon capture will be discussed in some detail, the materials and processes which make this possible, prospects for the utilization of the captured carbon dioxide and an outline of the activity so far in the form of projects and industries. Before finishing with a concluding chapter, the crucial question of the costs and economics of carbon capture are discussed in Chapter 14.

I hope that in a small way this book can contribute to the realization that to stop global warming carbon capture is a necessity and that no more time should be lost in turning it loose on power plants and on the carbon dioxide in the air.

I here wish to thank professor Frans van Lunteren of Leiden University for offering me a guest research position at Leiden University, which provided me with access to the university library and other university facilities. I am also grateful to Professor Brian R. Martin, former head of the physics department at University College London, for a reading of the manuscript and a number of useful remarks.

L.J. Reinders
October 2023

Contents

Acronyms and Abbreviations

ACTL	Alberta Carbon Trunk Line
BECCS	Bio-energy with Carbon Capture and Storage
CAES	Compressed-air energy storage
CCC	Committee on Climate Change (UK)
CCS	Carbon Capture and Storage
CCU	Carbon Capture and Utilization
CCUS	Carbon Capture, Utilization and Storage
CREA	Centre for Research on Energy and Clean Air
CDM	Clean Development Mechanism
CDR	Carbon-Dioxide Removal
CER	Certified Emission Reduction
CLC	Chemical Looping Combustion
CO_2	Carbon Dioxide
CO_2-EOR	Enhanced oil recovery using CO_2
COP	Conference of Parties
COP	Coefficient of Performance
CSP	Concentrated Solar Power
DAC	Direct Air Capture
DACCS	Direct Air Carbon Capture and Storage
DRI	Direct Reduced Iron
ECBM	Enhanced Coalbed Methane
EEA	European Economic Area
EEA	European Environment Agency
EGR	Enhanced Gas Recovery
EIA	Energy Information Administration
EOR	Enhanced Oil Recovery
EROEI	Energy Return on Energy Invested
ESA	European Space Agency
ESGR	Enhanced Shale Gas Recovery
ESR	Effort Sharing Regulation
ETS	Emissions Trading System
EU	European Union
GEM	Global Energy Monitor
GHG	Greenhouse Gases
GCI	Global CO_2 Initiative

Gt	gigaton (billions of tons)
ICAO	International Civil Aviation Organization
ICEF	Innovation for Cool Earth Forum
IEA	International Energy Agency
IEEFA	Institute for Energy Economics and Financial Analysis
IGCC	Integrated gasification combined cycle
IPCC	Intergovernmental Panel on Climate Change
LCOE	Levelized cost of electricity
LULUCF	Land Use, Land Use Change and Forestry
MSR	Market Stability Reserve
Mt	megaton (millions of tons)
MWh	megawatt-hour (1,000 kWh)
NDC	Nationally Determined Contributions
NETL	National Energy Technology Laboratory
NECP	National Energy and Climate Plan
NOAA	National Oceanic and Atmospheric Administration
OECD	Organisation for Economic Co-operation and Development
ORC	Offshore Re-injection of CO_2
PV	Photovoltaic solar panels
UCG	Underground coal gasification
UNEP	United Nations Environment Programme
UNCED	United Nations Conference on Environment and Development
UNFCCC	United Nations Framework Convention on Climate Change
toe	ton of oil equivalent
TFEC	Total Final Energy Consumption
TCRE	Transient Climate Response to Cumulative Carbon Emissions
TWh	terawatt-hour (billion kWh)
WMO	World Meteorological Organization

CHAPTER 1

Introduction

"The science is clear. No matter which IPCC pathway humanity will follow, holding the global average temperature increase below 1.5°C will require removing increasing amounts of CO_2 from the atmosphere."

The State of Carbon Dioxide Removal, Oxford University 2023

This book is about carbon: the backbone of life on Earth and the fourth most abundant element in the Universe (after hydrogen, helium and oxygen, with the first two together taking up 98%). The human body is comprised for 22.9% of carbon, taking second place after oxygen with 61.4% (for a great part bound with hydrogen into water), leaving rather little for all the other 19 elements that we are made of. Along with hydrogen (10%) and nitrogen (2.6%), oxygen and carbon account for 96% of the body's mass. Carbon keeps us upright and without it we would just be a puddle of dirty water. It plays an essential role in biology because of its ability to form many bonds—up to four per atom—in a seemingly endless variation of complex organic molecules. There is an immense number of organic compounds all containing at least one atom of carbon, with the known number of defined compounds being close to 10 million. These carbon compounds, in addition to water, are what we are made of and what we eat.

But that is just our bodies. Our economies, our homes, our means of transport are built on carbon too, on the fossil fuels buried in the Earth millions and millions of years ago. More specifically, on the energy released in the burning of these fuels. The transition to burning fossil fuels a few centuries ago entailed a radical change of our lives and economies. It freed mankind of the burdensome reliance on human and animal muscle power and ushered in an era of industrialization that saw a dramatic rise in living standards and consumption in large parts of the world. And now we, and especially the economies of industrialized countries, are completely dependent on them and, perhaps even more important, on the fact that the energy generated from these fossil fuels is cheap, as exemplified by the rude awakening due to rapidly rising fuel costs and panic reactions just before and after Russia's scandalous invasion of Ukraine in early 2022. No wonder that Alice J. Friedemann begins her book *Life After Fossil Fuels* (Friedemann 2021) with the words: "Even more than most of us realize, we are completely and utterly dependent on fossil fuels. Oil makes everything possible. Cement, steel, roads, cars, farm machinery, food, health care, and 500,000 products."

The great advantage of fossil fuels is their energy density, the amount of energy stored per unit volume. Especially natural gas and gasoline (derived from refining crude oil) are champions in this respect. In general, high quality fuels are gases, while low quality fuels are solids, with liquids in between. The fuel with the highest energy content is hydrogen gas. The drawback is that it cannot be mined anywhere (apart from so-called white hydrogen, see Glossary) and must be synthesized, either via electrolysis from water or from fossil fuels, consuming a lot of energy in the process.

Burning fossil fuels with the aim of getting at the energy buried in them is not bad per se; dumping the waste in the form of carbon dioxide and other hazardous pollutants into the atmosphere is bad, at any rate if it has, or with some certainty will have, profound negative effects. Like most things we need or love (eating, drinking, smoking, travelling) they can cause problems, in most cases due to consuming them in excess or wastefully. This is also true for carbon which is entwined with one of the most serious problems facing us today: man-made global climate change; the slow but relentless heating of the planet. Carbon dioxide is a particularly problematic greenhouse gas in this respect, owing to its longevity in the atmosphere, its lifespan being between 300 and 1,000 years (Ozkan et al. 2022), and the sheer quantity of the anthropogenic production of this gas. If there had been no humans on Earth or if we had not developed into an intelligent species, or not into such an unpleasant one that would so utterly and destructively dominate the planet as we do now, the climate would probably be slowly moving back to the next ice age. This will now not happen, which is about the only positive note in this matter.

Too many of us have indulged too much in carbon in the past, and still do, to such an extent that some, especially a certain type of radical environmentalist, have started to hate carbon, or at any rate the industry that has supplied us (and them) with carbon to our heart's delight in the last few centuries.[1] Someone or something must be the culprit, must be blamed for the bad habits we developed, in the same way as the pathetic chain smoker blames his lung cancer on the tobacco industry, in spite of all the warnings these spineless individuals ignored during their lifetime. Like them, we were just led astray and had we known how bad it would turn out to be, we would certainly have chosen a different path, wouldn't we? A good drama cannot do without antagonists, the good guys and the bad guys. So, instead of owning up to complicity in the 'crime', they are now out for the destruction of the fossil-fuel industry, not because they care so much about the climate or the environment, as their actions will probably make the problem worse, but mostly to get rid of their own feelings of guilt or some other hang-ups of no concern to anybody apart from themselves. But, as David Wallace-Wells remarks, "the burden of responsibility is too great to be shouldered by a few, however comforting it is to think that all that is

---

[1] An example of such a completely unreasonable and senseless attitude is presented in the publication *The Big Con: How Big Polluters are advancing a "net zero" climate agenda to delay, deceive, and deny*, which brands all net-zero proposals by industry and others as a big con, a scheme invented by the fossil-fuel industry to delay, deceive and deny, without proposing a single realistic plan.

needed is for a few villains to fall", although he rightly adds to this later in his book that "oil-backed denial [of climate change] will likely be seen as among the most heinous conspiracies against human health and well-being as have been perpetrated in the modern world" (Wallace-West 2019, p. 30 and 149).

Others cast their net wider and blame capitalism for all our climate woes,[2] i.e., the economic system in which private individuals or businesses own capital goods and all economic decisions are taken in the free market, notwithstanding the fact that they do not know of a viable alternative and have greedily enjoyed its sweet fruits, and continue to do so. Neither do they seem to feel an irrepressible urge to move to a paradise like Cuba or Venezuela, where this despicable system is not in place, although the burning of fossil fuels is going on there as well. Scapegoating capitalism and/or industry feels comfortable as it lays the blame somewhere else, on malicious outside forces or forces made to look malicious for the argument. In the meantime of course we just continue to enjoy the benefits these forces bring, in spite of the fact that "capitalism is now obsolete, since it is no longer compatible either with our survival as a species or our welfare as individual human beings" (Foster and Clark 2015). I wish it were true. Capitalism comes in many hues and colours; many Americans would call various versions in place in Europe pure socialism. We have known the capitalist system at least for 70 years now, but the version of the 1950s and 1960s is far different from the one we have now. It is hard to say whether it was better then, but it was at any rate much more regulated. Stiflingly so, many arch capitalists argued all this time, and eventually they got their way. In the course of the last half of the previous century most regulations were abolished. Think for instance of the banking regulations in the US and other parts of the world, the last of which were done away with in the 1990s, resulting a decade later in the 2008 financial crisis. The almost completely free *global* market, as we have now, is not capable of choosing general (*global*) human wellbeing over individual profit. We need market interventions to counter climate change and *global* warming, but the consequences of market interventions are notoriously difficult to predict and hardly possible where markets are larger than nations. And lacking a truly *globally* dominating power, a *global* market can only be regulated by *global* governance, if at all. Only *global* governance can deal with *global* problems. As Berners-Lee formulates it: "we need the self-adjusting nature of markets and we need the global overview that global governance can provide" (Berners-Lee 2019, p. 129). Unfortunately we don't have that and will not get it any time soon either, if ever.

---

2 E.g., Naomi Klein in her influential book *This Changes Everything* (Klein 2014, p. 59) blames free market ideology for suffocating the potential for climate action and speaks of "people responsible for the crisis" (apparently knowing who they are or at least having selected a group to place the blame on) and "a panicked, megalomaniacal elite". Introspection is clearly not her greatest asset. She also acknowledges that "contemporary, hyper-globalized capitalism" has not created the climate crisis, and actually only acerbates it (p. 159). That then, I suppose, does away with the need to abolish capitalism, to get rid of private ownership. If the relentless pursuit of profit is not the cause of climate change, there is no logic in the view that it should be changed for combating climate change. It should only be redirected and taught to see profit opportunities in decarbonizing the economy, by CCS for instance.

In some developed countries environmental organizations have successfully used the Paris Agreement in climate litigation, forcing national governments and an oil company (Shell in the Netherlands) to strengthen climate action. Although there is nothing wrong with pushing governments and oil companies towards taking more action in combatting climate change, a wry aftertaste of these litigation successes is that they were in the wrong courts, chasing after the facts and just child's play.[3] A huge number of climate litigation cases has by now been brought, as shown by the website http://climatecasechart.com/. It will be no surprise, I suppose, than none of these cases is instituted in Russia, China or other autocratically governed countries. And this while by far the biggest oil companies are national oil companies, like the Chinese Sinopec, the National Iranian Oil Company and the biggest of them all the Saudi Arabian Aramco. What we call 'Big Oil' is actually quite small. These companies control only 11% of global oil, roughly the same as Russia. The OPEC countries produce one third of global oil and possess even more than that in reserves.[4] These national oil companies are also active in Western countries and could be sued there as well, but for some reason are left untouched by the monopolists of climate action. Litigation in their country of origin is not an option and, if it were, would certainly not be successful as the rule of law in these countries is mostly equal to the rule of the clique in power. The action against Shell has not been followed up by legal action against other oil companies. Why not have a go at Lukoil or Gazprom, for instance? Only Western governments and Western industrial conglomerates are targeted, not a single critical or other remark is heard about (state) actors from Russia, China or other (authoritarian) regimes, who undoubtedly love to see such "useful idiots" in our midst serving their interests. This is for sure not the way to tackle the world's most serious environmental crisis.

To blame capitalism or some businesses is just a distraction, a way for environmental organizations to score pseudo-victories that impress the gullible part of the public and keep donor money flowing, but have little or no impact on the environment and do not at all contribute to climate-change mitigation. It is also an act of self-flagellation, something that is apparent from many books from climate activists and people portraying themselves as scientists. They now mostly indulge in predicting and describing horror scenarios, but all this doomsaying gets so tiring after a while, doesn't it? Haven't they been wrong with all these doom stories for close to fifty years by now? Who in the West wants to hear or read time and again that the Western world and lifestyle are to blame for all the climate mess we are in; the new white man's burden? The developing world apparently does not agree

---

[3] I make an exception for cases like the one recently (August 2023) decided in Montana where a judge ruled that a Montana policy that prohibits state agencies from evaluating the environmental effects of greenhouse-gas emissions when permitting energy projects was unconstitutional. The Montana constitution explicitly states that citizens have a right to "a clean and healthful environment". So, although state officials called the decision absurd and are planning an appeal, it seems a clear-cut case. Even without the explicit formulation of such a right in the constitution, such a government policy is utterly shameful. (*Inside Climate News*, 15 August 2023).

[4] US shale oil accounts for 12% and others, like Norway and other independent producers, for the rest (one third).

and is still desperately trying to catch up as soon as possible with the materialistic standards of these selfish Western 'bastards'. After all, doesn't developing mean to come as closely as possible to the Western way of life? It seems that nobody in these countries or anywhere else, including the environmental movement in developed countries, has a viable alternative to the high-consumption-based economic model the West has pioneered.

The civilisation built in the West in the century and a half since the industrial revolution and apparently the envy of the world, is a civilisation that by its nature lives in disharmony with Nature, exploits Nature and does this almost invariably in a negative way. It destroys Nature, does not respect it, it hardly ever respects human life. Let us look at a small but not insignificant example. It is of course quite sensible and logical to take a car and drive to a certain location where you have some business or for another good reason. It can serve a useful purpose to use a car for this. But, in our envied 'modern' society people drive a car, mostly alone, for 50–100 kms each day to get to work in the morning and in the evening to get back. It does not require a lot of thinking to realize that this is a very useless, wasteful and exploitive way to spend your time and money. It is lost time and lost money but it is one of the most common things to do. When you talk to people who do this, they may even agree with you that it is silly, but you will not convince them to mend their ways. They even get subsidized for doing it! Their employer is allowed to pay them a tax-free allowance at a certain rate per kilometre that essentially covers all their expenses for the car, including the driving they do for other purposes than commuting, which is necessarily not that much as the car is mostly standing idle at the workplace. A campaign in the streets to change hearts and minds will not easily be successful, I am afraid. It is easy to point the accusing finger at oil and gas companies, but it is unfair, as the people themselves refuse to organize their lives such that they do not need all this fossil fuel, and it draws the attention away from what really must be done. They must mend their ways, the same as with smoking, where the only solution is for people to stop doing it. We, the people, all of us, especially those living in developed countries, have been the largest contributors to the climate crisis, by buying the products of these companies in huge amounts and by loving the freedom, warmth and wealth it gave us. We have not been forced to buy ever larger polluting cars, use patio heaters to heat part of the outdoors in order to be able to barbecue to far into the winter, buy heaps of clothes we hardly wear, or go on holidays in far-away countries for no other purpose than to walk around half naked, showing our ugly bloated bodies to those who have the misfortune to have to serve us. Society has not been created by oil and gas companies, nor is it ruled by them. Or am I naïve when I still think that society is ruled by governments and parliaments, and that it matters when we cast our votes? The oil and gas industry has merely exploited the opportunities society offered them, happily jumped on the bandwagon of ever continuing growth. Nobody stopped Henry Ford when he started building cars for the 'common man', for his own workers. On the contrary, he was encouraged and everybody liked his ideas. Roads were built so that the common man could drive to their heart's delight, and the oil and gas companies saw possibilities for selling ever more of their products. And now we have decided that they are to blame, others too, but mainly them. It just won't wash.

The solution put forward by many climate writers is in most cases rather simple. Get rid of the villain of the piece, the fossil-fuel industry, switch to 100% renewable energy *tomorrow* (with emphasis on tomorrow) and all will be fine. Some continue to drive their car as much as before, but now in electric vehicles, like Saul Griffith (Griffith 2021), continuing the rape of Nature, while others want to return to the living standards of the 1970s. The latter is apparently Naomi Klein's idea of a solution, apart from the fact that she also wants to get rid of capitalism without specifying what should replace it (Klein 2014, p. 91). However, she did not explicitly say whose 1970 living standards she had in mind but I suppose she meant those in the Western world. But what about the Chinese, Indians and other people who still lived very poorly in the 1970s and whose numbers were considerably fewer at that time?[5] They will literally have to live and develop off the wind and the sun, supplemented perhaps with some tinned sardines we can still afford to dole out once we have settled back into a 1970s style of life. And we also have to do this with the carbon-dioxide emissions of the 1970s, I guess. For the Western world that is not such a problem as emissions today in the US and EU, for instance, are at or even below the level of the 1970s,[6] but for the developing world the picture is rather different. Today both China and India emit more than ten times as much as in 1970 (the world as a whole 2.3 times as much). Is this increase, mainly originating from coal, due to what is commonly meant by the fossil-fuel industry, i.e., Shell, BP, ExxonMobil and the rest of the lot? Or perhaps not? Has the wrong villain been targeted? It is always pleasant to see a simplified picture, one or at most two villains, a nostalgic craving for an idealized past and some other blissful ingredients, but reality is rather different and much tougher to handle.

But whatever it is and whoever is to blame, due to all this carbon, man-made climate change is now upon us. We have a truly global problem on our hands that knows no geographical boundaries, since the carbon dioxide emitted anywhere quickly spreads over the entire Earth and remains in the atmosphere for centuries. Hence decarbonization must also be global, but when looking for solutions it should be realized that for a considerable part of the people on this planet global warming and climate change are not the most urgent problems. Just plain survival and human misery are and not just misery caused by natural disasters and underdevelopment, but at least partly by the sheer callous behaviour of the developed world towards their fellow human beings born in less favourable circumstances.

At the same time and greatly helped by the cheap energy provided by fossil fuels, the global population has grown tremendously from just shy of 1 billion in 1800 at the dawn of the Industrial Revolution to close to 8 billion today, an eightfold increase in two centuries. The hockey-stick graph (Mann et al. 1998) is very popular with environmentalists to show how the Earth has warmed up in the last 10,000 years or so but, if there ever was a hockey-stick graph, then look at global population growth over that same period. The lives of almost all of these 8 billion people entirely depend

---

[5]   Total world population in 1970 was less than half of what it is today.

[6]   If this fact surprises you, see https://ourworldindata.org/co2-emissions, or Fig. 2.2; per capita the situation is even more striking with the US emitting less than in 1950, the EU less than in 1961 and the UK less than in 1930!

on fossil fuels. If the fossil-fuel industry were to vanish tomorrow, most of us would probably be dead within half a year.[7] In the unlikely event that the cows and sheep were to live longer, their belching might still increase greenhouse-gas concentrations for a while beyond that date. The world is addicted to and needs to be weaned from the drug gradually; it would not survive a state of cold turkey. The patient is seriously ill and we have to slowly and carefully nurse them back to a healthy state. One of the most important symptoms is a far too high concentration of carbon dioxide in the air, which due to the continued and even still increasing emissions of this gas can only get worse.

There obviously is a direct connection between the increase in the global population and the growing use in energy from burning large amounts of biomass (some 3 trillion trees;[8] half of the total) followed by fossil fuels. Everybody wants to burn fossil fuels to get at cheap energy. The concomitant emission of carbon dioxide into the atmosphere has been taken for granted and/or has just been ignored, a practice that has been going on for far too long. We have overused carbon, profited too much, become intoxicated by it and burdened the climate we live in with colossal amounts of carbon dioxide and other greenhouse gases that it can no longer accommodate. As James Lovelock (1919–2022) has written: "We have grown in number to the point where our presence is perceptibly disabling the planet like a disease" (Lovelock 2007, p. xv). The number of 8 billion we have reached now is clearly not sustainable.

We are approaching the ecological limits of the Earth, as there are too many people and all these people demand too much energy from carbon.[9] Decades of human exploitation and use, not necessarily misuse as some are now trying to portray it, have damaged our forests, our fisheries, our atmosphere, our soil, our watersheds and our farmlands. The time of reckoning has come. We are standing on a new threshold, a transition to a world that is supposed to be mainly powered by renewable and/or sustainable energy. At least that is what we hope. But, in whatever way it will pan out, this transition will be painful as we have to drastically change our way of life, adjusting to one in which less energy will be available to each of us.

Having said all this it should also be recognized that fossil fuels have transformed the world, are responsible for a century of unprecedented technological innovation, globalization and advancement of knowledge. Fossil fuels have so far been the only energy source able to provide cheap, plentiful and, most importantly, reliable energy for billions of people. Without the discovery of coal as a means of generating energy, there would not have been an industrial revolution and life on Earth would look rather different from what it is today, in all probability very miserable indeed. The rapid increase in prosperity in China from 1985 to 2015, in which period close

---

[7] https://medium.com/a-balanced-transition/fossil-fuel-divestment-our-painfully-ironic-gift-to-the-fossil-fuel-industry-28ea60f57277.

[8] https://a.plant-for-the-planet.org/trillion-trees/.

[9] It is however also important to realize that not all the carbon that is bothering us so much nowadays originates from the burning of fossil fuels. Agriculture for instance contributes 23% to mankind's greenhouse-gas footprint, most of which has nothing to do with fossil fuels (Berners-Lee 2019, p. 22ff).

to 800 million people were lifted out of poverty, is the latest testament to how fossil fuels can enable a tremendous increase in human welfare.[10]

As David Wallace-Wells observes: "before fossil fuels, nobody lived better than their parents or grandparents or ancestors from 500 years before, except in the immediate aftermath of a great plague like the Black Death, which allowed the lucky survivors to gobble up the resources liberated by mass graves" (Wallace-Wells 2019, p. 115–116). Hard manual labour has turned into easy and enjoyable work with machines or into mental labour. George Orwell wrote in 1937 in *Road to Wigan Pier*: "Our civilization (…) is founded on coal, more completely than one realizes until one stops to think about it. The machines that keep us alive, and the machines that make machines, are all directly or indirectly dependent upon coal. In the metabolism of the Western world the coal-miner is second in importance only to the man who ploughs the soil." The profession of coal-miner has almost died out by now and sometimes it looks as if coal is all but being phased out in various places of the world. It apparently has outlived its usefulness; the drawbacks are now allegedly greater than the benefits, but that does not mean that we have abandoned carbon, far from it. There are gentler forms of packaged carbon, natural gas and oil, which can be unwrapped to get at their energy content. The last decade has seen a transition away from coal to natural gas, which is now in high demand, so much so that 2021 has seen a dramatic rise in the price of natural gas. The price rise was partly due to stalling investment in fossil fuels, a deficit that renewables were unable to make good. This then continued and worsened into 2022 with Russia's revolting invasion of Ukraine.[11] Prices for natural gas and electricity tripled in a matter of weeks and for the first time in perhaps more than a century reintroduced energy poverty,[12] a common phenomenon in the third world, in affluent European countries where low-income households living in poorly insulated houses are falling behind on the payments of their utility bills or can't afford to keep their homes warm. It is estimated that in Europe between 50 and 125 million people are unable to afford proper indoor thermal comfort.[13] The curse called down upon coal, which has led to a virtual phase-out of coal in Europe, has much to do with this, while the much touted alternatives of wind and solar energy cannot (yet) fill its place, and perhaps never will. But, whatever the cause or reason, we must in some way transform the fossil-fuel economy; there is no longer any doubt about that, if only for the fact that one day we will run out of fossil fuels and our appetite is clearly out of control given that half of the oil consumption throughout human history has occurred since 1980 (Jaccard 2005, p. 2).[14] But how to do this? It is not carbon's fault that we handle it so unwisely. The basic reason is that we want

---

[10] Jack Goodman, Has China lifted 100 million people out of poverty? *BBC Reality Check*; https://www.bbc.com/news/56213271.

[11] This also had a positive effect, in that it suddenly turned out to be possible to save energy, e.g. gas consumption in the EU in 2022 was consistently below the 2017–2021 average. However, the most surprising effect of this situation was in my view that it took Germany just half a year to completely replace Russian gas, making up 65% of its imports in 2021, by gas from other countries, in particular Norway and the Netherlands.

[12] See Glossary.

[13] https://ec.europa.eu/energy/eu-buildings-factsheets-topics-tree/energy-poverty_en.

[14] In a more recent book (Jaccard 2020, p. 142) Jaccard says that the main worry is actually that we won't run out of oil (and oil substitutes) fast enough.

its content cheaply, as cheaply as possible. Overconsumption, an ugly spectre in itself, and its even uglier and outright repulsive consequences are only possible when energy is cheap, dirt cheap! There is a lot wrong with the way fossil-fuel companies try to extract the dwindling reserves of fossil fuels, like the exploitation of tar sands, fracking for oil or gas, deep-water drilling, and mountaintop-removal coal mining, all carrying significantly higher risks than earlier practices, as the example of the BP Deepwater Horizon oil spill in the Gulf of Mexico in 2010 has shown. To make these technically difficult and therefore higher cost operations competitive and profitable, corners have to be cut with accidents occurring more frequently. In particular also the way people living in affected areas are treated makes disturbing reading. There are quite a few horror stories in this respect, showing the ugly face of capitalism and the failures of democracy as mostly companies based in Western democracies perpetrate unacceptable, rapacious and criminal practices both in their own countries and abroad (See Klein 2014, Chapter 9). Although no excuse of course, such horror stories can be told about almost any industry, e.g., the garment industry in Bangladesh or the Bhopal disaster in India. They are not particular or inherent to fossil-fuel companies, but to the relentless drive for profit, whereby money is valued over people's welfare and lives.

There is no way around these facts and all agree that something must be done, and even urgently so. But what? Apart from the constant and tedious spelling out of impending disaster and doom, climate scientists, activists and others do not seem to be able to come up with anything else than calling for an immediate end to fossil-fuel extraction and burning, essentially asking the world to commit suicide, to stop producing, growing food, stocking supermarkets, going to work, running trains, building wind turbines, manufacturing solar panels, maintaining electricity grids, going to war, etc. Some people, among whom James Lovelock in his book *The Revenge of Gaia*,[15] claim that nuclear power must save us from being cooked, but that train seems to have left the station. Others believe that all our energy needs can be satisfied by renewables like solar and wind, while at the same time portraying anything nuclear as being criminal and inspired by the devil (e.g., Shrader-Frechette 2011). In spite of all the hullabaloo and the great vistas painted about the penetration of such renewable sources, it is extremely unlikely that they will successfully challenge the dominance of fossil fuels in this century. They will make a useful contribution, but the pace at which they are being introduced is just too slow; they can hardly keep up with the global growth in energy demand. There are moreover a few fundamental problems to be solved, especially as regards storage, before renewables can take over from fossil fuels. At best we can expect some real breakthroughs in that respect in the second half of the century.

And what about the 500,000 products Alice Friedemann refers to? The best bet is that the dependence on fossil fuels will remain until at least all the conventional

---

[15] Lovelock joins others in painting a very gloomy future predicting that only about 200 million people, or about one thirtieth of world population, will survive if competent leaders manage to make a new home for them near the present-day Arctic. In his view, which he expressed in 2006, it is much too late to rely on sustainable development and the potential of renewable energy sources. Power from nuclear fission is the only medicine that is left to us.

oil, i.e., easily exploitable, and most of the unconventional oil, e.g., from tar sands and shale, has gone. It implies that we must find ways of tuning down the greenhouse effect by ensuring that the CO_2 released in burning the fossil fuels does not end up in the atmosphere. Carbon capture and storage (CCS) can provide the solution, it will buy us time, provide a transition period in which to revise our energy system. A transition period that could last a century or longer, as it must also involve cleaning up the atmosphere.

Our energy system must be thoroughly revised, if only for the fact that, due to the slow but relentless depletion of resources, the 'energy return on energy invested' (EROEI) is rapidly going down for fossil fuels (which by the way has nothing to do with climate change). It costs ever more energy to get the oil and other fossil fuels out of the ground. As soon as it costs more energy to get at the oil than it will provide when burned, it will no longer make sense to try to get it out. The discussion on this is not yet settled as the 'energy invested' critically depends on technology, methodology, and various assumptions. But it is certain that if the EROEI of oil falters, so will the EROEI of nuclear, batteries, hydrogen, wind and solar, for all these technologies depend on fossil-fuel based transport, mining and so on. For wind an EROEI of 18 is quoted, in other words, at the current price of oil and metals you get 18 times more energy out of the wind turbine than you invest in it during manufacture, installation, operation, and dismantling. It is important to realize though that this number does not include the investment needed for storing part of the generated electricity to solve the problem of the intermittency of these energy sources. But, metals (copper, uranium, nickel, etc.) too require ever more energy to be extracted. More and more rock has to be hauled and treated to get the same amount of metal as in earlier days, as the better quality ores have been used up first (see, e.g., Perissi et al. 2021). We simply cannot go on like this for very much longer. Energy will become an increasingly precious commodity and we, especially the developed world, cannot continue using the vast quantities of energy as we do now. Things will be totally messed up; the entire environment will be wrecked and there will be nothing to enjoy in a hundred years from today.

The planet has suddenly become a rather small and fragile place. We either need to drastically change our behaviour, going back to living styles that were common half a century or more ago, for instance, or seeking technological solutions to solve the problems. Voluntary changes in behaviour are very unlikely to occur and in any case not on the scale required. We had mildly similar problems before when faced with air pollution and acid rain in the 1970s and 1980s, due to sulphur and nitrous oxide emissions and particulates in the exhausts of vehicles and the smokestacks of power stations. They were solved by requiring refineries to reduce sulphur in fuels, and power stations to capture the pollutants before releasing their smoke into the air. The same approach can be tried to bring down carbon emissions. Enact policies that compel greenhouse-gas reductions through regulation and/or carbon pricing, and use the revenues to hoover up the carbon dioxide in the air (Jaccard 2020, p. 154ff.)

We had our chance in the past with nuclear power. With any technology there are always doomsayers on the lookout who love to continue to hammer at the associated risks as they are mostly incapable of understanding the benefits such a technology might bring. This is the main reason that nuclear power has been unable to become a real major power generator. Carbon-dioxide emissions would still have been a problem but much easier to manage. Medieval sects like the German Green Party,[16] now sharing power in Germany, are much to blame here.[17] By spreading lies and cheaply playing on the fears of the public for intangible and invisible, and therefore scary, radiation about which it does not understand anything whatsoever, they earned their place in the limelight and killed off nuclear power in Germany. In this respect the German Greens form a remarkable contrast with the Finnish Green Party which voted by a large majority at its 2022 party conference to adopt a pro-nuclear approach.[18] The green in Finland is apparently of a different hue than the German variety. Had Germany and the rest of the EU followed the French example in building nuclear power stations in the 1980s and 1990s,[19] European CO_2 emissions would now have been more than 30% less, the CO_2-level in the atmosphere would have been considerably lower as fewer fossil fuels would have been burned, the panicky scramble for the building of wind turbines and the laying of solar panels would not have been necessary, and thousands of premature deaths from respiratory illness would have been avoided. The same can be said for most of the rest of the world.

Because we want to continue with the way we use energy, we curse carbon in the false belief that we can put things right by switching rapidly and drastically to renewable energy sources. Well, the energy may be renewable or sustainable, the means by which these sources are exploited are not. According to the reasonable definition provided by Jaccard (Jaccard 2005, p. 11–12) a sustainable energy system must: (1) have good prospects for enduring indefinitely in terms of the type and level of energy services it provides, and (2) extraction, transformation, transport and consumption of energy must not have any adverse consequences for people and ecosystems.

In other words, in this view the two key properties of a sustainable energy system are *endurance* and *cleanliness*. A system based solely on fossil fuels can obviously never be sustainable, even if we could get rid of the CO_2 problem, e.g., by CCS. It fails as regards both endurance and cleanliness. Their extraction is far from clean, although practices are perhaps not as bad as they were half a century ago, nor is their use, whereby air pollution from coal consumption (and increasingly too from vehicle exhausts) in developing countries is an especially pressing problem. In the end they

---

[16] The official name is *Bündnis 90/Die Grünen*, a 1993 merger of the non-communist East German political alliance *Bündnis 90* and the West-German party *Die Grünen*. In the 2021 election the party managed to almost double the number of its seats in parliament.

[17] They reject nuclear power but condone the burning of lignite, the most polluting fossil fuel available, bearing responsibility for thousands of premature deaths and billions of damage.

[18] https://www.neimagazine.com/news/newsfinlands-green-party-supports-nuclear-power-9727725, accessed 26 May 2022.

[19] In 2021 France derived about 70% (about 400 TWh) of its electricity from nuclear energy. In 1974 the French government decided to rapidly expand the country's nuclear power capacity. As a result of this decision, France now claims a substantial level of energy independence and below average electricity costs in Europe. It also has an extremely low level of carbon dioxide emissions per capita from electricity generation, since over 80% of its electricity is from nuclear or hydro (*World Nuclear Association*).

will all run out (or become too expensive to extract), a time that may actually come quite soon.

One wonders though whether a fully sustainable energy system is at all possible? Certainly at the mammoth scale of the current system whereby in the past half century the global population has more than doubled and economic output increased by a factor of seven!

Our current energy system certainly does not fulfil the above conditions, but renewable systems too will have a hard time in this respect, although some of them (e.g., wind and solar) easily meet the first condition formulated above. The emphasis nowadays is on greenhouse-gas emissions, as catastrophe due to the resulting climate change looms large. There are however a number of other possible adverse effects that need to be considered. Renewables are often intermittent, of relatively low energy density and inconveniently located. This will cause land use conflicts (especially as regards biomass, which is also seen as a renewable resource), environmental impacts and high costs, while the wind turbines, air source heat pumps, batteries for electric vehicles, solar panels and insulated walls and roofs of houses all need to be constructed. Materials are needed for this, which for the great part can only be produced by burning fossil fuels and/or extracted by doing untold damage to the environment and creating heaps of waste. In the future these materials will become scarce in the same way as fossil fuels are becoming scarce now. Wars will perhaps no longer or not solely be fought over oil, but also over nickel, cadmium, lithium, copper, rare earth materials and the rest. From an energy-intensive society we will switch to a materials-intensive society (see, e.g., Xu et al. 2020) and the destruction of Nature will just continue, as man has always been doing in order to ensure the multiplication of the species, similar to a plague of locusts that eat the Earth bare.

But the way we look at an energy system, i.e., the combined processes of acquiring and using energy in a given society or economy, is fundamentally wrong. Apparently, we only or mainly tend to consider how cheaply it can provide energy. Things that are cheap are not appreciated, but easily taken for granted. It must be 'competitive' else it will never be 'commercially' viable.[20] The consequences for Nature and/or more broadly for planet Earth are just a minor consideration, if at all a concern. That, in my view, is the basic mistake, the basic reason for the wasteful use of energy, a mistake that we are bound to repeat with renewable sources of energy even if they succeed in solving the carbon-dioxide problem. We may seem to have solved one problem with a certain system or technology, to only discover another damaging consequence popping up. The renewable energy systems praised sky high today will show to be no different from fossil fuels in their destructive impact before the century is out.

All nations and all people contribute to global warming as all nations and all people use fossil fuels and emit carbon dioxide and other greenhouse gases. And these emissions do not respect borders. A ton of carbon dioxide reduced in Europe is just as valuable towards combatting human influence on the climate as a ton reduced

---

[20] Even saving us from the eventual climate change catastrophe must be economically viable, else just don't bother. It would be a nightmare beyond believe if we were to survive and then came to realize that it could have been done more economically, wouldn't it?

in India or Australia. Avoiding activities that cause climate change truly is a *global public good*, whose provision benefits all countries and no country can be excluded. No single nation can be blamed for causing climate change and no single country can just by itself prevent climate change, but some nations are more equal than others in this respect. If China and the US would reduce their emissions to zero, global emissions would go down by about 45%. It shows that some nations have a greater responsibility to take action than others, but that does not absolve other nations of their own responsibility.

In assessing the risks that climate change poses, we face profound questions about how we ought to weigh the respective harms it may inflict on current and future generations, and on humans and other species. We face, in addition, difficult questions about how to act in conditions of uncertainty, in which at least some of the consequences of climate change—and of various human interventions to adapt to or mitigate it—are difficult to fully predict. Even if we agree that mitigating climate change is morally required, room exists for disagreeing about the precise extent to which this ought to be done. Finally, once we determine which actions we ought to take to reduce or avoid climate change, we face the question of who ought to bear the costs of those actions, as well as the costs associated with any climate change which nevertheless comes to pass (Armstrong 2018). In this connection environmental law uses the euphemistic phrase "common but differentiated responsibilities and respective capabilities" to express that all states are responsible for addressing global environmental destruction, yet not all equally responsible. The principle was formalized in the United Nations Framework Convention on Climate Change (UNFCCC) of the 1992 Earth Summit in Rio de Janeiro. The text of the convention reads: "… the global nature of climate change calls for the widest possible cooperation by all countries and their participation in an effective and appropriate international response, in accordance with their common but differentiated responsibilities and respective capabilities and their social and economic conditions." In 2014 in Lima the principle was slightly changed into "common but differentiated responsibilities and respective capabilities, in light of different national circumstances."[21]

Since no nation seems prepared to voluntarily cut emissions, as it would entail voluntarily reducing its (potential for) economic growth, we need global agreements, a coordinated global effort to tackle the problem (Chichilnisky and Bal 2019, p. 35). If every agent, institution or nation acts independently in its own self-interest, effective mitigation of man's influence on the climate will not be achieved, as the Intergovernmental Panel on Climate Change (IPCC) notes in its Fifth Assessment Report (Kolstad et al. 2014, p. 177). But, never have all or almost all countries with different interests and agendas been able to agree on anything significant in the past, and there is little reason to assume they will do so now. Effective mitigation will not be achieved either if agents act without coordination to reduce their greenhouse-gas emissions, if they try to achieve net zero at any cost by the earliest possible date, spending gigantic amounts of money in their own race to net zero, while such money could be spent to much greater effect by 'picking the lower-hanging fruit' elsewhere.

---

[21] See also Klein (2014), p. 413 ff for a strong analysis.

In 2015 at the Paris climate conference, the Paris Agreement on the mitigation of climate change was negotiated. On 22 April 2016 (Earth Day), 174 countries signed the agreement in New York, and began adopting it within their own legal systems. The agreement achieved one key result, viz. to limit average global warming to 2°C above pre-industrial levels. A peculiar feature of this statement is that it does not say what these pre-industrial levels actually are (the use of the plural makes it even odder), especially in the light of the fact that it is also stated that the parties will 'pursue efforts' to limit temperature increase to 1.5°C. Apparently the pre-industrial temperature level or levels are known within 0.5 degree, otherwise it would not be possible to distinguish between 2 and 1.5 degrees above that level.[22] It calls for net-zero human-related (anthropogenic) greenhouse-gas emissions to be reached during the second half of the 21st century, but the consensus among scientists seems to be that, for the temperature rise to be limited to 1.5°C, net-zero emissions must be achieved by 2050. So, one of the extraordinary features of the debate is that these values of 1.5 and 2°C are not based on much, are not magical tipping points after which life on Earth is doomed,[23] have not even been defined properly, and still managed to assume the status of almost mythical truths. No scientist could have gotten away with such sloppily defined numbers in my, admittedly limited, experience as a scientist.

And indeed the debate seems to have shifted. All the talk is now about net zero, mostly by 2050. Net-zero pledges are all the rage, from corporations and countries alike.[24] Net zero means that there is a balance between the amount of greenhouse gas produced and the amount removed from the atmosphere. We reach net zero when the amount we add is no more than the amount taken away. To achieve this, emissions must be reduced as much as possible and any emissions that remain must be balanced out by removing an equivalent amount. This will be a tall order, but is not impossible as this book intends to show.[25]

In 2018 the IPCC issued a special report on the impacts of global warming of 1.5°C above pre-industrial levels (IPCC 2018). It also uses the annoyingly sloppy language of 'pre-industrial levels', as does the most recent IPCC Sixth Assessment Report.[26] Chapter 1 of the special report starts with the words: "Human influence on climate has been the dominant cause of observed warming since the mid-20th century, while global average surface temperature warmed by 0.85°C between

---

[22] See also about this Koonin 2021, p. 214.

[23] They are more like estimates of when critical ecosystems, like coral reefs, will become imperilled, according to Zeke Hausfather, director of Climate and Energy at the Breakthrough Institute, while Naomi Klein (Klein 2014, p. 12) calls the 2°C target a "highly political choice that has more to do with minimizing economic disruption than with protecting the greatest number of people." To the latter I would say that minimizing economic disruption is most likely the best way to protect the greatest number of people.

[24] See https://zerotracker.net/; according to the Science Based Targets (https://sciencebasedtargets.org/) initiative as of November 2022 1558 companies had formulated net-zero commitments.

[25] According to the IPCC, net zero and carbon neutrality are overlapping concepts. There is a slight difference. Carbon neutrality refers to a policy of not increasing carbon emissions and of achieving carbon reductions through offsets, while net zero means making changes to reduce carbon emissions to the lowest amount—and offsetting as a last resort.

[26] See the Glossary for a short discussion of the term pre-industrial.

1880[27] and 2012, as reported in the IPCC Fifth Assessment Report. Many regions of the world have already [experienced] greater regional-scale warming, with 20–40% of the global population having experienced over 1.5°C of warming in at least one season. Temperature rise to date has already resulted in profound alterations to human and natural systems, including increases in droughts, floods, and some other types of extreme weather; sea level rise; and biodiversity loss—these changes are causing unprecedented risks to vulnerable persons and populations. (…) Worldwide, numerous ecosystems are at risk of severe impacts, particularly warm-water tropical reefs and Arctic ecosystems."

A key finding of the report is that meeting a 1.5°C target is possible but would require "deep emissions reductions" and "rapid, far-reaching and unprecedented changes in all aspects of society." I don't know all societies on the globe of course, but the reader will probably agree with me that such "far-reaching and unprecedented changes in all aspects of society" are still hardly noticeable, at least I did not spot any in the Netherlands or any other of the European countries I am somewhat familiar with.

If anything, it is more like business as usual, although there admittedly is a lot of wind-turbine construction and solar-panel laying. Pampered with subsidies, projects in those domains are now bringing in money for the vultures of capitalism (it is common to formulate this differently and say for instance that they create jobs). By pointing at the dire climate-change prospects for the world, any protests from ordinary people against these ugly monsters in their backyard can easily be silenced. These renewable energy sources (more about this term later) have become a boon to capitalist entrepreneurs, allowing them to earn vast amounts of money with little to no risk, a very attractive revenue model. They are now indeed generating a sizable share of electricity in some countries of Western Europe and elsewhere. This is good, but they don't lower the carbon concentration in the air. As things stand at the moment they actually increase it, only less rapidly than the outright burning of fossil fuels. So do electric cars. Most of the parts and materials are still produced by using energy generated from burning fossil fuels. Even the recent Covid-19 pandemic[28] has not instilled any further urgency in the matter, with most people hankering for a return to the 'former normal', i.e., to their former spendthrift lives.

Further findings from the report are that "limiting global warming to 1.5°C compared with 2°C would reduce challenging impacts on ecosystems, human health and wellbeing" and that a 2°C temperature increase would exacerbate extreme weather, rising sea levels and diminishing Arctic sea ice, coral bleaching, and loss of ecosystems, among other impacts.

---

[27] The mention of this firm date must be applauded, but is unfortunately a lone instant of clarity in an otherwise opaque picture.

[28] The pandemic had a positive effect on climate change: primary energy demand dropped nearly 4% in 2020 and global energy-related CO_2 emissions fell by 5.8%. The decline in emissions of almost 2,000 million tons (2 Gt) of CO_2 is without precedent in human history—broadly speaking, it is the equivalent of removing all of the EU's emissions from the global total. Demand for fossil fuels was hardest hit in 2020—especially oil, which plunged 8.6%, and coal, which dropped by 4%. Oil's annual decline was its largest ever, accounting for more than half of the drop in global emissions. Global emissions from oil use plummeted by well over 1.1 Gt CO2, down from around 11.4 Gt in 2019. https://www.iea.org/articles/global-energy-review-co2-emissions-in-2020.

The message of the IPCC report is alarming, and in this respect it should be emphasised that the IPCC is not prone to exaggeration. Far from it, it is a profoundly conservative organization that exercises restraint and weighs its every word before writing it down in a report. These reports are written by assessing published literature. The IPCC does not conduct its own scientific research. Defined language is used to characterize the degree of certainty in assessment conclusions. It is important to point out that the assessments made by the IPCC are based on well-established knowledge and evolving understanding. The IPCC is a messenger conveying to the world consensus results of the vast majority of scientists on the globe; facts the majority of scientists agree on and on which basis science can proceed.

IPCC reports certainly do not overestimate the impacts of climate change, but most likely underestimate them, they represent a bottom line everybody has been able to agree on. They are not mainstream science so much as conservative, lowest-common-denominator science, with solid credibility (Weart 2003, p. 162). Hence, nobody who is of sound mind can ignore what has been stated in the paragraph cited above. Climate change due to human influence is a fact; to deny it is no longer an option. Scientists are as certain that human-generated greenhouse-gas emissions are an important cause of climate change as that (cigarette) smoking is harmful to human health. From the geological record it is known that such greenhouse gases, principally CO_2, have had an effect on climate change in the past. Now humans are forcing the climate to change by the uncontrolled emission of greenhouse gases at a far more rapid pace than these changes occurred in the past—50–100 times faster than in the past several thousand years. Global warming is now everywhere on the planet, especially in the oceans which trap more than 90% of the extra heat caused by human carbon pollution (Romm 2018, p. 277; Cheng et al. 2022). The debate on the reality of global warming is pretty much over now. Ecosystems are deteriorating fast and, if we cannot arrest this deterioration, human wellbeing and prosperity will for sure also decline, if not already doing so. So, the interesting questions we have to answer are how bad a problem we have on our hands and what to do about it. The opinions on this vary widely.

In this respect one should be aware that there is a great danger in underestimating future impacts, as the IPCC report undoubtedly does. It can give people false reassurance. Planning a response for 'just' 1.5°C of global warming might well be inadequate or inappropriate when the warming turns out to be considerably higher. It is often better to prepare for the worst, especially as climate change issues have a long lifetime and decisions have to be made before there is any clarity about the actual impact of such changes. Once they are upon us, it is too late to react. An ounce of prevention is worth a pound of cure, as the saying goes. On the other hand, Broecker and Kunzig, both not known for their unreasonable utterances, write: "From Al Gore on down, environmentalists keen to spur action on global warming sometimes argue that it is a threat to Western civilization. The motive is reasonable, but the argument strikes us (i.e., Broecker and Kunzig) as both misleading and unhelpful–because Western civilization seems more resilient than that, and because such grandiose rhetoric converts many reasonable people into sceptics. There is no reason to overreach (…)" (Broecker and Kunzig 2008, p. 139). You see, there is endless room for arguing, to the utter enragement of the Greta Thunbergs of this world, who want action *now*, and rightly so!

So the pronouncements by the IPCC must be taken as going to happen *for sure*, but are most probably an underestimate of what is going to come. Figure 1.1 shows the picture as presented by NASA.[29] Although it is clear that the temperature has been rising in the last fifty years (by roughly 1°C since 1970), the picture before that is far from clear with a temperature decrease from 1880 to roughly 1910, then a rise to 1940 and a plateau from roughly 1940 to 1975.[30] The picture also makes clear that it matters what pre-industrial actually means, and that from 1910 to 2020 the temperature has risen by roughly 1.45°C.

Others give similar results, *Berkeley Earth*[31] estimates that global mean temperature in 2022 was 1.24 ± 0.03°C above the 1850 to 1900 average, which is frequently used as a benchmark for the pre-industrial period.[32] (For June 2023 it was 1.47 ± 0.09°C.) Both the land and ocean average temperatures relative to the average from 1850 to 1900 as reported by *Berkeley Earth* are shown in Fig. 1.2.[33]

Berkeley Earth predicts that at the current rate of progression, the increase in Earth's long-term average temperature will reach 1.5°C above the 1850–1900 average by around 2034 and 2°C around 2060. Heating of 1.5°C just cannot be prevented, whatever we do.

The figure shows that land averages have the tendency to increase more rapidly than ocean averages. This tendency is expected, based on our understanding of how increases in greenhouse-gas concentrations impact the Earth's climate. Even if emissions go (temporarily) down as with the Covid-19 pandemic, it still means an increase of the concentration of CO_2 in the atmosphere, as carbon dioxide from human activities accumulates. We will hit a new record in CO_2 concentration every year until we reach a situation of net-zero emissions, and the concentration will only go down when we start taking CO_2 out of the atmosphere. According to the data on the NASA website[34] the concentration of carbon dioxide in the atmosphere as of

---

[29] As reported in the text, the IPCC Fifth Assessment Report states a rise of 0.85 °C from 1880 to 2012. NASA has 0.86 for 2012. The NASA data shows the puzzling behaviour that from 2015 until 2022 the temperature has not risen, but seems to fluctuate between 0.85 (2018, 2021) and 1.02 (2016, 2020), not showing a real rise. This was also observed for the decade from 2005 to 2015. These seem to be small variations on a relentlessly upward trend in temperature, and do not change the fact that the last eight years have been unusually warm with global temperatures exceeding 1.2 °C above pre-industrial levels.

[30] There are explanations for these anomalies, e.g. the number of aerosols in the atmosphere, but so far I have not seen a convincing explanation for the entire curve or a model that can satisfactorily reproduce the behaviour in the picture from 1880 to the present day. Such an explanation is actually not of great interest either as the picture for the last half century is clear. Since data is much more reliable from 1970 onwards, it might be better to take 1970 as the base year, or the average of 1940-1970 as a benchmark, and stop using this vague term pre-industrial levels.

[31] *Berkeley Earth* (originally called the Berkeley Earth Surface Temperature project) is a well-respected, non-profit organisation based in Berkeley, California, that focuses on land temperature data analysis for climate science; http://berkeleyearth.org/.

[32] It is regrettable though that NASA uses a different benchmark, which makes comparison unnecessarily complicated.

[33] Berkeley Earth, *Global Temperature Report for 2022*, January 2023; https://berkeleyearth.org/global-temperature-report-for-2022/, accessed 12 January 2023.

[34] https://climate.nasa.gov/, accessed 9 January 2023.

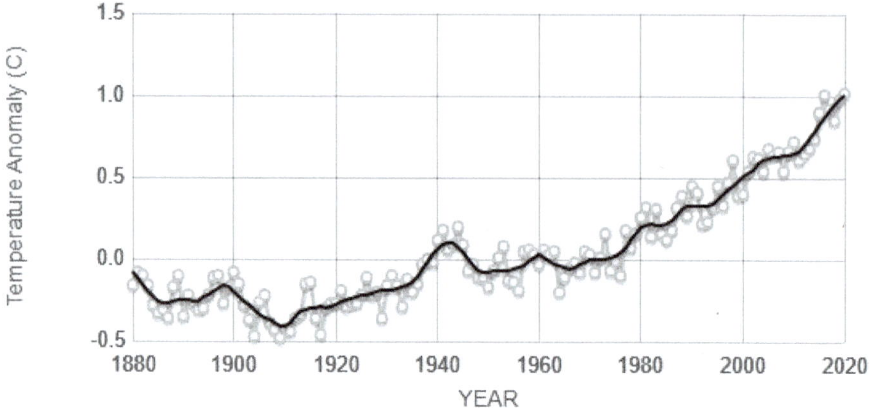

Fig. 1.1. Temperature anomaly from 1880 to the present day. The graph illustrates the change in global surface temperature relative to 1951–1980 average temperatures (https://climate.nasa.gov/).

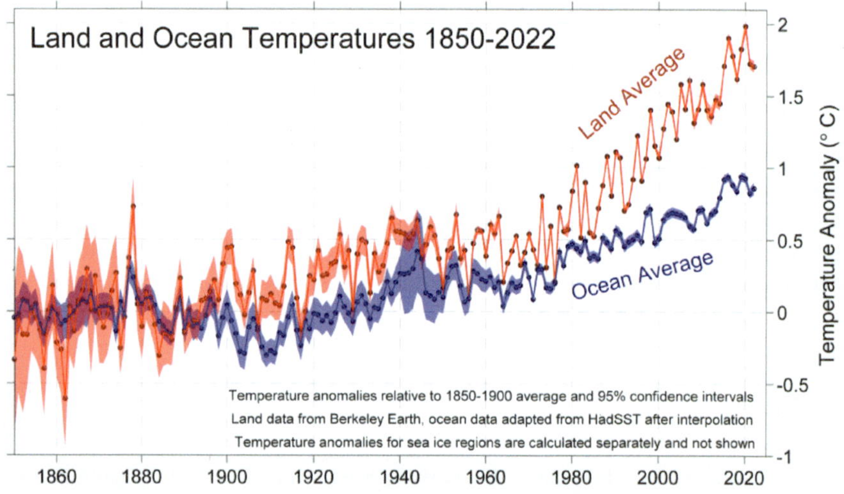

Fig. 1.2. Land and ocean average temperatures as reported by *Berkeley Earth*.

November 2022 is 420 ppm (parts pro million per volume), meaning that of every million particles in the air 420 (or 0.042%) are carbon dioxide molecules, while in 1750 the number was 280. To preserve the planet as we know it, atmospheric carbon dioxide must be drawn down, and urgently so! James Hansen (b. 1941), who was one of the first to study global temperatures in the early 1980s, and his collaborators concluded in 2008, when the level stood at 385 ppm, that "if humanity wishes to preserve a planet similar to that on which civilization developed and to which life on Earth is adapted, paleoclimate evidence and ongoing climate change suggest that CO_2 will need to be reduced from its current 385 ppm to at most 350 ppm, but likely less than that" (Hansen et al. 2008). Indeed, if possible to around 300, but 350 would be fine for the time being.[35] As Mohamed Nasheed, the then president of the Maldives, a country greatly at risk from climate change, forcefully put forth in his speech at the 2020 Klimaforum09 at the Copenhagen climate conference in December 2009: "Three-Five-O saves the coral reefs. Three-Five-O keeps the Arctic frozen. Three-Five-O ensures my country survives. Three-Five-O makes a better world possible" (Nasheed 2012, p. 285–288).

In view of the data presented above and in view of the fact that the temperature, barring a collective suicide of more than half of the global population, is for sure not going down in the coming decades, it is rather odd that at the COP26 in Glasgow people were behaving as if a rise of 1.5°C could still be avoided. Time is too short for that. So, within fifteen years after the Paris conference, mankind will already have utterly failed to pursue the efforts it promised to make.[36]

All this is reason for the International Energy Agency (IEA) to euphemistically observe in 2021 that the pathway to net-zero emissions by 2050 is a narrow one (IEA 2021a). The IEA is much too optimistic in suggesting that net zero by 2050 would give the world an even chance to limit global warming to 1.5°C in this century.[37] Net zero by 2050 is still possible, although unlikely as the political will to take the necessary action for this is lacking, but it does not give the world an even chance to limit global warming to 1.5°C, unless massive efforts are made to reduce the CO_2 concentration in the atmosphere, e.g., by Direct Air Capture in the second half of the century. Commitments made to date fall far short of what is required by the IEA's pathway. Even if the net-zero pledges made today[38] were successfully fulfilled, it would still leave annual emissions of around 22 billion tons of CO_2 worldwide in 2050 (more than total global emissions in 1990). Staying on the IEA pathway requires immediate and massive deployment of all available clean and efficient energy technologies, not for limiting global warming to 1.5°C, but just for achieving net zero for the world by 2050. It includes no new unabated (i.e., without carbon capture and storage) coal plants to be approved for development, the phase-out of unabated coal in developed economies by 2030 and phase-out of all unabated coal plants by 2040. That is just not going to happen, whether we like it or not. In the net-zero emissions pathway which the IEA presents in its report, the world economy in 2030 is some 40% larger than today but uses 7% less energy. Hence, increasing energy efficiency is an essential part of the scheme, resulting in annual energy intensity improvements of on average 4% to 2030—about three-times the average rate achieved over the last two decades. The pathway calls for rapidly scaling up solar and wind in this decade,

---

[35] This number is also the reason why the international environmental organization and website founded by Bill McKibben and others in 2007 is called 350.org; https://350.org/.

[36] The data can however be presented still differently. Lindsey and Dahlman on *Climate.gov* presented the warming data relative to the 20th-century average from 1880–2020, and concluded that "averaged across land and ocean, the 2020 surface temperature was 0.98°C warmer than the 20th century average of 13.9°C and 1.19°C warmer than the pre-industrial period (1880–1900)". But, of course part of the rise has then already been included in the average used.

[37] It says: this is consistent with limiting the global temperature rise to 1.5°C without a temperature overshoot (with a 50% probability).

[38] These pledges have grown rapidly over the last year and now cover around 70% of global emissions of CO_2, but most are not yet underpinned by near-term policies and measures.

in combination with hydropower and nuclear, the two largest sources of low-carbon electricity today, for providing an essential foundation for the transition, and this while nuclear has all but been banned in most developed countries. Most of the global emissions reductions through 2030 will come from technologies readily available today. But in 2050, almost half of the reductions must come from technologies that are currently at the demonstration or prototype stage. It mentions direct air capture and storage, together with advanced batteries and hydrogen electrolysers, as one of the biggest innovation opportunities. By 2030, global capture must amount to 1.6 Gt CO_2 per year, rising to 7.6 Gt[39] CO_2 in 2050. I doubt that this will be enough, but it is nevertheless gratifying to see that the IEA has squarely chosen for carbon capture, both by Direct Air Capture and CCS, as a necessary tool to combat climate change.

A few years earlier in its 2018 *Special Report on Global Warming of 1.5°C*, the IPCC had already formulated four main pathway scenarios to 1.5°C. Three of the four pathways include CCS as well as bio-energy with CCS (BECCS) as mitigation options. One of the conclusions of the report is that all pathways that limit global warming to 1.5°C with limited or no overshoot project the use of such techniques on the order of 100–1,000 $GtCO_2$ over the 21st century. In April 2022 Workgroup III put a more definite number on it by writing in its part of the IPCC's Sixth Assessment Report that the pathways that limit warming to 1.5°C by 2100 with a likelihood of 50% or more require global cumulative net-negative CO_2 emissions of 380 $GtCO_2$ in the second half of the century, and a rapid acceleration of other mitigation efforts across all sectors after 2030.

So, both the IEA and the IPCC show that the problems of climate change and global warming cannot be solved without Carbon Capture and Storage (CCS), Carbon Capture Utilization and Storage (CCUS), Bio-Energy with Carbon Capture and Storage (BECCS) and Direct Air Capture (DAC). All these technologies will be needed as it is also becoming increasingly clear that the most desirable goal of phasing out fossil fuels before 2050 or 2100 will be unattainable.

Hence, there are three paths open to us to resolve the climate problem: (1) continue with fossil fuels and capture and sequester the resulting carbon emissions, while at the same time reducing the concentration of carbon dioxide in the air; (2) revive nuclear power; and (3) turn to renewables. All three should be pursued with vigour. There is no single solution.

In the rest of this book we concentrate on the first solution and discuss carbon capture in greater detail, both capture from the flue gases of point sources like power stations, and the in the end indispensable technique of Direct Air Capture, scrubbing or hoovering some of the carbon dioxide directly out of the air. Before doing that we will first consider the carbon picture in more detail in the next chapter to get a feel for the enormous task we are faced with.

---

[39] One Gt (gigaton) is one billion tons; one Mt (megaton) is one million tons.

CHAPTER 2

The Global Carbon Picture

Introduction

Projections to 2100 as regards global warming have been neatly summarised in Fig. 2.1, originating from *Climate Action Tracker* (CAT),[1] a collaboration of Climate Analytics and the NewClimate Institute, an independent scientific project that tracks government climate action and measures it against the aims of the Paris Agreement.

The figure shows that with the November 2022 policies in force to curb greenhouse-gas emissions, global warming will reach 2.7°C (with a range from 2.2 to 3.4°C) by 2100, while the pledges made to date[2] for 2030 and the associated targets will limit global warming to 2.0°C (with a range from 1.6 to 2.5°C). It is also indicated in the figure that in 2022 warming already stood at 1.2°C, making 1.5°C in 2100 indeed the most optimistic of optimistic scenarios.

The scenario of 2°C warming by 2100 assumes the full implementation of all announced targets, including net-zero emissions by the US (by 2050),[3] China (before 2060) and a number of other countries, including the EU (by 2050), jointly responsible for 73% of global emissions.

The most recent Emissions Gap Report by the United Nations Environmental Programme, aptly called "The Heat is On", confirms the CAT picture and states in this respect that "[t]here is a fifty-fifty chance that global warming will exceed 1.5°C in the next two decades, and unless there are immediate, rapid and large-scale reductions in greenhouse-gas emissions, limiting warming to 1.5°C or even 2°C by the end of the century will be beyond reach."[4] The report suggests that with current policies the world is on course to warm around 2.7°C by the end of the century, which also agrees with the temperature foreseen by *Climate Action Tracker*.

---

[1] https://climateactiontracker.org/; *Climate Action Tracker*, Warming Projections Global Update, November 2022.

[2] The Biden administration has made the big promise to cut emissions by 50% to 52%, compared with 2005 levels, by 2030 (signed into law in August 2022 in the Inflation Reduction Act); the big question is whether the promise will survive his administration. The US's contribution to global emissions is about 15%, so a cut by 50% would be of considerable help.

[3] The Biden Administration has set three major goals for the United States: 50% emissions reduction by 2030, 100% clean electricity by 2035 and net-zero carbon emissions by 2050.

[4] UNEP (2021), *The Heat Is On – A world of climate promises not yet delivered*, p. XVI.

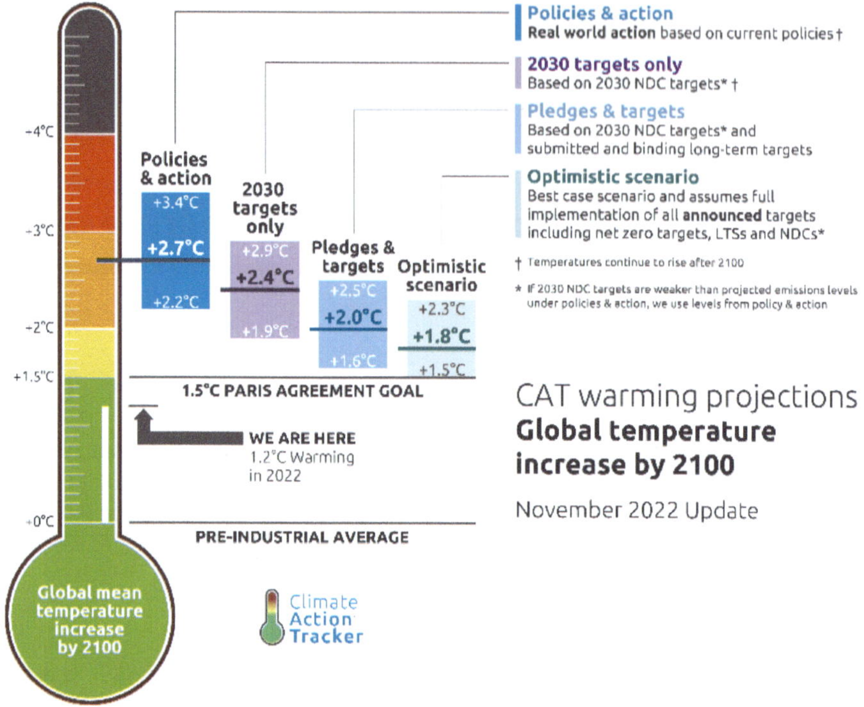

Fig. 2.1. The CAT Thermometer: *Climate Action Tracker* warming projections; global temperature increase by 2100 (November 2022 update). LTS stands for Long-Term Strategy and NDC for Nationally Determined Contribution, required to be submitted by the parties to the Paris Climate Agreement.

Although all this does not bode well for the goals set out in the Paris Agreement, it is in any case gratifying and encouraging that the horror scenarios of 4–6 degrees warming by the end of the century as sketched by some prophets of doom are fading out of the picture.[5] Hence "the most ferocious natural tempers" will likely not be unleashed upon mankind. In CAT's most optimistic scenario global warming by 2100 could be as low as $1.8 \pm 0.3°C$.[6] All these 'predictions' involve model calculations that depend on a number of assumptions, whereby the famous line written in 1976

---

[5] Like the journalist Mark Lynas in his 2020 book *Our Final Warning—Six Degrees of Climate Emergency* and Naomi Klein, *This Changes Everything—Capitalism vs the Climate* (Simon & Schuster, New York, 2014), p. 13–15 and 175. Perhaps James Watt, who unwittingly set us on the path to disaster, does not now have to fear of falling victim to the 'cancel culture'.

[6] It is remarkable that their projections, dating from May 2021, reported a most optimistic scenario of 2.0°C increase (ranging from 1.6°C to 2.6°C), while the worst case (i.e., a continuation of current policies) would result in an increase of 2.9 °C (with a range from 2.1 to 3.9°C). This implies that the situation has improved (a little). It is stated on the website, without further explanation, that although this temperature estimate has fallen since the September 2020 assessment (the same as the May 2021 one), major new policy developments are not the driving factor for this. So what is? What I also miss in their reports and on their otherwise splendid website is a definition of the 'pre-industrial average' temperature. Apparently they know this temperature so precisely that the difference between 1.8 and 2.0°C (just 0.2°C) is relevant.

by British statistician George Box (1919–2013) that "all models are wrong but some are useful" should be kept in mind.

Other analyses arrive at slightly different outcomes. In a fairly recent report (Sherwood et al. 2020), for instance, all lines of evidence were assessed and it was found that a large volume of evidence now consistently points to the view that substantial climate change (well in excess of 2°C warming) most probably cannot be avoided in a high-emission future scenario. It still cannot be ruled out that a doubling of carbon dioxide levels compared to the level at around 1850 would result in a temperature rise of more than 4.5°C, although this is unlikely. This conclusion is independent of climate models and in some sense still a positive result as it seems to roughly confirm the CAT scenario and exclude catastrophic warming predictions that go far beyond 4.5°C.

Nobody experiences the average global temperature though and the situation for individual countries can differ considerably.[7] Moreover the reported average covers both land and sea. As we have seen in the previous chapter, according to *Berkeley Earth* the global land average temperature rise has already hit 1.96°C in 2020, and the difference between land and ocean is more than 1°C (Fig. 1.2). Land areas generally show more than twice as much warming as the ocean. In 2021 global average land temperature went down to 1.70 ± 0.04°C and stayed at that level in 2022. Although slightly anomalous, this is still in agreement with the general spiky, zigzagging behaviour of the temperature, due to the El Niño and La Niña phenomena. The cyclic warming and cooling due to El Niño and La Niña is one of the largest sources of year-to-year internal variability in the global average temperature. *Berkeley Earth*'s current estimate is that 2023 is likely to be similar to 2022 or slightly warmer. With the current continuation of La Niña conditions, it is likely that 2023 remains relatively cool. Once El Niño related warming returns, a new record warm year is fairly likely.[8]

For individual countries *Berkeley Earth* reports that in 2020 the UK passed the 1.5°C mark, the US the 1.7°C mark, France and the Netherlands, among other countries, the 2°C mark, Germany and Denmark the 2.3°C mark, etc., ominously adding that in a scenario of stabilized carbon emissions and slow decline (equivalent to one of the IPCC's scenarios (SSP2-4.5), and as things stand now still an optimistic prospect) these countries are heading for respectively 3.5°C, 4.3°C, 4.3°C and 4.6°C in 2100. Russia already hit 2.7°C in 2020 and in that scenario will reach a staggering 5.6°C in 2100, while for China the corresponding numbers are 1.9°C in 2020 and 5.0°C in 2100. *Berkeley Earth* also reports that in 2021 the annual average temperature in China set a new record high, exceeding 2.0°C above pre-industrial levels for the first time.

If a tentative conclusion can be drawn from this, a cautious one may be that the picture *Climate Action Tracker* paints is rather rosy, while it still could turn out blood red, more in agreement with the prophets of doom.

---

[7] As recently shown by *Berkeley Earth* which gives data for individual countries, accessed 21 December 2021. Many countries had already crossed the 2°C mark by 2020.

[8] Berkeley Earth, *Global Temperature Report for 2022*, January 2023. The relative coolness of 2023 does not seem to be borne out by the experience so far during this year.

Carbon (dioxide) data

Global warming is caused by the emission of carbon dioxide and other greenhouse gases into the atmosphere. Carbon dioxide (CO_2) is the most important greenhouse gas and responsible for more than 60% of global warming.[9] It is a colourless gas with a slightly pungent odour and biting taste. It does not burn, support combustion or sustain life, and hence has widely been considered innocent and incapable of wrongdoing. It is about 1.5 times heavier than air and occurs naturally in the Earth's atmosphere as a trace gas, with its current concentration being about 0.04% (420 ppm by volume), fluctuating slightly with the seasons. Normally stored as a liquid (1 litre of the liquid weighing a little over 1 kg), carbon dioxide exists only as a solid or gas at room conditions.

The most detailed analysis of greenhouse-gas emissions is annually made by the *Global Carbon Project*, a huge international cooperation involving 70 institutes and close to 100 authors. This is the best the world has to offer on the quantification of carbon emissions and their absorption by the various carbon sinks from 1850 to the present day. The latest version has just been published in *Earth System Science Data* (Friedlingstein et al. 2022).[10] In 2020, due to the Covid-19 pandemic, CO_2 emissions decreased by 1.9 $GtCO_2$, bringing global 2020 emissions to 34.8 ± 1.8 $GtCO_2$, comparable to the emissions level of 2012. So, what mankind failed to achieve after sluggishly and spinelessly trying for several decades, a simple spineless virus managed to accomplish in a few months.[11] But don't worry we bounced back. In 2021 fossil CO_2 emissions rebounded by about 6% to reach their highest ever annual level. Almost all regions showed an increase in CO_2 emissions in 2021, ranging from growth of more than 10% in Brazil and India, to less than 1% in Japan. Emissions in China rose by 5%, while the US and European Union both registered increases of around 7%. China was the only major economy to experience

---

[9] To keep things simple, I will only consider CO_2 and not discuss the other greenhouse gases in any detail. The contributions of such gases can be included and instead of tons of CO_2 or the concentration of CO_2 in the atmosphere, quantities expressed in CO_2-equivalent (CO_2-eq or CO_2e) are often used. See Glossary where this concept is explained in some detail. However as Zeke Hausfather says "this proliferation in terms and their subtly different meanings lead to a good deal of confusion among policy makers and scientists alike, to say nothing of the general public". Such confusion should of course be avoided. The situation is made even more complicated by the fact that aerosols that are also released into the atmosphere when burning fossil fuels have a cooling effect, slightly undoing the warming effect due to greenhouse gases. With all greenhouses gases included global emissions amounted to 52.4 $GtCO_2$e in 2021 without land-use change and to 59.1 $GtCO_2$e when including land-use change (2020 UNEP Emissions Gap Report).

[10] The data in the paper by Friedlingstein et al. is given in tons of carbon C. This is often done without precisely saying what is what and can easily cause confusion. The amount of carbon can be obtained from the CO_2 figure by multiplying by 12/44 (carbon, element 6 in the Periodic Table of Elements, has mass number 12, while oxygen, element 8, has mass number 16, hence CO_2 is 44/12 times heavier than C). In the text I have converted the figures to CO_2.

[11] The war in Ukraine could have had a similar effect in 2022. According to preliminary data, emissions in 2022 have decreased considerably in some EU countries, e.g., by 9% in the Netherlands, as energy prices went through the roof and spurred people to save energy. In 2023 this gain will be undone by the ferocious wild fires raging in Canada and elsewhere.

economic growth in both 2020 and 2021, and the emissions increase in China more than offset the aggregate decline in the rest of the world between 2019 and 2021.[12]

Figure 2.2 illustrates the situation for carbon-dioxide emissions from 1950 to 2021 for a few key entities.[13] The figure clearly shows the sharp downturn in emissions for the US, EU and India from 2019 to 2020 due to the Covid-19 pandemic, a downturn that is strikingly absent in emissions from China.

In 2021, China, the US, the EU-27, India, Russia (1.8 Gt) and Japan (1 Gt) remained the world's largest CO_2 emitters. Together they account for 49.2% of global population, 62.4% of global Gross Domestic Product, 66.4% of global fossil-fuel consumption and 67.8% of global fossil CO_2 emissions. All six increased their fossil CO_2 emissions in 2021 compared to 2020, a clear rebound from the Covid-19 year 2020, with India and Russia having the largest increases in relative terms (10.5% and 8.1%, respectively) (Crippa et al. 2022, p. 3).

As can be seen from Fig. 2.2, the developed countries (EU, US) are no longer the main problem, in spite of what we hear all the time.[14] They have successfully curbed their emissions by moving a considerable part to the third world[15] and are on a downward trend already from the early 2000s. Consequently, it would not be such a big deal when some EU-countries, or even the EU in its entirety, reach net zero only in 2055 or 2060 instead of 2050, when at that time China will probably be responsible for close to half of global carbon emissions.[16] On an annual basis China passed 12 Gt in emissions in March 2021 and is now already emitting more than the US and EU taken together. This picture is also recognized by *Climate Action Tracker* which classifies the domestic targets of the EU, US and a few other countries as 'almost sufficient', while using the label 'insufficient' for China's and India's domestic targets, and 'highly insufficient' for Russia's. Striking in their assessment is that the UK is the only major economy whose domestic target is judged "1.5°C Paris Agreement compatible."

---

[12] IEA, Global Energy Review: CO_2 Emissions in 2021 (March 2022); https://www.iea.org/reports/global-energy-review-co$_2$-emissions-in-2021-2; numbers can differ between various publications as emissions due to land-use change and/or cement production may or may not be included.

[13] Most emissions data in this chapter, and also in other chapters of this book, are taken from the website *Our World in Data* or from the *Global Carbon Project*. Data is reported by the countries themselves which have no incentive to be honest and transparent in this respect. On 8 November 2021 the *Washington Post* reported that many countries give incomplete information about their emissions to the United Nations. According to the paper, annually at least 8.5 Gt and possibly even 13.3 Gt remains unreported. If true, even if true by half, this would be an unmitigated disaster but, since the concentration of carbon dioxide in the air is very accurately known, as well as how much is added to it every year, it would also mean that the carbon sinks absorb more carbon than previously thought. For, after all, these unreported emissions are apparently not in the air and must have gone somewhere.

[14] This situation has emerged in spite of the fact that more than 70% of the energy in the EU is still coming from fossil fuels. The 2019 figures are 71.4% fossil (12.6% coal, 36.4% oil, 22.4% natural gas), 13.1% nuclear and 15.3% renewable. The situation in other countries is even worse, e.g., US 83.3% fossil and China 85% of which 57.6% coal. See also Le Quéré et al. 2019.

[15] By globalization and free trade as Naomi Klein forcefully argues in her book *This Changes Everything*, and she definitely has a very genuine point here.

[16] See *Climate Action Tracker*. *Carbon Brief* reported that China's carbon-dioxide emissions have grown at their fastest pace in more than a decade (12 GtCO$_2$ in the year ending March 2021), increasing by 15% year-on-year in the first quarter of 2021.

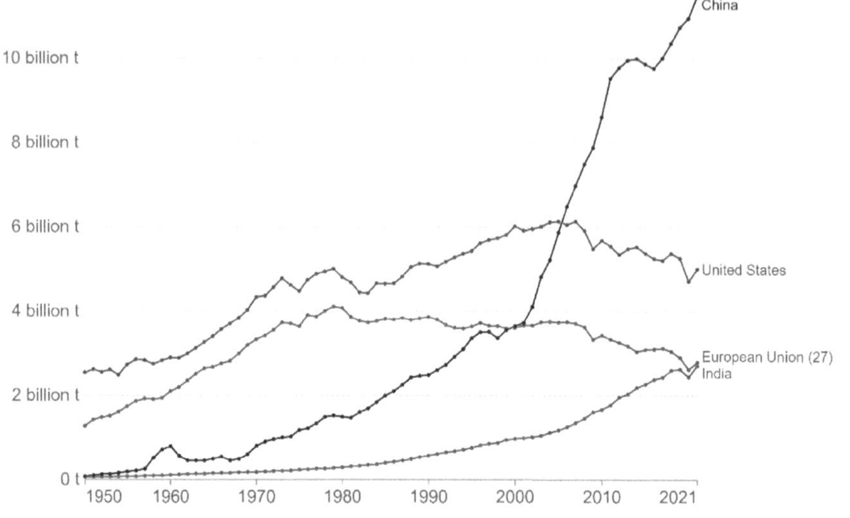

Fig. 2.2. Carbon dioxide emissions since 1950 by China, US, EU (including Britain), and India.

As Fig. 2.2 shows, the emissions picture for China shows a phenomenal rise from 2000 to 2010 (which is also the cause of its current severe air pollution problems), after which the increase slows down appreciably to take off again in 2016. In 2019 China's annual emissions exceeded those of all developed countries[17] combined and its emissions in 2020 are estimated to be equivalent to total annual emissions of nearly 180 of the world's lowest-emitting countries,[18] where the total number of countries in the world is only 195. In 1990 China's emissions were less than a quarter of emissions by developed countries, but over the past three decades they have more than tripled, reaching greenhouse-gas emissions of over 14 Gt of CO_2-equivalent[19] in 2019. Some 12 years ago somebody (Keim 2009) said: if China will not play, the rest of us will go under the waters with China. Well, the figures show that China does

---

[17] Including Australia, Canada, Chile, all 27 current EU member states, Iceland, Israel, Japan, Korea, Mexico, New Zealand, Norway, Switzerland, Turkey, the US and the UK. These countries house about the same number of people as China, but their combined GDP is vastly bigger.

[18] https://rhg.com/research/preliminary-2020-greenhouse-gas-emissions-estimates-for-china/.

[19] Not just CO_2, but also including other greenhouse gases (the so-called six Kyoto gases), and inclusive of land-use and forests and international bunker fuels. See Glossary.

not really want to play[20] and now the situation is such that both China and India must play, and if they don't we will all go under with them. India is still lagging behind a little but is rapidly catching up; in 2019 it edged out the EU-27 for third place with 6.6% of global emissions.[21] China (27%) and India (6.6%) are now responsible for more than one third of global emissions, close to twice the US (11%) and EU (6.4%) taken together, while their combined economies in terms of GDP are about half.

Net zero

Although China is indeed by far the worst polluter, that does not mean that the rest of the world is clearly on track to achieve net zero in the near future. Only two countries (Bhutan and Surinam; both emitting about 2 Mt of CO_2 per year) have currently achieved net zero, while just 12 countries[22] plus the EU have laid down a net-zero date in law (which is 2050 for all, except Germany which has 2045).[23] Three countries (Ireland (also bound to the EU target), Chile and Fuji) have proposed legislation to set 2050 as net-zero date. A further 53 countries have laid down a date, in most cases 2050,[24] in a policy document, as a pledge or statement of intent, including China (2060), US (2050), Russia (2060) and India (2070). A policy document can however easily be superseded by a subsequent policy document by the same or another politician, so such a pledge is not worth much, although still somewhat more than for the remaining 101 countries which are only discussing a date, apparently in all cases 2050.[25]

All pathways that are likely to limit warming to below 2°C relative to pre-industrial levels, having already given up on the desirable 1.5°C scenario,[26]

---

[20] Its pledge to achieve net zero by 2060 has generated a lot of hype on carbon neutrality in the country. However, as Bloomberg reported (https://www.bloomberg.com/news/articles/2021-12-20/china-s-climate-change-goals-are-creating-a-greenwashing-wave), this just amounts to a lot of greenwashing (a form of advertising or marketing spin in which green PR and green marketing are deceptively used to persuade the public that an organization's products, aims and policies are environmentally friendly). Lacking supervision from independent groups as in Europe, there are in China no effective mechanisms to expose such practices.

[21] https://rhg.com/research/chinas-emissions-surpass-developed-countries/.

[22] New Zealand, Canada, Japan, South Korea, the UK and seven EU countries, including France and Germany, representing 15% of CO_2 emissions in 2020.

[23] The EU currently has a target of reducing its net domestic greenhouse-gas emissions by at least 55% by 2030 (the "Fit for 55" package) compared to 1990 levels (hence total CO_2 emissions down to 1.74 Gt, the level of 1956!) and to become climate neutral (net-zero greenhouse-gas emissions) by 2050.

[24] Finland goes for 2035, and Austria and Iceland for 2040.

[25] From Energy & Climate Intelligence Unit; https://eciu.net/netzerotracker, can also be found at https://zerotracker.net/ (accessed 9 November 2021).

[26] As Elizabeth Kolbert says in her delightful book *Under a White Sky* (New York, 2021, pp. 148–149): "To stave off 1.5°C [emissions would] have to drop most of the way to zero within a single decade. This would entail for starters: revamping agricultural systems, transforming manufacturing, scrapping gasoline- and diesel-powered vehicles, and replacing most of the world's power plants." All this is obviously not going to happen, and would still only be the start of things we would have to do. She continues by saying that "[c]arbon dioxide removal offers a way to change the math. Extract large amounts of CO_2 from the atmosphere and 'negative emissions' could, conceivably, balance out the positive variety. It might even be feasible to cross the threshold of catastrophe and then suck enough carbon out of the air to keep calamity at bay, a situation that's become known as 'overshoot'."

require substantial emissions reductions over the next few decades and near zero emissions of CO_2 and other long-lived greenhouse gases by the end of the century. As noted in Chapter 1, the IEA still envisages a global pathway to net-zero emissions by 2050 that would be consistent with global warming to 1.5°C without a temperature overshoot (with a 50% probability). For this the following has to happen (IEA 2021a): "In the Net-Zero Emissions (NZE) scenario, global CO_2 emissions related to energy generation and industrial processes fall by nearly 40% between 2020 and 2030 and to net zero in 2050. Universal access to sustainable energy is achieved by 2030. There is a 75% reduction in methane emissions from fossil fuel use by 2030. These changes take place while the global economy more than doubles through to 2050 and the global population increases by 2 billion.

Total energy supply falls by 7% between 2020 and 2030 in the NZE [scenario] and remains at around this level to 2050. Solar PV and wind become the leading sources of electricity globally before 2030 and together they provide nearly 70% of global generation in 2050. The traditional use of bioenergy is phased out by 2030.

Coal demand declines by 90% to less than 600 million tce[27] in 2050, oil declines by 75% to 24 million barrels per day, and natural gas declines by 55% to 1750 billion cubic metres. The fossil fuels that remain in 2050 are used in the production of non-energy goods where the carbon is embodied in the product (like plastics), in power plants with CCUS (Carbon Capture Utilization and Storage), and in sectors where low-emissions technology options are scarce.

Energy efficiency, wind and solar provide around half of emissions savings to 2030 in the NZE [scenario]. They continue to deliver emissions reductions beyond 2030, but the period to 2050 sees increasing electrification, hydrogen use and CCUS deployment, for which not all technologies are available on the market today, and these provide more than half of emissions savings between 2030 and 2050. "In 2050, there is 1.9 Gt of CO_2 removal in the NZE [scenario] and 520 million tons of low-carbon hydrogen demand. Behavioural changes by citizens and businesses avoid 1.7 Gt CO_2 emissions in 2030, curb energy demand growth, and facilitate clean energy transitions."

It is clear from this summary that, even if all have their noses pointed in the same direction, it will be an incredibly tall order to pull this off. The 40% carbon emissions reduction to be achieved by 2030 implies that emissions will have to go down to about 20 Gt, the level of the late 1980s. It is predicted that for China alone CO_2 emissions will grow to 18 Gt by 2030, if it does not take substantive measures to control them, which it is very reluctant to do as is apparent from the country's reaction to the publication of the first part of the IPCC's Sixth Assessment Report in August 2021. It has now pledged that its emissions will peak by 2030 and to achieve net-zero in 2060 (ten years later than in the above pathway), but so far this has only remained a pledge and no new Nationally Determined Contribution (NDC), as required under the 2015 Paris Agreement, has been submitted. It is virtually impossible that this ambitious goal will be met (Wu and Zhu 2021).

Universal access to sustainable energy by 2030 is not very likely either with its global contribution currently being only 13.5% (2021) of total energy and 38%

---

[27] 1 tce, based on the amount of energy released by burning one ton of coal, is equal to 8.14 MWh.

(2021) of electricity, most of it from hydropower. According to the *Global Electricity Review 2022*,[28] the contribution of solar and wind in 2021 was still only 10.3% of electricity production, and just a few per cent of total energy. In spite of an impressive record rise in solar and wind electricity generation, only 29% of the global rise in electricity demand in 2021 was met by wind and solar, remaining demand increase being met by fossil fuels. So long as renewables cannot even meet the increase in demand they will never succeed in providing 70% in 2030. Coal lost its leading role in 2019 but will probably regain top spot in 2022, in part due to the war in Ukraine but also since the growth of clean electricity sources is stalling.

A fall in total energy use between now and 2030 is also very unlikely as the world is demanding more energy and a scenario in which energy use will grow considerably is much more likely.[29] The US Energy Information Administration (EIA) actually predicts a 50% increase in world energy use by 2050, with most energy still coming from fossil-fuel sources.[30] The latter implies that any outcome close to the IEA's NZE scenario, as the IEA itself also realizes, can only be achieved by deploying carbon removal and carbon-capture techniques. There is actually no compelling reason why *unabated* coal, i.e., without carbon capture and storage (CCS), could not be phased out completely by 2050 with the simultaneous much higher use of *abated* coal. It would require no more than fitting coal-fired power plants, new ones and units that still have a long lifespan, with CCS units. The same applies for natural gas. The IEA says in this respect: "The NZE [scenario] taps into all opportunities to decarbonize the energy sector, across all fuels and all technologies. (…) A failure to develop CCUS for fossil fuels could delay or prevent the development of CCUS for process emissions from cement production and carbon-removal technologies, making it much harder to achieve net-zero emissions by 2050." Not much harder, just completely impossible. The carbon does not leave the atmosphere by itself, it has to be taken out or prevented from going in. Else, the scary 4.5°C temperature rise as mentioned at the beginning of this chapter could still become a reality. It is important to realize though that the IEA does not see any realistic possibility of achieving net zero by 2050 without an important role for CCS, CCUS and/or Direct Carbon Removal.[31]

---

[28] Global Electricity Review 2022, https://ember-climate.org/app/uploads/2022/03/Report-GER22.pdf.

[29] According to projections from Goldman Sachs total global GDP will rise from roughly $100 trillion today to more than double this amount to about $225 trillion in 2050. It is hard to see how this can happen with less energy.

[30] https://www.eia.gov/todayinenergy/detail.php?id=49876.

[31] In this connection, the 2019 decision by the *National Contact Point for the OECD Guidelines for Multinational Enterprises* in a law suit brought in the Netherlands (BankTrack et al. versus ING Bank, see http://climatecasechart.com/) states: "The IEA scenarios for example see a big role for CCS (Carbon Capture and Storage), which is considered by the notifying parties as hindering progress towards emission reductions." The notifying parties are BankTrack et al. and this statement reflects the hostility of the environmentalist movement to the deployment of carbon capture technology, as in their view such technology will prolong the use of fossil fuels. In the same decision the Dutch government is called upon "to request the International Energy Agency to develop as soon as possible two 1.5 degrees scenarios, one with and one without CCS." Whether the Dutch government has indeed followed up this request, I don't know, but it is now clear that the IEA is of the opinion that a scenario *without* CCS is not feasible. I have not seen any signs of the environmentalist movement giving up its destructive and ill-considered opposition to CCS, an opposition that has already done much damage.

Efficiency and energy intensity

We have already noted above that China and India are responsible for more than one third of global emissions, close to twice the combined emissions by the US and EU, while their combined economies in terms of GDP are about half. This implies that in China and India much more energy is required for generating one unit of GDP. There is no obvious reason why that should be the case, apart from inefficient energy use.

It implies that efficiency improvements could also help China (and the world) a lot. China's coal plants are notoriously inefficient, just one seventh of plants in Japan (Li and Wan 2019), although other completely opposite voices are also heard.[32] A good measure to see how efficient energy is used is to calculate the number of kWh needed to produce $1,000 of GDP (the ratio of energy use to economic output), also called the energy intensity. This has been done in Table 2.1 for the most important economies in the world, including Switzerland as it is a real champion in this respect and should be made the benchmark. The table shows that in 2019 China used 2,674 kWh for each $1,000 worth of GDP, against a little over 1,000 kWh for other major economies. Of course, I realize that the comparison is not always straightforward and fair as some countries have a lot of factory production which consumes more energy than banking services, for instance. China is the factory of the world and therefore will need more energy than Switzerland which has relatively

---

[32] CAP Report 2017, *Everything You Think You Know About Coal in China Is Wrong*. This report blatantly claims that Chinese coal power plants are more efficient than the ones in the US and that China is "on track to overdeliver on the emissions reduction commitments it put forward under the Paris climate agreement", overlooking the fact that China is not reducing any emissions at all and that American coal power with 246 operating coal plants is just 23% of the 1087 Chinese plants in operation. Between 2010 and the first quarter of 2019, US power companies announced the retirement of more than 546 coal-fired power units, totalling about 102 gigawatts (GW) of generating capacity. Chinese plants are emitting 4.5 Gt of CO_2, American ones 1.1 Gt. China has plans to build 95 more, the US not a single one and existing ones are closing at a fast rate since the 2010s because of cheaper and cleaner natural gas and renewables. In 2017, there were still 359 coal-powered units at the electrical utilities across the US; in 2019 there were 241, showing that they shut down at a rate of dozens per year. It is not surprising that the 100 most efficient Chinese plants, which were built in the last few years, are more efficient than the 100 most efficient American ones, which are considerably older with many at the end of their normal lifespan. (Data from *Global Energy Monitor*, July 2021, and the *Energy Information Administration*. See also Emily Grubert (Grubert 2020), who has calculated that retiring all US fossil fuel plants by 2035 would strand only about 15% of fossil capacity-years and 20% of job-years.) There is a general tendency, all over the world, it seems, to be protective towards China and bash the US and EU. This is also apparent from the Vox website which likewise made outrageous untrue claims about China's coal plants (https://www.vox.com/energy-and-environment/2017/5/15/15634538/china-coal-cleaner), but also grudgingly has to admit that the US actually reduced emissions, but complains that was easy for the US by just switching from coal to natural gas, in addition to increasing wind and solar power and increasing fuel economy in light cars. This is true but the reductions are no less real. They all easily fall for the empty promises made by the Chinese autocrats. Another outrageous example, dating from April 2012, is an article on the website *ember-climate.org* by Rob Elsworth, who calls himself a climate change and international development policy specialist with expertise in mitigation, adaptation, climate finance, development and emissions trading policies. The paragraphs of his article, that earlier appeared in the *Guardian*, start with: Chinese authorities are acutely aware…, China is also increasingly conscious…, China is drawing on lessons…, China is advancing rapidly. China bootlicker should perhaps be added to his special skills. The results of China's awareness, lessons it drew and actions are visible in Fig. 2.2. I am sure that the autocrats in Beijing would love to cuddle Elsworth, and perhaps they actually do as such 'useful idiots' are too precious to let become befuddled with the truth.

Table 2.1. Carbon-dioxide emissions related to GDP and energy use for various countries.

Country	GDP in billion US dollars 2020 (nominal)	GHG-emissions 2016 in Gt	CO_2-emissions 2020 in Gt	CO_2 in tons per $1,000 of GDP	Total energy usage in TWh 2019	Energy usage in kWh per $1,000 of GDP	CO_2 in tons per capita 2020
US	20 932.75	5.83	4.71	0.225	26 291	1 256	14.24
EU-27	15 167.00	3.16	2.60	0.171	16 910	1 115	5.84
China	14 722.84	11.58[36]	10.67	0.725	39 361	2 674	7.41
Japan	5 048.69	1.26	1.03	0.204	5 187	1 028	8.15
Germany	3 803.01	0.81	0.64	0.168	3 650	960	7.69
UK	2 710.97	0.46	0.33	0.122	2 178	803	4.85
India	2 708.77	3.24	2.44	0.901	9 461	3 492	1.77
France	2 598.91	0.33	0.28	0.108	2 689	1 035	4.24
Canada	1 643.41	0.78	0.54	0.329	3 948	2 203	14.20
Korea	1 630.87	0.66	0.60	0.368	3 436	2 107	11.66
Russia	1 473.58	2.39	1.58	1.072	8 279	5 617	10.81
Australia	1 359.33	0.52	0.39	0.287	1 780	1 310	15.37
Switzerland	747.43	0.05	0.03	0.043	314	420	3.73

little heavy industry. But should the difference be that large? Manufacturing makes up roughly 25% of China's GDP, against 20% in Japan and 18% in Germany.[33]

As can be seen, energy usage per $1,000 of GDP varies wildly with Russia far ahead, more than 10 times as much as Switzerland[34] and more than 5 times as much as the EU. India and China also do rather poorly, showing that these countries can gain a lot from energy efficiency measures, which of course will automatically curb CO_2 emissions. Canada (an odd case as its electricity sector is almost 100% carbon free) and South Korea could also benefit from some improvement. Especially Canada is much less energy efficient than for instance Sweden, even if the differences in their industrial bases are accounted for (Jaccard 2005, p. 20). There is no single obvious reason why the differences should be so large as the table shows, other than just being callous (Russia) as regards energy use.

If all were as efficient in energy use as Switzerland, there would be no climate change problem whatsoever! Emissions would amount to just 5.4 Gt, the level of the world in the mid 1950s! And although Switzerland may be an odd case, this cannot be said of Germany or the UK which both score much, much better than China, Russia or India. For China too it should be possible to bring the energy intensity (the quantity of energy required per unit output or activity) down[35] close to German/UK levels, and they owe it to the world!

---

[33] https://www.visualcapitalist.com/visualizing-the-94-trillion-world-economy-in-one-chart/.

[34] CO_2 emissions in Switzerland are low as the country generates very little electricity from fossil fuels, just 5.4% in 2021 with 28.5% being nuclear and 66.1% renewable (mainly hydro).

[35] This however is not as simple and straightforward as this simple presentation suggests, as the monetary value of an output also changes for other reasons, such as quality improvements, which increase GDP, without changing the energy input into a good.

[36] 2020: 14.4 Gt (preliminary data Rhodium Group; hence a 24% rise in just 4 years).

But energy efficiency alone is not the panacea for our worries and can even have adverse effects. People in rich countries certainly use more energy per capita than people in poor countries, precisely because they have all these energy efficient gadgets at their disposal. Cars and heating systems are more efficient than 40 years ago, we drive more and, unable to withstand a normal summer, we acquire air conditioners, wine coolers and all kinds of other things we don't really need but can afford because energy is dirt cheap.

Carbon budget and historic emissions

Apart from current emissions as discussed above, there are also the sensitive issue of the carbon budget (the upper limit of carbon that still can be emitted before a specific global average temperature is exceeded) and the thorny issue of legacy emissions (cumulative past emissions, mainly by the developed, industrialized nations). If the carbon budget is large enough it does not matter much what has been put into the air in the past and we do not have to worry about legacy emissions. There still will be enough room for developing countries and others to burn fossil fuels to their heart's delight. It will be no surprise that is not the case, reason why people are demanding that fossil fuels be left in the ground.

The fact that there is such a limit before climate change becomes catastrophic is obvious, but to reliably calculate the budget left is quite a different matter.

According to the latest analysis by the *Global Carbon Project* (Friedlingstein et al. 2022, p. 1920) the remaining carbon budget for a 50% likelihood to limit global warming to 1.5°C has shrunk to 420 $GtCO_2$ and for remaining below 2°C to 1270 $GtCO_2$, equivalent to 11.5 and 35 years from the beginning of 2022, assuming 2021 emissions levels (36.4 Gt CO_2) in all those years. As can easily be calculated, these figures mean that the *Global Carbon Project* is of the opinion that emissions of 1270 − 420 = 850 Gt of CO_2 will cause 0.5°C of global warming, hence 170 Gt per 0.1°C of warming. They also imply that according to the *Global Carbon Project* current global warming is 1.25°C, half of which is due to emissions since 1980, and that at the current rate of emissions it will take about 5 years to cause 0.1°C of global warming.

We know however that, barring any unforeseen events, such as a continuing Covid-19 pandemic or a new virus outbreak, emissions will actually go up, implying that the 11.5 and 35 years mentioned will be considerably shortened. Global temperature has risen by 0.08°C per decade since 1880, but the rate of warming over the past 40 years at 0.18°C per decade is more than twice of that.[37] So, it is beyond doubt that within less than two decades a global average temperature rise of 1.5°C will have been realized. Only a sustained global economic collapse can prevent us from reaching this milestone. The 2°C mark may take a few decades more, but we can quite confidently conclude that there is very little carbon budget left, and that

---

[37] Climate Change: Global Temperature: https://www.climate.gov/news-features/understanding-climate/climate-change-global-temperature#:~:text=Earth's%20temperature%20has%20risen%20by,C)%20per%20decade%20since%201981.

concentrations of carbon dioxide and other greenhouse gases in the atmosphere have already reached dangerous levels.

The numbers given by the *Global Carbon Project* are however no gospel and the website *350.org*[38] gives a substantially lower number for the carbon budget, stating that we can still emit only 565 Gt of CO_2 to stay below 2°C of warming—anything more than that risks catastrophe for life on earth. At the current rate, which will increase for at least another decade or so, this budget will be exhausted in 16 years, by 2037, well before 2050.

The IPCC in its Sixth Assessment Report gives a value of just 400 Gt of CO_2 still available from January 2020 for the global carbon budget to keep global warming within 1.5°C with 67% confidence. This is 50 tons for each member of the eight billion world population. Who is allowed to emit them? Are the old 'rogues' (the countries that have emitted a lot in the past) still entitled to a share of this? It seems a futile exercise to discuss such issues, since before the discussion has really started the budget will already have been exhausted.

It should also be noted though that not all fossil fuels are the same, coal being the worst offender. So, what in any case should be done as soon as possible is to phase out coal power plants (or, if phasing out is no option, (retro)fit them with CCS installations) and to stop the use of gasoline in transportation. Natural gas, which is by far the most benign fossil fuel, could then be tolerated a little longer or, even better, the carbon emitted in its combustion should be captured and stored.

The *350.org* website also says that "burning the fossil fuels that corporations[39] now have in their reserves would result in emitting 2,795 gigatons of carbon dioxide—five times the safe amount", while a recent report by *Carbon Tracker* called *Unburnable Carbon* gives a number of 3,700 Gt. If all these reserves were burned, which apart from coal we could easily manage in this century, another 1,400 Gt of CO_2, or even 1,850 Gt, would be left in the atmosphere, assuming that the oceans and terrestrial biomass continue to provide the service of absorbing about half of them.[40] This is a bit less than what has been emitted and left in the atmosphere so far and would increase the current CO_2 concentration in the atmosphere by more than 200 ppm. The oceans absorb some of the carbon at the expense of ocean acidification, which is as bad as global warming but gets much less attention. Acidification reduces sea life's ability to produce skeletons and shells, which in the end might lead to a catastrophic and irreversible collapse of marine life.

It is clear from this discussion that emissions will soon exceed any of the remaining budgets estimated. To comply with the budget limits, the CO_2 already in the atmosphere will need to be captured on a huge scale and stored in products, the environment or underground. A 2015 study (Gasser et al. 2015) calculated that

---

[38] https://math.350.org/.

[39] It is not clear what is meant by this. The big Western oil companies (aka Big Oil) possess very little fossil fuel reserves, just 6% of global oil and gas reserves. The rest is owned by state actors, who of course also do business through corporations (Saudi Aramco, Gazprom, etc.). They however do the bidding of the mostly autocratic rulers in these countries and it is not clear whether they are included in the 'corporations' meant here.

[40] It is not clear whether they are still doing this, see, e.g., Johanna Miller (Miller 2021). There is evidence that part of the Amazon has flipped from carbon sink to carbon source.

carbon budgets can only be met by capturing CO_2. So, a cure is available, but carbon capture capacity in any form is at present vastly deficient for achieving any of this. The doctor has prescribed a cure for the terminally ill patient but others beyond his control have decided that the cure is worse than the disease and that it is better to let the patient die, while refusing to see that they will die with him.

In Chapter 1 we have already quoted the IPCC as saying that, to limit global warming to 1.5°C with limited or no overshoot, on the order of 100–1,000 $GtCO_2$ must be removed from the atmosphere over the 21st century. This must compensate for residual emissions and, in most cases, achieve net negative emissions to return global warming to 1.5°C following a peak. The IPCC intends to do this by carbon-dioxide removal (CDR). For a definition of this term see the Glossary. It encompasses a wide array of approaches that remove carbon dioxide from the atmosphere, including Direct Air Capture and other forms of Direct Carbon Removal, coupled to durable storage, as well as biomass carbon removal and storage. It does not include CCS from point-source emitters like power stations, which is considered a way of preventing CO_2 from entering the atmosphere. Apparently the IPCC does not see a large role for CCS from point-source emitters to prevent CO_2 from entering the atmosphere, although it is technically easier and every molecule prevented from going into the air does not have to be taken out later.

For the less strict goal of 2°C warming, carbon capture is needed too. The IPCC has only one scenario (they call it a 'Representative Concentration Pathway' (RCP)) which limits warming to 2°C: "RCP2.6 is representative of a scenario that aims to keep global warming likely below 2°C above pre-industrial temperatures. The majority of models indicate that scenarios meeting global warming levels similar to RCP2.6 are characterized by substantial net negative emissions by 2100, on average around 2 $GtCO_2$/yr."

The fact that there is hardly any carbon budget left makes legacy emissions the more poignant. The developed industrialized states have freely used the atmospheric commons in the past hundred and fifty years, and, so it is argued, used more than their fair share of the Earth's ability to absorb greenhouse-gas emissions that cause climate change without paying much attention to the consequences. By doing so they have robbed developing nations of the possibility to follow the same path to prosperity, as the carbon budget is shrinking rapidly. Such use in the distant past of more than a century ago is still important today as CO_2 stays in the atmosphere for a couple of centuries. Hence emissions 100 years ago are still very relevant today and current warming is determined by the cumulative total of CO_2 emissions over time. The latest IPCC report is actually quite specific about this and states that 1,000 Gt of CO_2 is responsible for warming of 0.45 ± 0.18°C, which it calls the transient climate response to cumulative carbon emissions (TCRE; the ratio of globally averaged surface temperature increase and cumulative CO_2 emissions). This is in the same ballpark as the estimate by the *Global Carbon Project*, which is 0.5°C for 850 Gt.

How large are these legacy emissions? To get some numbers we again turn to the annual analysis made by the *Global Carbon Project* (Friedlingstein et al. 2022, p. 1940). It shows that CO_2 emissions from land-use change have not changed much during the years from 1960 to 2019, while emissions from fossil fuels have tripled in that period. Cumulative anthropogenic CO_2 emissions for 1850–2020 totalled

2,420 ± 240 $GtCO_2$, of which almost 70% (1,670 Gt) occurred since 1960 and more than 30% (750 Gt) since 2000. These numbers include emissions from fossil fuels and cement production, as well as from land use, land-use change and forestry (LULUCF).[41] Total anthropogenic emissions more than doubled over the last 60 years, from 17.9 ± 2.6 Gt/yr for the 1960s to an average of 38.9 ± 2.9 Gt/yr during 2011–2020.

In 2020 total anthropogenic CO_2 emissions from fossil fuels plus land use amounted to 37.2 ± 2.9 Gt, while for 2021 these emissions are projected to be around 39.3 Gt. These figures differ from those that follow from Fig. 2.2 as in that figure, as with all data from *Our World in Data,* emissions from land-use change are not included.

During the historical period 1850–2020, 30% of historical emissions were from land-use change and 70% from fossil emissions. However, fossil emissions have grown significantly since 1960 while land-use changes have not, and consequently the contributions of land-use change to total anthropogenic emissions were smaller during recent periods (17% during the period 1960–2020 and 10% during 2011–2020).

The atmospheric CO_2 concentration was approximately 285 parts per million[42] (ppm) in 1850, reaching 300 ppm in the 1910s, 350 ppm in the late 1980s, and 412.44 ± 0.1 ppm in 2020.[43] The mass of carbon in the atmosphere increased by 48% from about 2,200 $GtCO_2$ in 1850 to 3,200 $GtCO_2$ in 2020. The observed stability of the airborne fraction over the 1960–2020 period indicates that the ocean and land carbon sinks have been removing on average about 55% of anthropogenic emissions. The growth rate in atmospheric CO_2 concentration was 2.37 ± 0.08 ppm in 2020, very close to the 2011–2020 average.

Cumulated since 1850, the ocean sink adds up to 620 ± 130 $GtCO_2$, with two thirds of this amount being taken up since 1960 at a rate of about 7 Gt CO_2/year. Over the historical period, the ocean sink increased in pace with the anthropogenic emissions increase; and shows no sign yet of slowing down. Since 1850, the ocean has removed 26% of total anthropogenic emissions. For the terrestrial carbon sink it applies that, cumulated since 1850, it has absorbed 715 ± 165 $GtCO_2$, 30% of total anthropogenic emissions. In this case too, the sink increased in pace with the anthropogenic emissions increase. So, total emissions during the period 1850–2020, amounting to 2,420 ± 240 $GtCO_2$, were partitioned[44] among the atmosphere (1,020 ± 18 Gt; 42%), ocean (650 ± 130 Gt; 27%), and land (750 ± 165 Gt; 31%).

---

[41] Land use is the total of arrangements, activities and inputs applied to a parcel of land. The term land use is also used in the sense of the social and economic purposes for which land is managed. Land-use change is the change from one land use category to another. It mainly concerns deforestation. Land use activities can result in emissions of greenhouse gases to the atmosphere or removal of such gases from the atmosphere. In the past land-use change often involved deforestation while nowadays reforestation is very much in vogue.

[42] Equivalent to 0.0285%.

[43] This is 3–4 ppm less than reported by NASA.

[44] In the report an imbalance of 3% (about 100 Gt) is reported which we have equally partitioned over the three sinks.

The cumulative land sink is almost equal to the cumulative land-use emissions, making global land nearly neutral over the whole 1850–2020 period.

From this it follows that atmospheric carbon from legacy emissions amounts to about 1,000 Gt of CO_2. In ppm this is about 130 ppm, implying that 1 ppm corresponds to 7.7 Gt of CO_2. It also applies that with the current rate of emissions more than 2 ppm are added each year to the carbon-dioxide concentration in the atmosphere.

The *Our World in Data* website has calculated how much of these legacy emissions can be attributed to individual entities or nations. The results are shown in Fig. 2.3.

The starting point in compiling this figure is that total cumulative fossil CO_2 emissions from 1750–2020 amount to 1,700 Gt,[45] 75% of which were emitted by the following 7 entities: US (417 Gt or 25%), EU-27 (290 Gt or 17%), China (236 Gt or 14%), Russia (115 Gt or 7%), UK (78 Gt or 5%), Japan (66 Gt or 4%), and India (54 Gt or 3%). The next largest contribution at 42 Gt (or 2.5%) is from International Transport, which cannot be assigned to a specific country.

Carbon Brief[46] also includes land-use emissions and attributes 20% (484 Gt) to the US, 11% to China and 7% to Russia. Due to the inclusion of emissions from land use the UK (3%) and Japan drop down the ranking, while Brazil with 5% and Indonesia with 4% move up. This reflects that early industrialized countries have

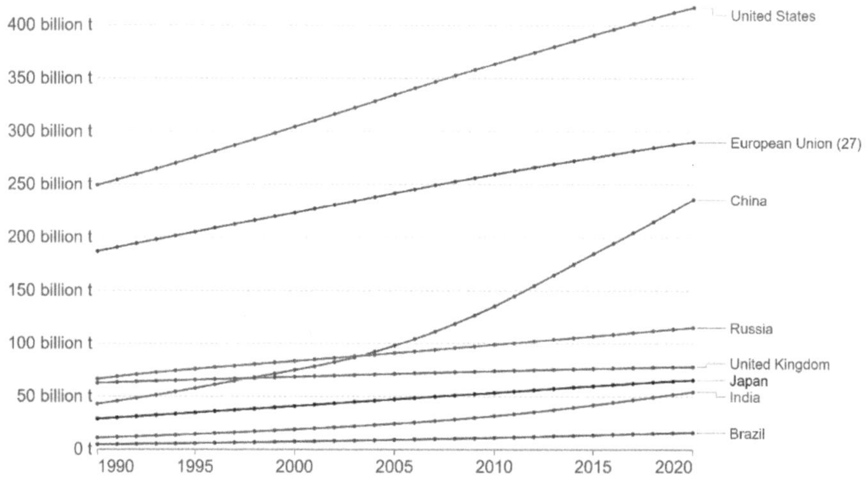

Fig. 2.3. Cumulative CO_2 emissions for a number of entities.

---

[45] This includes CO_2 from fossil fuels and cement production; land-use change is not included. It agrees well with 70% of 2,420 Gt quoted by the *Global Carbon Project.*

[46] https://www.carbonbrief.org/analysis-which-countries-are-historically-responsible-for-climate-change.

made considerable changes to their land use in the 19th century, while younger nations are still doing this now.

All this confirms the conclusion already made above that, forgetting about land-use change emissions (about equal to the absorption by biomass on land) and accounting for absorption by the oceans (this carbon is after all no longer in the air although it still does harm), the atmospheric carbon from relevant historic emissions so far amounts to roughly 1,000 Gt of CO_2, with 75% of the responsibility for them attributed as stated above (US 25%, EU-27 17%, China 14%, Russia 7%, UK 5%, Japan 4% and India 3%).

Taking $100 per ton as a guide, this would add up to a cumulative debt of these countries to the world of $100,000 billion, about equal to current total global GDP. Do they indeed owe such a debt to the world or to other countries? It is easy to answer this question with a resounding 'yes' as some people do (see, e.g., Naomi Klein (Klein 2011)), although China and India are normally excluded as their sins are too young, as it were, and there does not appear to be any clear answer to the question what a fair share of climate sink capacity would have been historically. It is equally easy to demand that the countries, mostly rich ones, that are responsible for these legacy emissions should pay a certain number of billions per year to countries, mostly poor or developing countries, that allegedly suffer the most from climate change. But such a simple approach will never be accepted and is in any case too harsh and unjust, if only for the fact that present generations cannot be held responsible *as a matter of course* for what earlier generations have done. This certainly applies if these earlier generations were for a large part unaware of the fact that their behaviour was or could be harmful, let us say until 1990. On the other hand, it cannot be denied either that the current climate crisis makes it difficult, if not impossible, for developing countries to lift their populations out of poverty, at any rate when trying to do so in an environmentally responsible way, which very few do. As we saw in the preceding chapter, all this should be solved by all countries "in accordance with their common but differentiated responsibilities and respective capabilities and their social and economic conditions." Nice words, which have however not yet resulted in agreement on any form of payment for legacy emissions from rich countries to developing ones. The ethics question is huge and has been written about by many authors. It is too complex and too delicate to deal with in a few paragraphs and therefore goes beyond the scope of this book. That the developed world has a special duty to mitigate the consequences of climate change for the developing world goes however without saying and has also been recognized in the Paris Agreement, although not or not sufficiently acted upon.

Forgetting about legacy emissions, it must be possible to establish national 'fair shares' for countries in the costs of a global mobilization for climate mitigation relief. This should include the cost of carbon capture at existing and future fossil-fuel power plants as well as the cost of Direct Air Capture. In this way developing countries could be provided with the means of further development through the use of fossil fuels. An interesting development in this respect is the formulation of so-called Greenhouse Development Rights that seek to transparently calculate such fair shares and to develop at the same time a framework for reconciling the twin crises of climate change and development.[47]

How tricky blame apportioning can be is shown by a recent study (Callahan and Mankin 2022) which has assessed the economic impacts caused by individual countries to other individual countries through their contributions to global warming. The research claims to draw direct connections between cumulative greenhouse-gas emissions per nation to losses and gains in the gross domestic product in 143 countries for which data is available. Previous studies provided estimates of the total global level of economic loss due to such emissions but were unable to determine the warming attributable to individual nations, undermining efforts to hold emitting countries accountable for legal damages because of the uncertainties involved. For some unclear and rather incomprehensible reason the study only considers emissions for the 25-year period from 1990 to 2014, leaving the period before 1990 out of the discussion, which as can be imagined will be a relief to some developed countries as it lets them off the hook.

By creating an analytical framework that links emissions from individual countries to the losses and gains in every other country, this new research hopes to help resolve questions of climate liability and national accountability to inform climate policy. The study discredits the idea that climate mitigation is simply a 'collective action problem', where no one country acting alone can have an effect on the impacts of global warming.

According to the study, emissions from the U.S. and China alone are responsible for global income losses of over \$1.8 trillion each in the above-mentioned 25-year period. Economic losses caused by Russia, India, and Brazil individually exceed \$500 billion each for the same years. The \$6 trillion in cumulative losses attributable to the five countries equals to about 11% of annual global GDP within the study period. It is perhaps worth noting here that for the US and China \$1.8 trillion is less than 10% of these countries' GDP, while the more than \$500 billion for Russia, India and Brazil is 20–30% of their GDP. Table 2.2 shows (calculated from data given by *Our World in Data*) that in de period from 1990–2014 Brazil 'only' emitted 8.6 Gt CO_2, just 6% of emissions by the US in that period, while Brazil is said to have caused \$500 billion in damage equal to more than 25% of the damage caused by the US. This is incomprehensible, and throws doubt on the entire approach adopted in the study. A gigaton of Brazilian carbon dioxide is far more damaging than a gigaton of any of the other countries included in the study.

The picture this study presents is also at variance with total cumulative emissions over the entire period from 1880 to the present day as presented in Fig. 2.3 and Table 2.2, where the EU-27 and the UK (both not mentioned in the study) are shown to be responsible for a greater share in total emissions than for instance India and Brazil, the latter two having actually emitted very little. The fact that in this study China is awarded the top spot, jointly with the US, which of course is due to only including emissions from the 1990–2014 period, comes across as a glaring injustice

---

Table 2.2. Cumulative carbon-dioxide emissions by a number of entities in various periods.

	to 1990	to 2020	1990–2020	to 2014	1990–2014
China	43.0	235.6	192.6	174.6	131.6
US	249.2	416.7	167.5	385.6	136.4
EU-27	186.6	290.1	103.5	272.3	85.7
Russia	66.4	115.3	48.9	105.5	39.1
India	11.0	54.4	43.4	39.7	28.7
Japan	28.8	65.6	36.8	58.7	29.9
UK	62.7	78.2	15.5	75.9	13.2
Brazil	4.7	16.2	11.5	13.3	8.6

as China's cumulative emissions over the entire period from 1880 are much less, as stated above, than the totals emitted by the US and the EU-27. Moreover, through these emissions China has managed to lift its people out of poverty. Should it be punished for having done so? This and the inclusion of India and Brazil among the top damage-causing nations in this study appear to place a further burden on developing countries. The results of the study suggest that, in planning their development, developing countries must take into consideration to what extent such development may cause damage to other countries.

CHAPTER 3

The 100% Renewable-Energy Option Explored

Introduction

In the last few decades energy generation from renewable sources, like wind turbines and solar panels, has made great strides and is now touted as the way to mitigate climate change and global warming. This is without doubt a task of prime importance but will not by any means be the only function of renewables. Nobody can ignore the fact that stocks of fossil fuels are finite. The time of peak oil, peak natural gas or peak coal may not yet have arrived, but that time will surely come and, at any rate for gas and oil, is just a few generations away, meaning that at the end of the day only five or six generations of humans will have been able to enjoy the abundance of cheap energy from fossil fuels.

It is actually quite accurately known how much oil, gas and coal there still is to be had (at the current level of technology).[1] It is estimated that at current consumption levels we still have enough oil for 47 years, natural gas for 52 years and coal for 133 years.[2] Not everybody agrees with this assessment. In his latest book Marc Jaccard (Jaccard 2020, p. 138 ff.), for instance, argues that so far we have only exhausted just 5% of the available fossil fuels and that the end of oil is nowhere in sight. If anything, "our worry is that we won't run out of oil fast enough". But, whatever the case, we don't want to ever run out of oil, gas or coal. We must diversify our energy sources, so that we can continue to use fossil fuels for the multitude of purposes and products we need them for. As Richard Somerville notes "many climate activists who talk about weaning the world quickly from fossil fuels rarely understand the details of how crops become food and how steel and cement are used in cities and highways, or how widespread plastics have become and how central fossil fuels are to these processes" (Somerville 2022). Think of diesel, for instance. Without diesel, civilization would end within a week, as Alice Friedemann makes clear (Friedemann 2021, p. 28). Although we will not soon run out, we should think

---

[1] See the discussion in Chapter 1 on EROEI (energy return on energy invested).

[2] https://www.worldometers.info/oil/; https://www.worldometers.info/gas/, https://www.worldometers.info/coal/.

now of what can replace it. And if we cannot find a replacement,[3] we should save the remaining fossil fuels for making diesel.

In view of all this, in some sense, climate change comes at a good point and the drive towards affordable renewable/sustainable energy generation ensures that in future we will still have oil for all these other uses. But are these renewables up to the job we want them to do? Can they provide what fossil fuels have given us in the past and are still doing now, i.e., a reliable and affordable energy system?

And actually, why would anyone want a 100% renewable-energy system in the first place, at gigantic costs and by 2050 at the latest? There is no need for that. A net-zero or close to net-zero system by that date is just as good. Emitting CO_2 is not a problem if we capture it before it enters the atmosphere and store it out of harms way. It is far cheaper than doing away with the entire fossil-fuel infrastructure and replace it by a vast renewable-energy infrastructure. Moreover, the only way to tackle legacy carbon emissions is by Direct Air Capture, which has in any case to be deployed on a vast scale. Let us count our blessings. My view is that renewable sources have come along in the last few decades as *additional* sources of energy that should make our lives easier and generate vast amounts of extra energy that can partly be used to transform the old system and repair the planet.

By renewable energy sources we mean in particular wind turbines, solar panels (solar photovoltaics (PV)) and hydropower, perhaps supplemented, I hesitantly add, with biomass. I say hesitantly as it should be realized that renewable does not necessarily mean environmentally benign, as biomass definitely fails in that respect. Actually none of them are, even hydropower, normally considered one of the cleanest forms of energy, may result in stream flows being altered, large numbers of people being displaced, thousands of hectares flooded, etc. The created water reservoirs are even sources of greenhouse-gas emissions by releasing carbon dioxide and methane (CH_4) from submerged and decaying trees and shrubs (Smil 2017, p. 196). Hence the appetite for developing hydropower where that is still possible, particularly in Africa and Asia, is rather subdued. Moreover, the transition to renewables will involve a switch from an energy-intensive to a materials-intensive society, which will involve a lot of (environmentally harmful) mining, not necessarily an improvement over current practices, being very energy intensive and creating heaps of waste.

The great asset of these energy sources is of course that they do not produce any carbon dioxide, reason why they play a crucial role in the decarbonization of global energy usage and in mitigating the impact of climate change. But we should not be blinded by this fact and throw the baby out with the bathwater.

In the end the alleged ultimate is when all energy economy-wide is supplied by renewable sources only, and it is not difficult to devise scenarios in which such a renewable-energy system is possible and in the last 15 to 20 years a lot of effort was spent in just doing that.[4] The question is however not whether such a 100% renewable-energy economy is possible, after all much is possible but still does not

---

[3] There has been promising recent progress in developing a modified conventional diesel engine that runs on a mix of hydrogen and diesel, cutting CO_2 emissions by 85% (Liu et al. 2022), but a practical engine is still decades away.

[4] Research into 100% renewable-energy systems is much older and was already in vogue in the 1970s following the oil-price crises of that time. See, e.g., Sørensen 1975.

happen. More important is whether the concocted scenarios are realistic. Other words used in this connection are feasible and/or viable. Ideas are easy, execution is everything, as a wise man once said.

In this respect it is important to realize that an energy system that provides abundant and affordable energy in a reliable manner is vital to a modern society. The system we have in place now fulfils these criteria quite well and, if it is going to be changed, it must be dead certain that what we get in return is at least as good as the old system. There is no margin for error here. To show this and convince others that this will indeed be the case is far from easy. The jury is still out and no consensus has so far been reached in the (scientific) literature. Actually the disagreement in the literature is rather large. In this chapter we will look at this debate in some detail.

A 100% renewable-energy economy first involves the clean-up of the electricity sector, i.e., generate all electricity we use, roughly 20% of total energy, from renewable sources, and then continue with the total electrification of everything else that can be electrified, whereby all energy consumed, including for heating, transport and industrial processes, will in the end be electric. It would mean only electric, biofuel or hydrogen-driven cars, trucks, trains and planes, the closure of all nuclear and fossil-fuel power plants that are deployed to generate electricity or to provide heat. The latter would of course also remove the need to equip such plants with CCS installations, leaving only Direct Air Capture as a necessary means to reduce the concentration of carbon dioxide in the atmosphere. It should be noted that a 100% renewable-energy system is different from a net-zero system as the latter does not necessarily require the total phase-out of fossil fuels. Most 100%-renewable scenarios foresee the use of large quantities of biomass, which is falsely considered a renewable resource. The most recent IPCC report (AR6 published in February 2022) states that it requires large areas of land which can conflict with the need to produce food and protect biodiversity. Such a system does not even have to be net zero, as only large biomass plants are generally equipped with CCS (BECCS) and the small ones are allowed to continue to emit CO_2 (and other harmful substances).

It is neither sufficient to show in an abstract calculation that the renewable sources mentioned are capable of generating enough energy, i.e., enough kWh, by simply adding up the output of all wind turbines and solar panels. That creates energy but not an energy system. Electric energy cannot be used for everything. It is for instance hard to make steel from ore with electricity. So, steelmakers are looking at green hydrogen or electrochemistry to reduce iron oxides to iron instead of the cokes used in traditional processes. Renewable electricity must then first be used to produce hydrogen by electrolyzing water. So, alongside the wind turbines and solar panels a hydrogen infrastructure has to be created. Another example is home heating which is currently often done by burning natural gas in a furnace at home, supplied by gas networks covering the entire country. Burning 3,000 m³ of natural gas to keep your home warm in winter is equivalent to using about 30,000 kWh in electricity if you would extract it straight from the grid. That would be very expensive, so the solution might be to use one or several air source heat pumps as intermediaries, which might reduce electricity demand by a factor of three or more, depending on the coefficient of performance (COP) of your air source heat pump. This electricity can then be extracted from the grid or generated (in summer) by solar panels on your roof.

In such a system you are supplying solar electricity to the grid when there is little demand and withdraw electricity from the grid in winter when there is little supply from solar panels. Electricity grids have to be greatly extended to guarantee that the generated power can be transmitted whenever and wherever power is consumed, while the gas networks will become redundant. Now and then on very sunny summer days the news media proudly announce that more electricity has been generated from solar panels than used; a mixed blessing as it implies that the generated power has essentially become worthless, causing the price of electricity to drop below zero, a consequence of the intermittency of solar power and the lack of affordable storage. It shows that the hard part of a carbon-free energy system is not making carbon-free electricity, but matching the electricity generated to demand every minute, of every day, of every season.

To avoid such undesirable situations requires an energy infrastructure that works rather differently from the current system which balances supply and demand almost instantly. As the amounts of electricity will be vastly bigger than in the current system, the capacity of the grid needs to be greatly expanded. Switching to a 100% renewable-energy system does not merely involve building large numbers of wind turbines and laying acres and acres of solar panels but demands much more. As Vaclav Smil notes there is a fundamental difference between systems that derive 20–40% of electricity from renewable intermittent sources and a national electricity supply that relies completely on them. The latter would require either mass-scale long-term electricity storage to back up for the intermittency or extensive grids of high-voltage lines to transmit electricity across time zones and from sunny and windy regions to urban and industrial concentrations. (Smil 2022, p. 31.) We have neither the storage capacity yet nor the grids; this still apart from the question whether these renewable sources could produce enough electricity to replace all electricity for current and future uses.

Current status of renewable energy generation

In spite of all the hullabaloo about wind and solar, their current contribution to the electricity sector is actually rather disappointing. According to the IEA Global Energy Review 2021 (IEA 2021b, p. 23), the share of low-carbon sources in global electricity generation has risen from 25.7% in 1971 to 39.5% in 2021, just 14% in 50 years, while total electricity demand in that period has roughly increased by a factor of 3. The largest contribution comes from hydropower, while the share of wind and solar, the renewables that are mainly at issue here, was just 10.1% in 2021, about the same as the share of nuclear, as can be seen in Fig. 3.1. The contribution from wind grew from 1.6% in 2010 to 6.7% in 2021, and from solar from 0.1% in 2010 to 3.4% in 2021; sizeable increases but total output is still rather small. As a comparison, the output from coal was 35.4% in 2021, compared to 40% in 1971.

The Global Status Report 2021 on Renewables, which covers the entire energy sector, not only electricity generation, sums up the situation as follows: "As in past years, the highest share of renewable energy use was in the electricity sector (26% renewables); however, electrical end uses accounted for only 17% of total final energy consumption (TFEC). The transport sector, meanwhile, accounted for an estimated

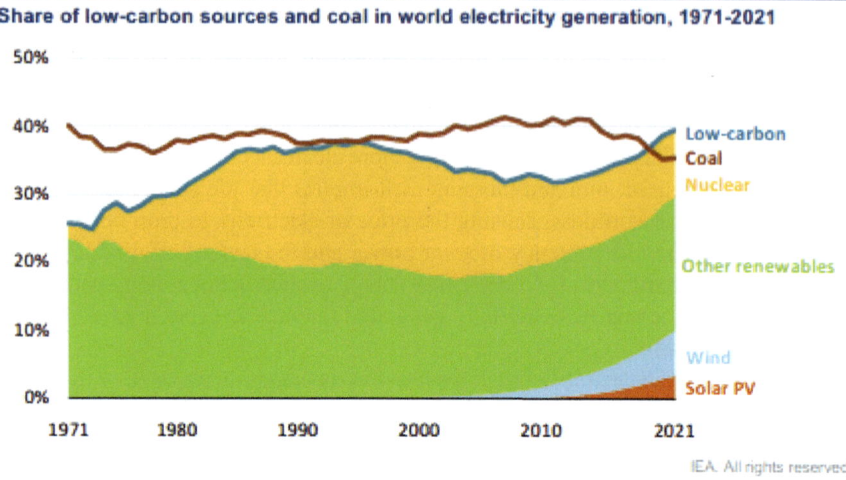

Fig. 3.1. Share of low-carbon sources and coal in world electricity generation, 1971–2021; other renewables are mostly hydropower and biomass (IEA 2021b, p. 23).

32% of TFEC and had the lowest share of renewables (3.3%). The remaining thermal energy uses, which include space and water heating, space cooling, and industrial process heat, represented more than half (51%) of TFEC; of this, renewables supplied some 11%. As of 2019, modern renewable energy (excluding the traditional use of biomass) accounted for an estimated 11.2% of TFEC, up from 8.7% a decade earlier. Despite tremendous growth in some renewable energy sectors, the share of renewables has increased only moderately each year. This is due to rising global energy demand, continuing consumption of and investment in new fossil fuels, and declining traditional use of biomass (which has led to a shift towards fossil fuels)" (REN21 2021, p. 15).

This picture is not encouraging, in spite of the record increases in energy from wind and solar, and not what one expects from a rapid movement towards a 100% renewable-energy economy. There is very little good news for renewables, with only a dozen or so countries producing over 15% of their electricity from solar PV and wind power, and the contribution of solar and wind to total global energy use just growing by 2.5% in the last decade. Total and utter failure! At this rate we will not even get there by the end of the century, let alone by 2050; the contribution from low-carbon sources can only just keep up with the increase in electricity demand. It raises serious doubts about the feasibility of a 100% renewable-energy economy. We even don't seem to get it off the ground soon enough for the electricity sector.

Intermittency

In a 100% renewable-energy system all energy needed for any purpose is supplied from renewable sources, some of which, in particular wind and solar, are variable, or intermittent as it is often called. Variable means that they can't be dispatched or turned on and off at will. This contrary to fossil-fuelled power plants which provide enough flexibility for balancing the grid's load, as their output is controllable. The output of renewables is dictated by the availability of the natural resource (wind and sun) and it cannot be increased or decreased to meet demand. When generated, the energy must be absorbed into the supply, transmitted along the line, or stored for later use. If not, it's wasted. The consequence is that, as renewables replace energy generation from fossil fuels, the grid loses flexibility and the challenge of balancing energy supply and demand intensifies.

Renewable sources have their variability in common with energy demand, but the two variabilities are not in step with each other. Wind fluctuations in a single country can span the range from 5% to 80% of installed capacity. To solve this intermittency we need to realize large-scale affordable storage of the surplus electricity generated from wind turbines and solar panels when plenty of the resource is available, i.e., when the sun shines and/or the wind blows. Storage alone is however not sufficient. Storage systems have to be able to store enough energy to last through the periods that no energy is generated, but also have to be able to deliver that energy fast enough to meet electricity demand. Once you know both the energy storage capacity (e.g., in megawatt-hours) and the output power (e.g., in megawatts), you can simply divide these numbers to find out how long the backup power will last.

The amount of storage needed depends on the penetration of renewables in the energy system. No energy storage is needed for penetrations lower than about 25%; the renewable energy is just added to the grid which is mainly supplied by power stations that can be turned down or off when demand is low. For penetrations up to about 80%, a relatively small storage capacity is needed. When the penetration of renewables approaches 100%, a very large increase in storage capacity is required. For the UK the situation has been analyzed in a recent paper (Cárdenas et al. 2021a)[5] and the wind/solar mix that minimizes the size of the storage required for a 100% overall renewable penetration is 84% wind + 16% solar. In that case a storage capacity of 43.2 TWh is needed (about 15% of total annual electricity production). This must be considered, at any rate according to the study, the storage capacity the UK needs to decarbonize its electricity supply. An investment of about £165.3 billion (7% of GDP) is quoted as being needed for this based on current costs of bulk energy storage technologies. That is a lot of money and, although storage technology will undoubtedly improve, it seems a far cheaper and easier option to fit a few of the existing gas-fired power stations with CCS and keep them in operation to generate the electricity needed to smooth out the erratic supply from renewable sources.

This storage problem is essentially the great unsolved problem of the energy transition, apart from the fact that solar and wind come on line much too slowly, which in turn may actually be due to the fact that the storage problem remains unsolved. I am confident that this problem will be solved in future but the solution may still have to wait a few decades.

There are a few other strategies to provide flexibility to the grid, such as interconnecting different grids, demand-side management, and supply response, which however are expected to play a smaller role than storage.

---

[5]   Remarkable is that in the paper it is assumed as a fact that a 100% renewable-based electric supply can be achieved.

Review of 100% renewable-energy systems

According to a 2019 review (Hansen et al. 2019) of 181 scientific studies of 100% renewable-energy (RE) systems published in peer-reviewed journals from 2004 to 2018, the "great majority of all publications highlights the technical feasibility and economic viability of 100% RE systems." It is however not clear from the review what this means and whether the definitions of feasibility and viability in all these studies are similar, and what they entail. For the UK for instance a 100% renewable-electricity grid, i.e., not yet a 100% renewable-energy system for the entire economy, is feasible but at high cost (Cárdenas et al. 2021a, b). The majority of the reviewed studies finds that 100% RE is possible from a technical perspective, which of course is not such a great deal, with only few publications arguing against this. Many publications focus on the electricity sector only and conclude that 100% renewable energy is possible within this sector, but that it is an altogether different kettle of fish when other sectors, such as transport and heating, must also be electrified, while other studies find that it is technically achievable for all sectors in a long-term perspective. More than half of all articles cover Europe, followed by the US and Australia. This while the really big problems in achieving decarbonization by the middle of this century lie not with these countries but with the developing world; China, India and the still considerable part of the world that does not have (electric) energy available at the simple turn of a switch. If it is hard for Europe, i.e., the EU, to duly decarbonize the energy sector, it will be much harder still and perhaps impossible for the developing world.

Few studies investigate the transport sector and, in even rarer cases, all transport modes are considered (road, rail, marine, aviation). Especially rail, when electric, is not that hard to decarbonize. Moreover, the heating and cooling sectors gain little attention despite the fact that these sectors in some cases exceed the final energy demands of the electricity sector.

People who advocate the possibility (which as said is easy), feasibility (less easy) or viability (perhaps impossible) of 100% renewable energy by a date not too far in the future, and at any rate no later than by 2050, are of course the darlings of the environmentalists, and with good reason for they give us the prospect of a new Garden of Eden. We have a predilection for being lured by rosy prospects and so-called breakthroughs, and many scientists are happy to provide them, or present their results as such, as it gives them the attention they are craving for and further money to march on to further breakthroughs, even if it amounts to publishing science with serious shortcomings by hiding, in obscure wording (scientific jargon) or otherwise, the 'ifs' that have to be fulfilled.

A system of completely carbon-free energy sources is of course preferable over a patched-up solution like carbon capture and storage. Unfortunately we do not live in a world in which this seems possible in the short term. Carbon capture and storage would make it possible to keep a number of fossil-fuelled power plants, equipped with CCS, in operation, if only as backup, and ensure that the transition to a decarbonized economy can be carried out smoothly and in a cost-effective manner.[6] To ask for 100% renewable energy on a global scale as early as 2030 is clearly impossible and even 2050 does not seem to be feasible. There may be a few individual countries endowed with large hydropower or geothermal capability where this might be possible, but not on a global scale. Sweden perhaps already in 2030 (if it also keeps its nuclear power capacity),[7] or certain regions, like the EU by 2050, if we manage to move some more pollution to the third world and exclude some sectors, like aviation, although I believe it will already be an astounding feat if the EU manages to achieve net zero by 2050, and there is no need for it to go further. Sweden was one of the only nine EU countries, together comprising 50 million people, a little over 10% of total EU population, that generated more than half of their electricity (so just electricity, not total energy) from renewable sources in 2022, with Luxembourg and Denmark in top spot with over 80%. It is not realistic to assume that in particular the developing world, but also more developed countries, like China, will switch to a 100% renewable-energy system for generating all energy, including manufacturing processes, heating and transport by 2050.

Many bold claims have been made about the opportunities of renewable-energy sources. An example is Fthenakis et al. (2009) who as long ago as 2009 analyzed the technical, geographical, and economic feasibility for solar energy to supply the energy needs of the US and concluded that "it is clearly feasible to replace the present fossil fuel energy infrastructure in the US with solar power and other renewables, and reduce CO_2 emissions to a level commensurate with the most aggressive climate-change goals." Further claims are that "[b]ased on expected improvements of established, commercially available PV (*photovoltaic solar panels*), CSP (*concentrated solar power*), and CAES[8] (*Compressed-Air Energy Storage*) technologies, we show that solar energy has the technical, geographical, and economic potential to supply 69% of the total electricity needs and 35% of the total (electricity and fuel) energy needs of the US by 2050. When we extend our scenario to 2100, solar energy supplies over 90%, and together with other renewables, 100% of the total US energy demand with a corresponding 92% reduction in energy-related carbon dioxide emissions compared to the 2005 levels." This is quite a momentous conclusion and there are many of those. Their analysis concerns the US, is mainly concerned with the cost price of electricity from solar, sees a relatively small role for wind energy, which today is considered the most important renewable resource, and wants especially CAES to solve the intermittency problem. The paper is at variance with the UK studies referred to above. I am certain that if they would write the paper today, the conclusion might perhaps look similar but with solar and wind having swopped places.

---

[6] It should be noted though that powering such power plants up and down in line with the lulls in solar and wind power generation will reduce their lifetimes caused by material stress and creep from thermal shock. It will also produce more carbon dioxide (which however will of course be captured).

[7] Its only problem is oil, responsible in 2021 for 23.5% of energy consumption. So, the Swedes must be forced to exchange their petrol car for an electric one. In 2022 Sweden generated 68.4% of its electricity from renewable sources.

[8] See Glossary for more details about CAES and its feasibility.

Feasibility and viability of proposed 100% renewable systems

To discuss all this in greater detail requires first of all a definition of (technical) feasibility and (economic, social, political) viability. Here I will follow a paper by Heard et al. (2017) which reviewed the feasibility of 100% renewable-electricity systems in a thorough and careful manner. It is one of the few studies mentioned in the aforementioned 2019 review on 100% renewable energy systems that argued against the technical feasibility of a 100% renewable energy system. They defined feasibility as meaning that it is technically possible to construct such a 100% renewable energy system and that it works. A demonstration of feasibility requires that evidence is presented for a proposed system to work with current or near-current technology at a specified reliability. It does not yet mean that it is also viable in an economic and/ or social sense, which would mean that it is possible within environmental and social constraints and at reasonable cost.

In the following I quote freely from this paper, whose abstract states: "While many modelled scenarios have been published claiming to show that a 100% renewable-electricity system is achievable, there is no empirical or historical evidence that demonstrates that such systems are in fact feasible. Of the studies published to date, 24 have forecast regional, national or global energy requirements at sufficient detail to be considered potentially credible." The studies are evaluated against four objective feasibility criteria for reliable electricity systems needed to meet electricity demand this century. These criteria are: (1) consistency with mainstream energy-demand forecasts; (2) simulating supply to meet demand reliably at hourly, half-hourly, and five-minute timescales, with resilience to extreme climate events; (3) identifying necessary transmission and distribution requirements; and (4) maintaining the provision of essential ancillary services. Ancillary services include for instance frequency regulation and voltage control, which keep the grid stable and have typically been supplied by fossil-fuel power plants. These four criteria are not particularly earthshattering and are also fulfilled by the energy system that is currently in place. The authors were concerned with evidence for the strict technical feasibility of proposed 100%-renewable *electricity* systems only, i.e., that it will work, but only as regards the electricity sector. The striking outcome of the study is that none of the 24 studies provides convincing evidence that these basic feasibility criteria can be met, which simply implies that the energy systems proposed in these studies are probably inferior to the one we have now. The authors rightly demand that strong empirical evidence of feasibility must be demonstrated for any study that attempts to construct or model a low-carbon energy future based on any combination of low-carbon technology.

The only developed nation today with electricity from 100% renewable sources is Iceland, thanks to a unique endowment of geothermal aquifers, abundant hydropower, and a population of only 387,000 people. It has just 1.8 MW of operational wind-power capacity and solar power is not really an option due to the low solar irradiation on the island. There are countries that are close to generating 100% renewable electricity, e.g., Paraguay (99%), Norway (97%), Uruguay (95%), Costa Rica (93%), Brazil (76%) and Canada (62%). Almost all of the electricity in these countries is hydroelectricity, which is renewable for sure, but its presence in these countries has nothing or not much to do with the energy transition we are currently faced with. There is virtually no contribution from solar and wind which must be the motor for the energy transition in the greater part of the world. None of these nations can be put forward as evidence that an energy transition based on renewables is possible, as done in a critique of the study under discussion here (Brown et al. 2018).[9] European nations like Denmark and Germany which have made considerable progress towards such a transition and are lauded for their efforts in renewable energy deployment, i.e., solar and wind, are still far from a 100% renewable electricity system, as the following data bears out. According to the IEA,[10] the share of oil, gas and coal in Denmark's total energy production in 2020 was 56.7% (37.7, 14, and 5% respectively), with renewables providing the rest of which 32.8% came from biofuels and waste and a rather disappointing 10.7% from wind and solar. The electricity sector fares much better with more than 80% generated from biofuels en waste (23%), and wind and solar (60%), with solar providing only 4%, the main share coming from wind. In the same year 2020, Germany[11] generated just 6% of its energy from wind and solar, 11.2% from biofuels and waste, 0.6% from hydro, 6% from nuclear and a surprisingly large 76.2% from fossil fuels. Electricity is generated for 43.6% from fossil fuels; for 56.5% from renewables, including nuclear, with wind (22.6%), solar (8.7%) and hydro (4.3%) accounting for 35.6%, waste and biofuels for 9.8%. No wonder that Germany is still emitting more CO_2 per capita (7.69 ton) than the EU average (5.84 ton), and Denmark at 4.52 ton/per capita not that much less. In contrast, for China the numbers for 2019 are that 87.5% of total energy production comes from fossil fuels, mostly coal (61.1%), just 10% more than the German contribution from fossil fuels, 2.7% from nuclear and 9.7% from wind and solar (2.8%), biofuels and waste (3.7%) and hydro (3.2%). With 7.41 tons in CO_2 emissions per capita in 2020 a Chinese emits less than a German![12] So, who should be lauded here? The only laudable aspect probably is that the emissions figure for China is likely to go up and the one for Germany down. A recent study from the Oxford Institute for Energy Studies (Dickel 2022) claims to show that for Germany an all-renewables, predominantly electric approach to achieving net zero by 2045 will not work, nor will it maintain reliable energy supply. In addition, the study concludes that the inclusion of CCS for power plants and blue hydrogen[13] are essential for achieving Germany's decarbonization goals. And all this before the war in Ukraine threw a spanner into the works.

Two of the criteria mentioned above, (1) and (2), are that electricity demand must be projected realistically over the future time interval of interest, and that electricity supply is reliable, close to 100% as is common in most advanced countries. All reliable projections for future energy use, in line with the trend of the last fifty years in which energy use by mankind has grown by an average 2.4% per year, entail an increase in electricity demand due to an increasing global population and greater

---

[9] The paper makes the further mistake that it equates a 100% renewable-energy system with a net-zero system.

[10] https://www.iea.org/countries/denmark.

[11] https://www.iea.org/countries/germany.

[12] https://www.iea.org/countries/china.

[13] See Glossary for the meaning of blue hydrogen.

parts of this population being connected to electricity networks, e.g., the close to one billion people who currently have no access to electricity at all. Although it is clear that we cannot continue for much longer on the path of ever continuing growth, it is unrealistic to assume that countries in Africa and Asia with rapidly growing populations and at present a very low level of energy usage will consume less energy by 2050 or so, as some proponents of 100% renewable energy such as Jacobson and co-workers (Jacobson and Delucchi 2011; Jacobson et al. 2017a) do. Below we will say more about their work, which has attracted a considerable amount of attention, but now just note that for Nigeria, for instance, a country that emits the puny amount of 0.64 ton of CO_2 per capita[14] and has no need at all to switch to a 100% renewable-energy system, their scenario requires a reduction in energy demand of almost 40% by 2050 compared to the current level, for India and Indonesia even more than 40%. This while energy use in these countries will have to vastly increase to lift the population out of poverty. As Heard et al. remark: "any future global scenario that presents static or reduced demand in either primary energy or electricity is unrealistic, and is inconsistent with almost all other future energy projections. Such an outcome would be at odds with the increase in global population, ongoing economic development for the non-OECD majority, and the firmly established link between industrialization and increased energy consumption" (Heard et al. 2017).

In the critique by Brown et al. mentioned above it has been rightly noted that primary energy is not the right quantity to use here. Secondary energy, the energy that is actually used, is about 30% less than primary energy due to conversion losses. When switching from fossil fuels to renewables primary energy consumption goes automatically down as there are no or less conversion losses. In converting fossil fuels to electricity more than 50% is lost, which is not the case for renewables. This is however only the case when fossil fuels are used for generating electricity. For home heating by burning natural gas in individual furnaces there is very little to no waste heat; to replace this type of heating by electricity is more wasteful, as in electricity transport 15% is lost per thousand kilometres of power transmission line. To resolve the intermittency problem, part of the energy from renewables must be stored which also involves considerable losses. For instance, the CAES technology mentioned above stores electricity in compressed air which can then later again be reconverted into electricity with a loss of about 40–50%. A second point is that renewables-based systems avoid the significant energy use for mining, transporting and refining fossil fuels. The latter is only partly true as renewables require more materials and these must be mined. A third argument that living standards can be maintained while increasing energy efficiency is not convincing either as it equally applies to energy derived from fossil energy and living standards in the developing world must actually be increased, maintained is not sufficient. Finally, most 100% renewable-energy systems rely to a considerable extent on biomass for which the primary energy argument still holds. In generating energy, biomass does not differ from fossil fuels. If looked at more carefully, it is actually worse.

---

[14] Our World in Data, 2021 figure. Energy use in Nigeria is only 2,500 kWh per capita, while for the US it is 76,600 kWh.

Reliability

The second important point is that an electrical power system must provide reliable electricity to its customers as economically as possible. Such reliability depends on both adequacy and security. Adequacy refers to sufficient generation of electricity by the system to satisfy demand at any time, and security concerns the ability of the system to adequately respond to multiple types of disturbance in the quality of power supply. These concepts together define a reliability standard. High reliability (> 99.9% of customer demand being met in a year) is a common requirement of modern electricity supply; outages occur but are very rare. The reliance on variable, climate-dependent sources of electricity generation, such as wind and solar generation, provides additional challenges for managing system reliability. The wind turbines and solar panels may have high reliability in terms of being in working order, yet they have low and intermittent availability of the resource itself, i.e., the wind does not always blow or too hard (especially relevant for offshore wind parks) and the sun does not always shine (and if it does we don't need to switch on the lights). These are the elephants in the room which some people refuse to see by just concentrating on the positive side of the story and on a desired outcome for which applause can be earned. Any proposed supply system must demonstrate that the supply will meet any foreseeable demand in real time at a very high reliability standard and with a sufficient reserve margin for unscheduled outages like breakdowns. In addition, transmission networks to transport electricity from generators to distribution networks must be shown to be reliable and critical ancillary services must be in place to ensure power quality and the reliable operation of the network.

Based on these criteria, none of the 100% renewable-electricity studies examined provided a convincing demonstration of feasibility. Most studies were found to be seriously flawed.

A few scenarios attempted to maintain final energy demand at values consistent with mainstream projections. These scenarios assumed a transition of up to 100% of total energy from either direct electrification or electrolytic hydrogen production, with reliance on flexibility of demand and/or widespread storage of energy using a range of technologies (most of which—beyond pumped hydro—are unproven at large scales, either technologically and/or economically).

The importance of system simulation

According to Heard et al. (2017), whole-system simulations are vital. After all, there must be absolute certainty that the new system will work before embarking on introducing it. The absence of such simulations from nine of the reviewed studies suggests that many authors and organizations have either not grasped or not explicitly tackled the challenge of ensuring reliable supply from variable sources. For example, in the WWF scenario (Singer 2011) it is assumed that by 2050 the share of energy from variable renewable sources could increase to 60% via all of the following: (i) grid-capacity improvements, (ii) demand-side management, (iii) storage, and (iv) conversion of energy surpluses into storable hydrogen. This series of assumptions for managing a system dominated by supply-driven sources is largely repeated in the Greenpeace scenario (Teske et al. 2015). In neither case evidence from system simulation is provided for how this might work.

The system-simulation approaches applied so far mostly cannot demonstrate the feasibility and reliability of 100% renewable-energy systems. The only study that simulated below half-hourly[15] reliability carried out by Jacobson et al. in 2015 (Jacobson et al. 2015) offers a system simulation for the continental United States. The results show a perfect match between supply and demand based on a renewable-energy scenario that assumed (i) expansion in the use of thermal stored energy, (ii) total electrification of the United States' whole-of-economy energy needs, (iii) nation-wide dependence on underground thermal-energy storage for space and water heating based on a system that has not yet been commissioned, and (iv) flexibility in demand ranging from 50% to 95% across different energy sectors, including some industrial applications. It claims that no natural gas, biofuels, nuclear power, or stationary batteries are needed, but according to Heard et al. depends strongly on extraordinary assumptions relating to electrification, energy storage and flexibility in demand. As such, the scenario is unrealistic, violating the first criterion formulated above, viz. consistency with mainstream energy-demand forecasts.

For countries like New Zealand, with large endowments of hydro and geothermal resources and a small population (4.5 million people), a 100% renewable-electricity system might be possible at reasonable cost. Other studies reinforce the notion that integration of variable renewable energy sources into existing grids can be cost-effective up to penetrations of around 20%, after which integration costs escalate rapidly. If not much hydropower is available, alternative dispatchable generation supplies are required to meet the balance of supply, and of course if you allow these to come from biomass or fossil fuels, as Brown et al. (2018) apparently do, any eventuality can be met, provided you keep these biomass or fossil-fuel plants ready to start at all times. In addition to feasibility issues, the heavy reliance on exploitation of hydroelectricity and biomass raises concerns regarding environmental sustainability and social justice. Other studies, including the WWF and Greenpeace scenarios mentioned above, found biomass to be essential to ensure reliability of 100%-renewable scenarios, supplying from as little as 2% to as high as 70% of the electricity from biomass. The conclusion of the review is that efforts to date seem to have substantially underestimated the challenge and delayed the identification and implementation of effective and comprehensive decarbonization pathways.

In view of the enormous discrepancies in the literature, Heard et al. were perfectly right in raising concerns about the 100%-renewable scenarios presented. The widespread assumptions of deep cuts in (primary) energy consumption defy historical experience, are generally inconsistent with realistic projections, and would likely raise problems for developing countries in meeting goals of poverty alleviation, as we noted above. Economic growth and poverty reduction in developing countries is crucially dependent on energy availability. A reduction in (primary) energy use is an unlikely pathway to achieve these humanitarian goals.

---

[15] Brown et al. (2018) argued that for most scenarios and areas hourly simulations would suffice to deal with short-term weather fluctuations, where Heard et al. argued that nine studies had no simulations at all, and only five studies demonstrated sub-hour reliability, which is clearly insufficient.

In applying so many assumptions to deliver changes far beyond historical precedents, the failure in any or several of these assumptions regarding energy efficiency, electrification or flexible load would nullify the proposed supply system. As such, these systems present a fragile pathway, being conceived to power scenarios that do not exist and likely never will. The evidence from these studies for the proposition of 100% renewable electricity must therefore be heavily discounted, modified or discarded.

Storage

A common assumption is that advances in storage technologies will resolve issues of reliability both at sub-hourly timescales and in situations of low seasonal availability of renewable resources. Storage of sufficient power to cover a 3 day lull in wind power is not possible with existing technology, at least in countries with large wind capacity and low hydro capacity.[16] Robert Hargraves (Hargraves 2021) calculates that storage to cover 1 day of global energy use requires 1 cubic mile of batteries, or 36 billion Tesla Powerwalls at 13.5 kWh each, which have to be built at a rate of 1,000 per second for 10 years at a total cost of $250 trillion (about $7,000 a piece, i.e., $518 per kWh of storage). It is true that battery prices are dropping but still far too little to be affordable.[17] The lifespan of batteries is an additional limiting factor. It is reasonable to assume that in future a greater range of cost-effective energy-storage options will be available. Such solutions will undoubtedly assist in achieving reliability standards in systems with greater penetration of variable renewable generation. However, whether such breakthroughs will enable the (as yet unknown) scale of storage required for 100% renewable remains unknown. To bet the future on such breakthroughs is arguably risky and policy makers must realize that dependence on storage is entirely an artefact of deliberately constraining the options for dispatchable low-carbon generation, for there is no reasonable argument against using some fossil-fuel powered plants with CCS as backup for an energy system made up for a large part of renewable sources.

In this connection books like the one by Hargraves mentioned above and another one by Saul Griffith (Griffith 2021) argue in favour of electrifying everything. Although they are basically right and I agree that we must electrify as much as possible of our energy supply, they have not been able to convince me if only for the fact that they mainly focus on the US, which is not a major player in climate mitigation. That does not stop Griffith from painting a rosy future for a decarbonized America, after which the world will follow while part of the world is actually ahead of America, although admitting that it will not be easy but also insisting that it is doable without giving much evidence for the latter. Of course we must electrify as much as possible, I cannot agree more but we cannot rely too much on the magic wand of innovation to solve the fundamental problems that still exist with this approach. As Bill McKibben says "it's time to think with special clarity about the future" (McKibben 2011, p. 48).

---

[16] https://www.climate-and-hope.net/electricity-generation/reducing-carbon-emissions-with-wind-power?c=analysis-conclusions.

[17] See also https://www.eia.gov/analysis/studies/electricity/batterystorage/pdf/battery_storage_2021.pdf.

Instead of providing this clarity, Griffith's book is full of wishful thinking. Making the statement that "it's important that manufacturers double or triple battery cycle life, which will make the storage cost for each cycle mere pennies per kWh" is not enough to make it happen, no matter how desirable this may be. Or, "the energy game will change forever when the combined cost of rooftop solar and battery storage can beat the cost of the current grid." There is no need to write such truisms down in a book or anywhere else for that matter. Knowing that the global total of all electric vehicles in 2020 is 10 million of which only 1.78 million in the US, it is equally silly to write that "if all of America's 250 million vehicles were electrified, they would have the capacity of about 20 terawatt-hours (TWh) [of storage], enough by themselves to balance out the daily fluctuations of our new electrified world." The realization of this 'if' requires very audacious wishful thinking indeed. It is more likely that there never will be 250 million electric vehicles in America. At least I hope not. Electric vehicles are not going to make our world sustainable, let alone regenerative as some want. If people resort to statements as Griffith does, everyone of us can solve any problem within minutes. It says enough that currently more than 90% of our grid-connected storage is still pumped hydro, which involves the primitive practice of pumping water uphill (during periods of low-cost surplus off-peak electric power) to let it run back down later, when electricity prices are higher. Not particularly a technology to be proud of in the twenty-first century. But as the saying goes, beggars can't be choosers. Other so-called gravity batteries working on the same primitive principle are currently being developed and involve hoisting massive metal or concrete blocks upwards before gradually releasing them back to earth to power a series of electric generators. The Swiss company Energy Vault managed to lure gullible investors in throwing away $100 million by investing in this concept, called idiotic by some.[18]

The biomass issue

Perhaps the most worrying finding relates to the strong dependence of 100%-renewable scenarios on biomass, which must be seen as the revival of another non-innovative practice, akin to the burning of wood practised by mankind before large-scale coal burning came along, so a step or several steps backwards. The UK scenario is typical in this respect; even when assuming a 54% reduction in primary energy consumption, the contribution from biomass requires 4.1 million hectares of land to be committed to the growing of grasses, short-rotation forestry and coppice crops (17% of UK land area). It will be hard to find this land. Another study (Lund and Mathiesen 2009) describes how Denmark would need to reorganize farming from wheat to corn to produce the requisite biomass, in a scenario of 53% reduction in primary energy consumption from the year of reference. For Ireland, a biomass requirement equal to 60% of the total potential biomass resource in the country was calculated (Connoly et al. 2011). For Australia too a similar completely unfeasible scenario was worked out (Turner et al. 2013). The WWF scenario, already mentioned above,

---

[18] https://www.wattisduurzaam.nl/20976/energie-opslaan/reservoirs/topinvesteerder-kiept-110-mln-in-idiote-toren-voor-energieopslag/; https://medium.com/the-future-is-electric/absurd-gravity-storage-non-solution-energy-vault-being-built-once-a8b8b52e31aa.

demands up to 250 million hectares, equal to about one sixth of global cropland, for biomass production for energy, along with another 4.5 billion m^3 of biomass from existing production forests to meet the demand in their scenario (also including a 15% reduction in energy demand by 2050 compared to 2005). This is clearly not feasible, which the authors of the WWF report also seem to realize when they add "what is possible on paper, even after the most rigorous analysis, is a different matter in practice. We have yet to identify where this land is, and how it is being used at the moment" (Singer 2011, p. 61). One would have thought them to have done this before forcing a report down our throats.

A 100%-renewable scenario

As an illustration, I will discuss here one 100% renewable-energy study that has obtained a fair amount of attention and controversy, conducted under the leadership of Mark Z. Jacobson of Stanford University. Jacobson is also one of the founders of The Solutions Project,[19] an initiative that aims to show how every state in the USA, and indeed almost every country in the world, can transition to 100%-renewable energy (WWS: water, wind and solar). They have also managed to lure Naomi Klein into their camp who in her in many respects splendid book *This Changes Everything* takes it as a given that 100%-renewable energy is within our grasp, as proven by top-level research, by which she means research carried out by Jacobson and co-workers (e.g., Klein 2014, p. 452).

Since 2009 Jacobson and his co-worker Mark Delucchi (in the following J&D) have been proposing that the world should move to 100%-renewable energy, and just from WWS, *in all energy sectors*, i.e., not just electricity generation. Everything is electrified, including transportation and industry. They claim that with a few exceptions it can be achieved with currently commercially available technologies. Remarkable and laudable in their study is that they exclude the use of biomass globally, citing irreconcilable concerns relating to air pollution, land use and water use, which are indeed genuine concerns. This is in contrast to many other studies which, as noted above, feel that biomass is needed to meet the balance of supply.

In Part I of their study (Jacobson and Delucchi 2011) they "estimate that 3,800,000 5 MW wind turbines,[20] 49,000 300 MW concentrated solar plants, 40,000 300 MW solar PV power plants, 1.7 billion 3kW rooftop PV systems, 5350 100 MW geothermal power plants, 270 new 1300 MW hydroelectric power plants, 720,000 0.75 MW wave devices, and 490,000 1 MW tidal turbines can power a 2030 wind, water, and solar world that uses electricity and electrolytic hydrogen for all purposes. Such a WWS infrastructure should reduce world power demand by 30% and requires only about 0.41% and about 0.59% more of the world's land for footprint and spacing, respectively." As can be seen from the list of devices that need to be constructed, in addition to biomass, J&D also exclude both nuclear power and carbon capture.

---

[19]   https://thesolutionsproject.org/.
[20]   These alone would cost roughly $5 trillion in construction at $1.3 million per MW.

I am not going to try to refute any of the numbers they propose, I trust that they are all correct and that if you add up the energy generated by all these appliances there will be enough to power the entire world, although as said above that does not yet create an energy system.

But, let us just consider the 3,800,000 wind turbines with a nominal capacity[21] of 5 MW each that must be built before 2030. That is a lot of wind turbines with total capacity amounting to 19,000 GW. Actually, a 5 MW wind turbine is not a particularly big one, at any rate not for offshore ones, which nowadays are built up to 12 MW[22] (with a blade length up to 80 metres). So the numbers can even be improved upon and there is no doubt that offshore wind power offers tremendous potential. For onshore wind a 5 MW turbine is still a fairly big one (average capacity being 2.5–3 MW). But what are the facts about wind power? The J&D papers referred to here date from 2011 and in 2021 total global wind energy capacity amounted to 837 GW, up from 733 GW in 2020, while in 2011 this was still only 220 GW (IRENA 2021). In ten years time the capacity has increased by a respectable factor of 3.8. J&D demand 19,000 GW capacity by 2030, so we have to speed up by a factor of 25 to reach that goal. In the next ten years wind turbine capacity must increase annually by 1816 GW, more than twice total current global capacity. Compare this to newly installed wind capacity of 93 GW in 2020, a record year for wind energy and up from 63 GW in 2019,[23] but still a factor of 20 removed from what J&D demand. In 2021 construction slowed down and added capacity went up only slightly to 93.6 GW.[24] J&D would probably counter that no all-out effort is being made and that their suggestion requires such an all-out effort. This is certainly true, but at the same time it shows that their analysis is rather futile, a purely theoretical exercise with little relevance to the real world.

Investment in offshore wind alone reached $303 billion in 2020, where they would apparently want an *annual* investment of $6,000 billion, more than the combined economies of Germany and the UK! It is clear that what J&D want is more than one bridge too far, completely unfeasible, and this only concerns the story for wind turbines. As Clack et al. noted: "Given unlimited resources to build variable energy production facilities, while expanding the transmission grid and accompanying energy storage capacity enormously, one would eventually be able to meet any conceivable load" (Clack et al. 2017, p. 6722), but as said such schemes are no more than a rather futile theoretical exercise.

---

[21] Capacity should be distinguished from generated power; a wind turbine with a nominal capacity of 5 MW could in principle when working full time throughout the year generate $5 \times 24 \times 365 =$ 43,800 MWh. In actual fact the capacity factor of a wind turbine is less than 50%, implying that a 5 MW wind turbine will generate less than half this number. Fossil fuelled plants have capacity factors of 80–90%.

[22] At 12 MW GE's Haliade-X claims to be the most powerful offshore wind turbine in the world today; the Chinese company MingYang Smart Energy is developing the 16 MW MySE 16.0–242, which will provide 80 GWh per year very close to the theoretical Betz limit.

[23] GWEC Global Wind Report 2021, https://gwec.net/global-wind-report-2021/, 23,000 wind turbines were constricted in 2019.

[24] GWEC Global Wind Report 2022.

Building castles in the air can be quite enjoyable and seems to be a favourite pastime of many climate scientists, but such castles by definition lack a solid foundation and I agree with other critics of J&D's study that "our energy system is the backbone of our economy. We cannot accept an unreliable or unaffordable energy system. It is our belief that the analysis presented by Jacobson et al. substantially underestimates the costs and consequences of a transition to solely wind, water, and solar power by 2030 and in doing so provides a misleading assessment that is counterproductive for guiding sound, rational energy policy" (Gilbraith et al. 2013, p. 69).

In 2017 they more or less repeated their message (Jacobson et al. 2017a), providing a detailed and admittedly impressive analysis for 139 countries in the world, at the same time shifting the date of 100% renewable energy to 2050, as it became clear I suppose that 2030 as target date was not realistic. Later this was extended to 143 (Jacobson et al. 2019) and finally in 2022 to 145 countries (Jacobson et al. 2022). The revised roadmap now envisages 80% conversion by 2030 (ideally) and 100% by 2050. One of the things I wish to point out here is that, while the lifetime of a fossil-fuel power plant is 40–50 years, lifetimes of wind turbines, heat pumps and solar panels are less than or at most half, so during the time of the transition which J&D envisage to take place to 2050 these devices must continually be replaced; few of the devices installed before 2030 will still work properly by 2050. This will lead to recurring costs and create mountains of waste.

In the latest paper global energy demand will drop by 56.4% compared to the situation with continued reliance on fossil fuels.[25] This number for the drop in global energy demand is rather unstable, it seems, as in earlier papers they quote 30%, as said above, and 42.5% (Jacobson et al. 2017a). As already mentioned, such drop in energy demand is at variance with most projections for future energy demand. This decline has to come from three sources: (1) for 38.4% from the higher energy-to-work conversion efficiency of using electricity for heating, heat pumps, and electric motors, and using electrolytic hydrogen in hydrogen fuel cells for transportation, compared with using fossil fuels; (2) for 11.3% from the elimination of energy needed to mine, transport, and refine coal, oil, gas, biofuels, bioenergy, and uranium; and (3) from 6.64% from assumed additional energy-efficiency measures.

Whether this indeed can be achieved is not a priori obvious. For instance, in many countries homes and offices are heated by the beautiful and efficient system of individual gas-fired heaters, whereby a primary energy carrier is directly converted into heat with higher than 90% efficiency. When natural gas is phased out, air source heat pumps must become the main driver of home heating. Home heating will then become fully electric. Replacing the current system by heating with heat pumps, if at all possible, does not necessarily result in a saving in energy use, as the efficiency

---

[25] This they call BAU (Business as Usual). It is not clear what is actually meant by this scenario. Most energy predictions include some penetration of renewables in their business-as-usual scenarios, e.g., in 2021 the US Energy Information Administration envisaged energy demand to grow 47% by 2050, with renewables making up 27% of the 2050 global energy mix and liquid fuels 28% (a 36% increase in liquid fuel demand and a 165% increase by renewables from 2020 levels).

of heat pumps drops rapidly with temperature.[26] As far as electrolytic hydrogen is concerned, typical efficiencies of electrolysers lie between 75 and 80%. Further losses of between 5 and 35% result from the compression or cooling of hydrogen, to store and transport the gas in a proper way. As far as mining is concerned, the energy transition will involve a switch to a materials-intensive society and a lot of (environmentally harmful and energy-intensive) mining of materials, e.g., of nickel, cobalt and lithium for batteries in private cars. Oil flows, lithium does not.

In their 2022 update to 145 countries Jacobson et al. continue this unrealistic approach although an extensive list of uncertainties pertaining to their analysis is given. They still finish with the unrealistic conclusion that "Here, roadmaps to transition 145 countries to 100% clean, renewable WWS energy and storage across all energy sectors are developed. The full transition should occur no later than 2050, but ideally by 2035, with no less than 80% by 2030" (Jacobson et al. 2022, p. 3358).

Since I live in the Netherlands, that country has my special interest. J&D envisage that all energy needs of the country must be fulfilled by rooftop solar (3.4%), solar plants (43.1%), onshore wind (10.4%) and offshore wind (43.1%), hence just solar and wind; nothing else. The Solutions Project[27] claims that by improving energy efficiency and powering the grid with electricity from wind and sun overall energy demand reduces by a stupendous 79%.[28] At any rate that is one version. In the supplemental information to their 2017 paper a more realistic figure of 45% reduction is quoted, quite a difference. For both versions solar and wind are the only two energy sources and the division is roughly the same as given above.

But relying solely on wind and solar requires a gigantic transformation of the energy infrastructure and housing stock in the country at gigantic costs. According to the Dutch environmentalist group Urgenda, on average €35,000 per home will be needed for making it sustainable, to be spent on insulation, solar panels and air source heat pumps. This amount is probably too low,[29] but for the sake of the argument let us assume that it is approximately correct. For the roughly 7 million houses in the Netherlands that are not yet sustainable, this would require a total investment of €245 billion, more than 25% of the country's GDP. Saul Griffith in his *Electrify* book, referred to above, mentions first an amount of $70,000 (Griffith 2021, p. 111) and then later[30] $40,000 for doing in the US what the Urgenda scheme wants to do in the Netherlands, while in his 2023 book *No Miracles Needed* Jacobson himself, certainly

---

[26] The COP (coefficient of performance) quoted for air source heat pumps refers to heating water at temperatures to 35°C at an outdoor temperature above 7 degrees, while in temperate climates in homes equipped with radiators (instead of underfloor heating) temperatures to 60°C are needed when it is freezing outside. In such circumstances the efficiency of air source heat pumps rapidly diminishes and the COP can easily drop to 1 instead of the 4 advertised.

[27] https://thesolutionsproject.org/what-we-do/inspiring-action/why-clean-energy/#/map/countries/location/NLD.

[28] This can all be found at https://thesolutionsproject.org/why-clean-energy/. The value of 79% for the Netherlands comes from this website on which short summaries are published for all countries. In the extensive tables belonging to their paper in *Joule* it says 45%, I believe.

[29] In addition, it will not be possible to make all houses sustainable, as they are too old and cannot be insulated to the required standard for air source heat pumps to supply enough heat.

[30] On page 126 without explaining the difference; see also Somerville (2022).

not a man to overestimate such costs, estimates that between \$30,000 to \$115,000 are needed for the upfront costs to make a US home sustainable (Jacobson 2023, p. 108). With 140 million US homes this would require a gigantic amount of somewhere between 4,200 and 16,100 billion dollars!

Where a gas-fired heater with a lifetime of 15 to 20 years costs somewhere from €1500 to €3,000, solar panels and an air source heat pump easily cost from €10,000 to €30,000,[31] and such amounts will have to be forked out every 15-20 years, the lifetime of these devices!

Millions of houses, especially older and smaller ones in cities will not be suitable for this, even with maximum insulation, double glazing and suchlike. The noise of thousands of humming air source heat pumps will drive people crazy, if there actually is room to put them anywhere. In the infographic maps of results by country[32] J&D quote a total investment of \$249 billion for the Netherlands.[33] That is far too little, at least double will be needed. The amount quoted is just sufficient to make most of the existing housing stock sustainable and get them decoupled from the gas grid, if we accept Urgenda's assumptions. The overhaul of the electricity grid itself will require similar amounts to make it suitable for absorbing the electricity supplied from all the solar panels. This electricity is mostly generated in summer when demand is low and must be stored or used to produce hydrogen, for which a separate infrastructure must be built.

In the case of the Netherlands, 39% of all end-use energy in the J&D scenario is to come from utility solar panels, requiring 6.4% of the country's land area, which does not sound that much but amounts to more than 10% of all land now in use as farm land and is equal to 70% of total built-up area. It will not be easy to find that land, in addition to the land needed for the onshore wind turbines envisaged in the scheme. Such a scenario also requires a lot of storage if backup fossil-fuel plants are not allowed. How this should be realized is not discussed.

It also seems to be compulsory nowadays, no matter what kind of scheme you come up with, to create the magnificent vista of colossal numbers of jobs. J&D create no less than 52 million jobs worldwide, while only 27.7 million jobs get lost. It is never mentioned how much these jobs get paid and who pays for them but, if they get paid roughly the same as the people in the jobs that are lost, total expenditure for salaries will be double while a lot less energy is being delivered. Doesn't that make energy much more expensive?

Closing remarks

The criticism of Heard et al. of the kind of scenarios sketched here is very much to the point. Efficiency gains can certainly be made, but many analyses in this respect

---

[31] I just spent €13,050 on 20 430 Wp solar panels and €16,000 on a high-temperature Samsung 12kW air source heat pump. In the Dutch climate the solar panels will generate about 8,000 kWh per year, which should hopefully be sufficient to power the heat pump through the winter and keep my home comfortably warm.

[32] https://sites.google.com/stanford.edu/wws-roadmaps/home.

[33] Total investment for the US is supposed to be \$9 trillion (https://sites.google.com/stanford.edu/wws-roadmaps/home), while Griffith's number just for homes is already either \$5.6 or \$9.8 trillion.

overestimate technically achievable energy gains, intangible costs to consumers of adopting technologies that are not perfect substitutes of current equipment,[34] overlook cost decreases to energy supply technologies that make efficiency investments less profitable, and overlook new profit seeking practices and underestimate rebound effects (Jaccard 2005, p. 216). The latter involve the greater use of a more efficient appliance (e.g., a bigger TV-set since its energy use is only marginally higher than for a smaller one, or why not take two if they consume that little electricity), partly undoing the efficiency gain.

The value of energy is also reflected in the price. Lower energy prices encourage greater energy use. For instance, the steep rise in gas and electricity prices across Europe in 2022, due to the war in Ukraine, suddenly made it possible to lower heating levels in offices and homes, a trend that will no doubt continue and spread to other activities (car use perhaps) when this high price level persists.

J&D did not test their supply systems by doing simulations, instead they referenced other studies to assert that system reliability is possible. They did not apply simulation processes to their own, different proposed systems, nor did they address the uncertainties, challenges and limitations articulated in their supporting references or related critiques as pointed out by Heard et al. in the review discussed above. These and other errors in the methodologies of Jacobson and co-authors have been highlighted and severely criticized in the paper by Clack et al. (2017), already referred to above, who concluded that there are significant shortcomings in the analysis, that the work used invalid modelling tools, contained modelling errors, and made implausible and inadequately supported assumptions. Policy makers should treat with caution any visions of a rapid, reliable, and low-cost transition to entire energy systems that rely almost exclusively on wind, solar, and hydroelectric power. The dispute reached such a pitch that Jacobson filed, and later withdrew, a libel lawsuit against the publisher, an odd thing to do in a scientific dispute. The final result was that in 2020 the court ordered Jacobson to pay the legal fees of Clack and the publisher. J&D answered the criticism in a 2017 letter (Jacobson et al. 2017b) in which they not surprisingly claimed that the premise and all error claims by Clack et al. are wrong. In climate-change science it is very difficult to discern any willingness to learn from one another. You preferably only preach to the converted who loudly applaud your every utterance, no matter how silly. Their research was however also criticised by other authors as being insufficient to provide a reasonable defence of their claims (Gilbraith et al. 2013).

Conclusions

Of course it is always possible to envisage the construction of specific numbers of wind, solar and water facilities that can provide all the energy the world needs. That only requires a few simple division sums: take global energy consumption and divide by the output of a solar panel, wind turbine or whatever you fancy and the required

---

[34] Jaccard (2020) has several humorous stories to tell about experiences of his own and others with energy efficiency, showing that the notion that energy efficiency is easy and *profitable* is far from true.

number of such pieces of equipment rolls out. I put it here perhaps a little irreverently as it is of course not quite as simple as that, but I definitely get the impression when reading such papers that the authors often have little grasp of the complexity of a nation-wide or even regional energy system that absolutely must be reliable, i.e., *always* provide a reliable supply of energy. It should *reliably* meet instantaneous demand for electric power throughout the year, at peak demand, at night, and when the wind is not blowing. They do not seem to fully appreciate that it is far from trivial that we have such a system in place and that replacing such a system is not just a matter of putting up a lot of wind turbines and laying huge numbers of solar panels.

Remaining issues as regards feasibility lie in the largely ignored, yet essential requirements for expanded transmission and enhanced distribution systems (grid extensions and the like, already now an issue in many countries where grids already receive too much electricity from rooftop solar installations), both to transport electricity from more sources over greater distances, and to maintain stable system operations. An expansion of the scale required is no mere detail to be ignored, as is done in many studies.

As Heard et al. have shown, the case made for feasibility of a 100% renewable-energy system is inadequate for formulating responsible policies directed at responding to climate change. "Efforts to assess the viability of 100%-renewable systems, taking into account aspects such as financial cost, social acceptance, pace of roll-out, land use, and materials consumption, have substantially underestimated the challenge of excising fossil fuels from our energy supplies. The desire to push the 100%-renewable ideal without critical evaluation has ironically delayed the identification and implementation of effective and comprehensive decarbonization pathways; the early exclusion of other forms of technology, from plans to decarbonize the global electricity supply is unsupportable, and arguably reckless.

For the developing world, important progress in human development would be threatened under scenarios applying unrealistic assumptions regarding the scale of energy demand, assumptions that lack historical precedent and fall outside all mainstream forecasts" (Heard et al. 2017, p. 1130).

In his above-mentioned book *No Miracles Needed*, Jacobson (2023) relates all the successes that his 100% WWS movement has garnered in the US and elsewhere in the years from 2009 to 2022. He and his co-workers have drawn up 100% renewable roadmaps for all energy sectors for cities, countries and all US states. Striking is though that none of these roadmaps are incorporated into laws or resolutions for all energy sectors, but, if at all, just for electricity, which few people will dispute as not being feasible. The important point is that these roadmaps have apparently not been able to convince people of the feasibility of 100%-renewable energy for all energy sectors.

CHAPTER 4

The Need for Carbon Capture

Introduction

The burning of fossil fuels since the start of the industrial revolution has resulted in a very pronounced increase of carbon dioxide (and other greenhouse gases) in the atmosphere. Carbon dioxide spreads very quickly all over the world and, although in the vicinity of specific (point) sources, like power stations and industries, the concentration is temporarily much higher, it is basically the same everywhere. It is also very persistent and unlike many other pollutants remains in the air for centuries, which implies that, even if we were to stop burning fossil fuels tomorrow, climate change and global warming will still continue for quite a while, until the Earth system has settled at a new equilibrium with its corresponding higher temperature.

At the start of the industrial revolution the concentration of carbon dioxide in the atmosphere stood at around 280 ppm (parts per million per volume), meaning that of every million particles in the air 280 were molecules of carbon dioxide (0.028%). In February 2023 NASA reported a value of 419.5 ppm (an increase of 50% compared to the pre-industrial value), a value that goes up each year by 2–3 ppm, as can be seen in the famous Keeling curve presented in Fig. 4.1. At this rate of increase we are rapidly heading for a doubling of the concentration compared to the pre-industrial value. As stated earlier, already in 2008 James Hansen and co-workers urged that, if we wish to preserve the climate on the planet, the concentration of atmospheric carbon dioxide has to be brought down to at most 350 ppm, the level achieved in 1987, as Fig. 4.1 shows. We did not heed their call. It has now passed the 400 ppm level and is rapidly going up further with a growth rate that has been accelerating from on average 1 ppm per year in the thirty years from 1960 to 1990 to on average 2 ppm from 1990 to 2020, and 2.4 in the last decade.

The same paper warns: "If we stay our present course, using fossil fuels to feed a growing appetite for energy-intensive life styles, we will soon leave the climate of the Holocene, the world of prior human history. The eventual response to doubling pre-industrial atmospheric CO_2 likely would be a nearly ice-free planet, preceded by a period of chaotic change with continually changing shorelines." It will still take several centuries before the planet will be nearly ice free, but judging from the extreme weather events in July 2021, with severe floods in Western Europe and China, accompanied by unprecedented heat waves in North America, already predicted twenty years ago by the IPCC in its Third Assessment Report, and extensive wild

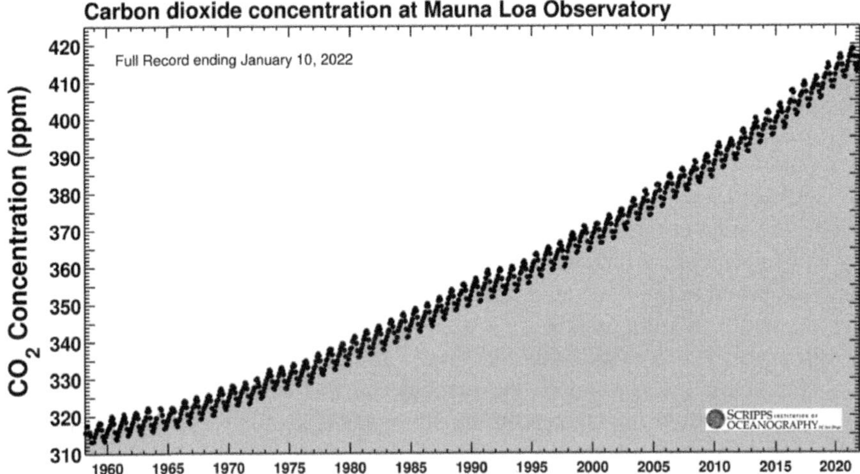

Fig. 4.1. Annual average carbon-dioxide level in ppm at NOAA's Mauna Loa Observatory on Hawaii (Scripps Keeling curve) from the start of observations in 1958 to the present day. Over the course of the year, values are higher in the Northern Hemisphere winter and lower in summer.

fires in Siberia and elsewhere, followed by an unprecedented number of heat waves and an extended period of drought in Europe in the summer of 2022, topped by disastrous floods in Pakistan engulfing one third of the country, this period of chaotic change has indeed already arrived. The only question is where it will strike next time.

So, as Hansen says, and he is now joined by many others, including organizations like the IPCC and the IEA, we must take out (part of) this surplus of CO_2 and, at the same time, reduce emissions, e.g., from point sources like power plants and fossil-fuel combusting industrial installations, to a level that natural carbon sinks can absorb and process. This latter type of emission reduction requires the capture of CO_2 from flue gases, compressing it, and injecting it deep underground in secure geological formations, and ensuring it remains stored there indefinitely. This technique of CO_2 removal is a way of slowing our CO_2 release into the atmosphere, and is a cost-effective alternative to developing means of CO_2 free-energy production. If we don't perform this type of capture, the first task of taking CO_2 out of the air will just be a waste of time and money. At the same time we must of course also continue with the transition to renewable energy sources but, in view of the world's dependence on fossil fuels for its energy supply, this transition needs time. Both carbon capture and storage (CCS) at point sources to prevent emissions from ending up in the atmosphere, and Direct Carbon Removal for taking part of the CO_2 out of the atmosphere, are invaluable in making such a transition possible. In this chapter I do not yet make a distinction between CCS from point sources like power plants and Direct Carbon Removal, although they are very different. This book is about both and the distinction will be made later. They are both forms of carbon capture. CCS from point sources is of course the most sensible thing to do as the concentration of CO_2 in flue gases is much higher than in ambient air but, if you fail to capture the carbon from those flue gases, it will end up in the air and Direct Carbon Removal is your only salvation.

In spite of all this, carbon capture in whatever form has no large fan club. Not many seem happy with this form of combating climate change. Those against the burning of fossil fuels are against carbon capture as they feel that it is putting the cart before the horse. The fossil-fuel industry does not like it either as it will increase the cost of generating energy from fossil fuels, although already in 2008 the then Shell CEO Jeroen van der Veer argued that the EU should combine its Emissions Trading System with developing CCS technology.[1] In general, industry is seldom against any measure or proposal so long as it applies to all and does not disturb the level playing field, so that any costs can be passed on further down the chain. It is they who provide the innovation and make it work, not the screaming NGOs and environmental pressure groups.

However, in our market- and profit-oriented approach to everything under the sun, carbon capture has indeed the fundamental and seemingly fatal flaw that it only costs money. Although carbon dioxide is not a simple waste product, carbon capture and its subsequent storage is mainly "investment and running costs, but no revenue," and at the scale we need it for making an impact the cost will be enormous. It can only abate the problem of carbon dioxide in the atmosphere and that's it. The sewage and clean drinking-water problem of the nineteenth century can be seen as an analogy. Dung and other solid and liquid filth were just dumped in the streets. A few cholera and typhus outbreaks were needed to impress on people that sewage had to be removed and treated properly. That is why we see waste water and sewage treatment plants dotted in the landscape. What we are doing for water, we also need to do for air; build an air treatment infrastructure to scrub the air clean of CO_2 (and of other things that we think should not be there). To reverse climate change we have to *lower* the concentration of greenhouse gases in the atmosphere, in particular the long-lived carbon dioxide. There is no other option! In view of the numbers of people milling about on the planet and the activities they engage in, we can no longer take our environment for granted and use it as a rubbish bin but must tend it with more care.

It is true though that there currently does not yet exist a full-blown industry that could carry out such a gigantic scrubbing and capture effort, the main reason for this being that carbon removal has not been a very active field of research and until a decade or so ago it was generally thought that carbon capture would be an impossible venture that could never be made cost effective. It got little or no support, while sustainable energy options were pampered with subsidies. Large and rapid progress could have been made if capture projects had just been allowed to run their course instead of being cancelled for dubious reasons at an early stage, as has happened several times. As we will see, there are some success stories which clearly show that especially CCS is not a bridge too far, but so long as the burning of fossil fuels and the concomitant emission of carbon dioxide into the atmosphere can go on without

---

[1] Van der Veer spoke at the 19th World Petroleum Congress in Madrid. He was of the opinion that it would be difficult for the EU to achieve its 20% emission reduction target, compared to 1990, without CCS. He was wrong. EU emissions in 2020 were 24% lower than in 1990, while the economy grew by 60% in that period and no CCS was needed for this. He underestimated the reductions that could be made in burning coal in the EU. It would have been much better though if the EU had actually heeded his call, and Van der Veer may still be right as far as the rest of the world is concerned, as it is hard to see how China and the developing world can phase out coal to the same extent as the EU in the next few decades.

paying a price for the pollution caused, there is little chance of widespread success as competition with something that is free is hopeless. We will discuss the necessity of a carbon tax or other levy mechanism in Chapter 6, but an international revenue mechanism, like a universal charge of as little as $1/tCO_2$, would be able to generate more than $30 billion annually that could be used to fund critical internationally sponsored and monitored R&D into zero-emission technologies and install CCS installations on new and older coal-power plants. In that simple way huge progress could already be made.

Wind turbines and solar panels also cost money of course but they provide energy, clean energy. They are viable alternatives to burning fossil fuels for energy, carbon capture isn't because it doesn't generate energy. There is no doubt that carbon-free sources of energy are vastly preferable over something that only costs money and can do little else than perhaps make the further burning of fossil fuels a little more palatable. It is a good argument but not a sufficient one, as it has become clear over the last twenty years or so, in spite of all the clamour to the contrary, that sustainable forms of energy alone will not be able to do the job in practice. See the previous chapter. In theory they can, but in practice there are only a few rich countries in the West, notably in the EU, which seem to be prepared to follow that course and completely cut out fossil fuels, at any rate coal, by the middle of the century. Other countries are not prepared to take that route, or are unable to. China, India and many other developing nations in Asia and Africa do not have that luxury. They just don't have the capital to invest in expensive clean energy generating options and will continue to rely mostly on much cheaper coal power plants. In this respect it is not surprising that at the recent COP26 climate conference in Glasgow India and China objected to writing the "phase-out of unabated coal power" in the final declaration and could only agree to a phase-down. There is always talk of getting out of fossil fuels, but if fossil fuels are the only thing you have as a developing nation there is not much chance that this will happen. Although I in general disagree with Alex Epstein, he rightly pointed out that "there is an incredibly strong correlation between fossil fuel use and life expectancy and between fossil fuel use and income, particularly in the rapidly developing parts of the world" (Epstein 2014, p. 13). Abatement by the installation of CCS facilities and paid for by the developed world is the way to go forward in this respect. If, as was said, "nearly 500 global financial services firms agreed to align $130 trillion—some 40% of the world's financial assets— with the climate goals set out in the Paris Agreement, including limiting warming to 1.5 degrees Celsius", is it then impossible to find $1 trillion or so for CCS installations for existing and projected coal power stations? China, for instance, has more than 1,000 coal power plants in operation, providing almost 60% of the energy it consumes. It cannot afford to scrap these plants. It will continue to use most of them for the duration of their normal operational lifespan and build new ones. They will not be prepared to forgo this option because of some noisy pampered Western NGOs. The developing countries have never enjoyed the luxury of cheap energy at the simple turn of a switch and will just ignore the warnings of these NGOs, rightly telling them to first put their own house in order and clean up the mess they have caused in the last hundred years or so, before telling them what to do. One of the ways to do this is by carbon capture and storage and paying for this in developing countries.

The IPCC has also realized the need for carbon capture and the scenarios in its special report on the 1.5 degree target are based on the removal of vast quantities of CO_2—730 billion tons by 2100 (10 Gt/y by 2050 and 20 Gt/y by the end of the century), in total almost 15 times current annual CO_2 emissions (Rogelj et al. 2018, p. 122). Various removal methods, which can all be grouped under the heading of Negative Emissions Technology (NET), have so far been considered with the most important ones being reforestation/afforestation[2] and, most importantly, the use of biomass for energy generation in combination with carbon capture and storage (BECCS).

Carbon-capture Approaches

There are six major natural and technical approaches to remove and sequester CO_2 (Ozkan et al. 2022):

1. Coastal blue carbon, the carbon captured by living coastal and marine organisms and stored in coastal ecosystems;

2. Terrestrial carbon removal and sequestration, the process through which CO_2 from the atmosphere is absorbed by trees and plants through photosynthesis and stored in soils and biomass;

3. Bioenergy with carbon capture and storage (BECCS);

4. Carbon mineralization;

5. Geological sequestration, capturing carbon dioxide from point sources like power plants and industrial installations and store it permanently in a geological formation (CCS);

6. Direct air capture (DAC), scrubbing the carbon dioxide out of the air and store it.

The first two are natural processes, part of the natural carbon sinks of the oceans and the biomass on Earth. They will not be further considered in this book, although we will later see some instances in which these processes can be accelerated. The third one, BECCS, plays a dominant role in global climate mitigating scenarios and is the dominant active carbon removal approach. BECCS involves the burning of biomass for generating energy and is touted almost everywhere as an acceptable form of carbon removal. Its components are called technologically mature, and the transport and storage of CO_2 captured in this way is called a known and commercially available technology (Fuss and Johnsson 2021), while other carbon-capture approaches that are identical from a technological point of view, are decried as too expensive, risky and unrealistic. If BECCS is acceptable, I see no reason why a similar sort of capture from the exhaust stacks of coal- and gas-fired power plants would not be acceptable. In view of the general arguments presented above, actually even more so. You can only reject this possibility if you cling to the unreasonable idea that burning fossil fuels *as such* is a cardinal sin and should be banned immediately, irrespective of costs and suffering. It is gratifying in this respect to see that the latest report by

---

[2] Capture from the atmosphere in wood biomass through photosynthesis by the planned expansion of forest areas on land that has not been covered by trees during the last 50 years.

Climate Action Tracker (Boehm et al. 2022, p. 135) argues in favour of the development of a robust portfolio of carbon removal approaches.

In the case of BECCS the biomass that is burned has captured carbon dioxide from the air when it grew, which without capture would of course escape in the combustion process. If captured and stored, the end result is that some of the carbon dioxide has been removed forever, while it has in the meantime served an extra heat-providing round. 'Negative emissions' are possible if the CO_2 stored is greater than the CO_2 emitted during biomass production, transport, conversion and utilization. These methods will however be nowhere near sufficient to remove the huge amounts required in the IPCC scenarios. We cannot do without other methods, including Direct Air Capture and CCS to achieve the goals set in them.

As BECCS does not involve the burning of coal, natural gas or oil, it is more benevolently gazed upon, although coal, natural gas and oil are of course just fossilized biomass and cost issues are identical. The more benevolent attitude is mainly based on the fallacy that biomass is a renewable resource, renewable in the sense that the burned biomass can be "naturally replenished on a human time scale". However, as has repeatedly been pointed out (Norton et al. 2019 and references therein), this is a gross misrepresentation of the atmosphere's CO_2 balance since it ignores the slowness of the photosynthesis process which requires several decades for trees to reach maturity.[3] It completely arbitrarily connects the concept of renewability to the time scale of a human life.

The dark side to BECCS is that the burning of biomass for power emits not only carbon dioxide (which is captured), but also many other pollutants, including harmful particulates. The burning of biomass, even without CCS, is generally accepted and even seen as a carbon-neutral way of generating energy. It is argued that the biomass burned is just a waste product that would otherwise be left on the land to rot and emit the CO_2 stored in it by the photosynthesis of the tree. This would be true if only waste were burned. However, in many cases whole forests are cut down and nothing is replanted. The trees are transported, sometimes after having been processed into pellets (resulting in CO_2 emissions), over thousands of kilometres in polluting ships (resulting in further CO_2 emissions) to be burned in another continent. The carbon dioxide emitted into the air has been captured by the trees over decades and, if not captured, is now released into the atmosphere in a flash.

Of course, there is no objection against burning (locally gathered) crop residues, waste wood from other uses and municipal solid waste (consisting of everyday items discarded by the public) for generating energy, if it is done with CCS. In practice, however, CCS is only applied to large biomass-burning power plants which need a constant supply of fuel to be economical. Most of the biomass plants in operation

---

[3] The simplicity of BECCS brought with it political and economic advantages and led to the inclusion of biomass in the European Commission's definition of renewable energy in its 2009 Renewable Energy Directive (RED; EC, 2009), being treated as 'part of the package of measures required to reduce greenhouse gas (GHG) emissions.' Far from reducing GHG emissions, replacing coal by biomass for electricity generation is likely to initially increase emissions of CO_2 per kWh of electricity as a result of the lower energy density of wood, emissions along the supply chain, and/or less efficient conversion of combustion heat to electricity. The initial negative impact is only reversed later if and when the biomass regrows. Woody biomass contains less energy than coal, so that CO_2 emissions for the same energy output are higher.

are small, do not deploy CCS and, since they also emit particulate matter, are more polluting than state-of-the-art coal- or gas-burning power plants. This must be considered totally unacceptable. It makes burning biomass even more reprehensible than coal or natural gas and this practice should be terminated as soon as possible.

An example of such unacceptable practices is the Drax power plant in the UK, on which Friends of the Earth reported[4] in 2021. It is an old coal-burning power plant that since 2012 received more than £4 billion in subsidies to convert four of its six boilers to burn wood pellets instead of coal, with another £2 billion still to be provided until 2027. Over 7 million tons of wood pellets are burned each year to produce electricity. This requires 14 million tons of green wood, because pellets must be dried and compressed before they can be burned as fuel. Most of the wood pellets are imported from the US, Canada and the Baltic states and Drax uses more than twice the amount of wood pellets to produce the same amount of usable energy as a modern plant. It says that 60–80% of its wood pellet feedstock comes from sawmill residues, tree thinnings, branches, tops, and bark, but according to Friends of the Earth more than one third of the wood burned originates from large whole trees. Drax opened in 1974 and has a terrible thermal efficiency of around 38%, meaning that for every 10 trees burned, 6 are wasted as uncaptured heat. Its carbon emissions, 13 Mt of CO_2 in 2020, are not captured and the amounts actually emitted are downplayed by opaque stories about carbon being captured by new growth. The fact of the matter is that Drax's activities increase harvesting of wood and even with regrowth there is a lower amount of carbon stored in the forest than if the forest was left to grow undisturbed.

When coal is burned in a coal power plant, air and coal are fed into a boiler and the combustion process produces steam that is used to spin a turbine to generate electricity. The flue gases from the combustion are subject to the conventional technologies that must currently be applied to reduce harmful emissions, such as selective catalytic reduction to reduce NO_x emissions, flue gas desulphurization to reduce SO_2 emissions, and electro-static precipitators or baghouse to reduce particulates (Jaccard 2005, p. 189). Carbon dioxide can easily be added to this. Acid rain caused by the sulphur and nitrogen compounds in exhaust gases and destroying forests in the 1980s (in Germany called *Das große Waldsterben*) were never a reason to cry for the abolishment of fossil fuels or petrol-driven cars. A technological fix was found, applied and apparently accepted by everybody. There is no reason why CO_2 could not be removed from the flue gases as well. Its capture from a coal plant's flue gases does not require the development of new technologies. Various technologies have been used in the past and could be developed further. A scrubbing technique can be integrated into new coal-fired power plants and even retrofitted to existing plants.

Already in 2016 the IEA wrote that "Retrofitting carbon capture and storage (CCS) on existing coal-fired power stations in the People's Republic of China (…) represents a major opportunity, with significant benefits for emission reductions. In total, some 310 gigawatts (GW) of existing coalfired power capacity meet a number of basic criteria for being suitable for a retrofit" (IEA 2016, p. 5). Apparently nobody picked up this message as so far not a single coal power station in China, or

---

[4] Friends of the Earth, *The future of Drax: old, inefficient, damaging and expensive* (2021).

anywhere else for that matter, has been equipped with CCS and since 2016 hundreds of new coal-fired power plants have been taken into operation in China alone. All the new ones should have been equipped with CCS as a matter of course.

IPCC Scenarios

In 2005 the IPCC issued a special report (IPCC 2005)[5] on carbon capture and storage in which a rather optimistic view on this technology was presented. In the IPCC's later Assessment Reports (AR5, AR6) CCS continued to be mentioned as a possible means to aid in climate mitigation, but only fleetingly and bunched together with a number of other mitigation approaches. As we have seen, carbon capture can be achieved by various means, including CCS, Direct Air Capture, afforestation, etc. In spite of the fact that in the ensuing 15 years climate problems have only worsened and become more urgent, there is still no large-scale deployment of CCS and neither did the IPCC in any way push for CCS in its subsequent reports. The 40-page Summary for Policy Makers, probably the only pages policy makers (and others) will ever read of its latest 4,000 page Assessment Report (AR6 2022), only includes two rather vague statements[6] about carbon-dioxide removal as having "the potential to remove CO_2 from the atmosphere and durably store it in reservoirs." It will not come as a surprise to anybody I suppose that removing has the potential to remove. CCS, the topic of its 2005 special report, is not mentioned at all in the entire summary, and only a few times in the rest of the report (mainly in combination with bio-energy (BECCS)). The above-mentioned 2005 IPCC report on CCS is not mentioned at all in AR6. The reason could be that it is rather old, but a more recent report does unfortunately not exist.

Another (*high-confidence*) statement from AR6 is that anthropogenic CO_2 removal leading to global net negative emissions would lower the atmospheric CO_2 concentration and reverse surface ocean acidification. That is again an obvious statement, but if it is true why is it not pushed more forcefully? And can global net negative emissions be achieved without CCS applied to the massive emissions of coal-fired power stations and other point-source emitters?

In Chapter 5 of the report the message is downplayed by noting that when more CO_2 emissions are removed than human activities emit (net negative CO_2 emissions), and atmospheric CO_2 declines, the land and ocean sinks initially continue to take up CO_2 from the atmosphere. This because carbon sinks, particularly the ocean, exhibit inertia and continue to respond to the prior trajectory of rising atmospheric CO_2 concentration. After some time land and ocean carbon reservoirs begin to release CO_2 to the atmosphere making carbon removal less effective. So, if a lot of CO_2, more than we put in, is removed from the atmosphere, the oceans and CO_2 land sinks will release some CO_2 to establish a new equilibrium. Well, I wish we were at the stage of removing more CO_2 from the atmosphere than we put in, as most of our problems would then be solved. The minor problem of the ocean releasing CO_2

---

[5]   Available at https://www.ipcc.ch/report /carbon-dioxide-capture-and-storage/.
[6]   Statements D.1.4 and D.1.5 on page SPM-35 (almost at the end of the Summary for Policymakers).

and the ensuing benefit of reversing acidification[7] can then easily be dealt with. It is rather odd that such a farfetched issue is mentioned at all, and even odder that it is mentioned as a drawback of removing carbon from the atmosphere. It is a repeat of a similar statement in IPCC AR5 (p. 546–547) which states that it is thus virtually certain that the removal of CO_2 will be partially offset by outgassing of CO_2 from the ocean and land ecosystems. Returning to pre-industrial CO_2 levels would therefore require that an amount of carbon equal to total anthropogenic CO_2 emissions be permanently sequestered, roughly equal to twice the excess of atmospheric CO_2 above pre-industrial levels. It is of course not necessary to bring the CO_2 concentration in the atmosphere back to pre-industrial levels (if we knew what these were). The planet has enough flexibility to handle some of the mischief we cause. Hence, down to 350 ppm would be sufficient, but this bogus argument is nevertheless repeatedly put forward in an attempt to put carbon capture in a poor light.

The IPCC statements are based on various studies, including one recently published in *Nature Climate Change* which confirms this asymmetry (Zickfeld et al. 2021): it is more effective to avoid CO_2 emissions than to take CO_2 out of the atmosphere, or as they put it "an emission of CO_2 into the atmosphere is more effective at raising atmospheric CO_2 than a CO_2 removal is at lowering atmospheric CO_2, indicating that the carbon cycle response is asymmetric." An extra amount of CO_2 removal is required to compensate for an emission of a given magnitude to attain the same atmospheric CO_2 concentration. This is due to a rebound effect caused by CO_2 being released by the terrestrial biosphere and the ocean in response to declining atmospheric CO_2 levels. In my view this is rather obvious as the CO_2 removal disturbs the (movement towards an) equilibrium between the carbon sinks and the atmosphere and the system will adjust in such a way that after some time the equilibrium is restored.

There is apparently a cabal of (environmental) organizations and possibly other shady actors whose goal from the outset has been that fossil fuels should be banned and have done their utmost best to put CCS (or carbon capture in general) in a poor light as it would in their view extend the use of fossil fuels. The latter is probably correct but no reason to let the world go to pieces by not taking action and trying to get rid of the cause of global warming. As recently as 1 June 2022 Greenpeace even calls it "the great carbon capture scam", a swindle by the oil companies to "pretend to help solve a problem, while making the problem worse, socialize the costs and liabilities, and privatize the profits."[8] Such organizations say to be concerned about global warming and to be busy with climate change all day but contribute very little, if anything at all, to solving the problems and repairing the damage. At any rate, harping on the same string of the need to stop using fossil fuels *now* while knowing full well that this is not going to happen does not help at all. I just cannot comprehend

---

[7]   It is somewhat misleading to talk about ocean acidification as the ocean water is naturally alkaline with a pH ranging from 7.8 to 8.5; acidic would be a pH below 7. There is no danger of the ocean pH to drop below 7; so far the pH has dropped by only 0.1-0.2 in the 200 years from 1800 to 2000. However, since the pH scale is logarithmic, this represents a significant increase in the H^+ ion concentration, and sets in motion a number of chemical reactions that greatly affect life in the oceans.

[8]   Rex Weyler, The Great Carbon Capture Scam (1 June 2022); https://www.greenpeace.org/international/story/54079/great-carbon-capture-scam/.

how someone who is genuinely concerned about climate change and the havoc the resulting global warming will wreak with the world can be obsessed by just part of the problem and ignore the already dangerously high levels of greenhouse gases in the atmosphere.

In this connection, quite extraordinary pronouncements are sometimes made in the literature on the attractiveness of CCS as a potential solution for climate-change mitigation. For instance, in a 2019 paper (Heyes and Urban 2019) on the economics of CCS, it is incomprehensibly stated that since CCS "is based on a suite of mature, well known technologies, most of the cost reductions have already occurred." Although it is true that carbon-capture techniques go back to the 1930s, not a single fossil-fuel fired power plant with a CCS installation and no large scale Direct Air Capture installations are currently, or have ever been, in operation, and the same paper states elsewhere that "experience with CCS (geological sequestration) remains limited." How can a technology that has hardly been used be called 'mature'? It would be similar to rejecting wind-turbine development in 2000 with the argument that windmills have been operating in various countries, e.g., the Netherlands for more than 500 years and therefore are a 'mature' technology with slim prospects of successful deployment. Indeed, in 2011 the European Technology Platform for Zero Emission Fossil Fuel Power Plants issued a report[9] in which it stated that CCS "is an emerging technology and historical experience with comparable processes shows that significant improvements are achievable—traditionally referred to as learning curves." One of the rather optimistic conclusions of the report is that "post 2020, CCS will be cost-competitive with other low-carbon energy technologies." Unfortunately the recommendations made in the report, especially the construction of demonstration projects, have not been followed up, resulting in the situation that this learning curve for CCS still has to begin.

The IPCC in its latest report (AR6) uses a core set of five scenarios to explore climate change over the 21st century. They are labelled SSP1-1.9, SSP1-2.6, SSP2-4.5, SSP3-7.0, and SSP5-8.5, with SSP standing for Shared Socioeconomic Pathway, and correspond to increasing emission levels, with SSP1-1.9 representing the low end of future emissions pathways, leading to warming just below 1.5°C in 2100 with limited temperature overshoot during the century, and SSP5-8.5 representing the very high warming end of future emissions pathways from the literature, the consequences of which are just too awful to contemplate. In this latter scenario the average global temperature will be a scorching 4.4°C higher by 2100. SSP1-1.9 is characterized by a rapid decline of net CO_2 emissions to zero by 2050 and net negative CO_2 emissions in the second half of this century (e.g., by employing carbon-removal technologies). The next best scenario SSP1-2.6 was designed to limit warming to below 2°C. Global CO_2 emissions are cut severely, but not as fast as in SSP1-1.9, reaching net-zero after 2050. The temperature rise stabilizes at around 1.8°C by the end of the century. Even in the most optimistic SSP1-1.9 scenario, the IPCC calculates[10] that in 2050 cumulative CO_2 emissions from 1850 will amount to about 3,000 Gt, about 600 Gt more than today, of which about half, 1,500 Gt, will remain in the air.

---

[9]   European Technology Platform for Zero Emission Fossil Fuel Power Plants, *The Costs of CO_2 Capture, Transport and Storage* (2011).

[10]   IPPC, AR6 (2021), Fig. SPM-10.

From these scenarios it is clear what we have to do. CCS cannot be a "false solution" or part of the problem, as some environmentalists are saying. On the contrary, it is an essential part of the solution. We need to embrace every available possibility to reduce both emissions *and* the concentration of greenhouse gases in the air if we want to control global warming. By 2050 the CO_2 concentration will most likely be close to or even over 500 ppm. It is no option to just gradually cut emissions to zero, if we were indeed able to do so, achieve net zero by 2050 and leave carbon dioxide in the air for the next two hundred years or so at such a high concentration,[11] meaning that climate change will just continue to get worse and worse until a hundred or hundred-fifty years from today a new balance will have established itself. So long as current emission levels continue, and so far we only have seen a year-by-year global rise (see Fig. 2.2), 500 ppm will be reached within 3–4 decades. The problem is already here and noticeable: the Earth is rapidly warming up and will continue to do so even if we were able to cut emissions of greenhouse gases to zero from tomorrow. Some people do not seem to realize this utterly trivial fact. Quite frequently, also in the main media, it is stated or suggested that greenhouse gases in the atmosphere can be lowered below the current level by reducing the emission rate.[12] So long as we continue to emit such gases above the absorption capacity of carbon sinks the concentration will inevitably rise, unless we employ other means to take out more than we put in. So, bringing emissions down to 1990 levels or achieve net zero by 2050 or another date as is the aim of many climate mitigation programmes nowadays will simply not be enough.[13] Global emissions in 1990 were more than 20 Gt of CO_2, too much to be absorbed by the carbon sinks we have. Moreover, for China and equally for other developing nations it will be much harder, if not impossible, to cut back to 1990 levels. As can be seen from Fig. 2.2, China, with a population three times as large as the US, would be required to cut back to half the US level. India was only emitting 600 Mt in 1990, a factor of ten less than its current emissions, which in view of its development path are bound to increase to the level of China today or even higher if no stringent mitigation measures are taken. Net zero by 2050 is of course much better than emitting at 1990 levels but even that will not be sufficient, as it neither removes any carbon that is already in the atmosphere or will still be emitted until that date.

As said, it is imperative to bring the concentration in the air down to roughly 350 ppm as a minimum, 300 ppm would be even better. About 70 to 100 ppm, or even 150 ppm if we postpone large-scale carbon scrubbing for another twenty years

---

[11] Although some seem to be satisfied with a situation in which CO_2 concentration is stabilised at 500 ppm by 2050. See, e.g., Del Grosso and Cavigelli 2012.

[12] E.g., Powers 2020 and very recently in *Inside Climate News* weekly newsletter which stated "The first week of COP26 wrapped up with several international climate commitments, including more than 100 nations pledging to cut global methane emissions by 30 percent or more between now and 2030 in an effort to quickly and significantly curb global warming." If implemented, this effort does not significantly curb global warming at all. Global warming will continue, at best a little more slowly! After all, concentrations of greenhouse gases will increase, not go down! The tub will at best overflow a little later!

[13] The Dutch Climate Act, for instance calls for a 49% reduction in greenhouse-gas emissions by 2030, compared to 1990 levels, and a 95% reduction by 2050. As part of the European Green Deal most recent EU proposals want a 55% reduction of emissions by 2030 compared to 1990 levels, and no net emissions of greenhouse gases by 2050.

or so, have to be removed from the atmosphere. In Chapter 2 we have seen that 1 ppm of atmospheric CO_2 is equivalent to 2.1 Gt of carbon or 7.7 Gt of carbon dioxide. This implies that we must take out the staggering amount of 700–1,000 Gt of CO_2 *and* at the same time vastly reduce emissions to make sure that the world will remain liveable.[14] This is about half of the total amount of carbon that has been belched into the air in the last 150 years, the other half having been graciously absorbed by the oceans and the biomass (trees, plants) on land. In this book we will show how this can be done (already with current technology), at high but affordable cost. No miracles and/or great feats of innovation are required, although there will be many eureka moments along the way as technologies get developed. We will describe the (old and emerging) technologies for this, and what is required from individual societies and the international community to achieve this. If we do not start with the deployment and further development of these techniques in the short term, the situation will only worsen and before the century is out we will have reached a point at which there will be no further option than deploying some of the really risky, never tested and scary geoengineering techniques[15] to reflect sunlight (albedo modification), like stratospheric aerosol injection[16] or to increase the ocean's ability to suck carbon from the air,[17] in a last drastic and desperate attempt to at least try to save part of the Earth. Such action would be similar, but then on a sustained level, to the short-lived effects of large scale volcanic eruptions.

Concluding remarks

So the starting point of the discussion are two observations. The first one, the concentration of atmospheric carbon dioxide is too high, has been set out above. The second one is that climate change is due for a great deal to the energy sector, the burning of fossil fuels for electricity. The power plant structure providing us with all this energy is worth \$45 to 55 trillion. Such amounts cannot be written off at the stroke of a pen, nor can they easily, if at all, be replaced by a sustainable energy infrastructure in the two to three decades we need to take action to avert catastrophic climate change. Countries would go bankrupt if we did, therefore it will not happen. CCS can help avoid this to happen. When presenting a realistic vision for the future, key real-world constraints that stand in the way of shifting away from current trends must be taken into account. We need to shift slowly and in the short term capture the carbon at source, in addition to starting to scrub CO_2 out of the atmosphere. Some people want to get rid of fossil fuels from tomorrow, but that is not a realistic aim. They are advised to first read Vaclav Smil's latest book[18] and then re-join the discussion. Fossil fuels are here to stay in one form or another for a considerable time to come.

---

[14] Subject to the outgassing from these carbon sinks as discussed above.

[15] Naomi Klein (Klein 2014, p. 282) is not certain, it seems, that for the geoengineering aficionados and some industries like ExxonMobil and General Motors "geoengineering isn't a Plan B should emission cuts fail, but rather a Plan A".

[16] See about this National Research Council, *Climate Intervention: Reflecting Sunlight to Cool Earth* (The National Academies Press, Washington D.C., 2015).

[17] https://www.science.org/content/article/panel-calls-2-5-billion-ocean-geoengineering-research?utm_campaign=news_daily_2021-12-08&et_rid=414026595&et_cid=4026336.

[18] Smil 2022; you may skip chapter 5, but please read the rest and pay special attention to Chapter 6.

The upheaval and scramble for fossil fuels since the start of the war in Ukraine brings this home in a rather nasty way, with European consumers suddenly being faced with sky-high energy prices and suffering from energy poverty, a situation they never had encountered before. Sustainable/renewable energy generation is a long term solution and, although an absolute necessity, for the time being at best a partial solution. In due course it will be able to cover a large part, perhaps all, of global electricity generation and through 2050 consumption of non-hydroelectric renewable energy will be the fastest growing energy source, but a considerable part of production will still come from in particular natural gas and especially in developing nations from coal power plants, as these countries often have no affordable alternative. The developed world should pay for the CCS installations for these plants. Not all greenhouse-gas emissions can be completely eliminated, for technological, economic or political reasons. No one in the debate doubts that there are unavoidable emission sources which will have to be balanced by sinks or by some form of carbon capture.

As Smit et al. (Smit, Reimer, Oldenburg and Bourg 2014, p. 141) say in their book, there is nothing magical about carbon capture and storage (CCS). The technology to capture and store CO_2 in geological formations is already available commercially. So why isn't carbon-capture technology more widespread? Currently, the costs of separating CO_2—in terms of required energy and capital costs, and the resulting increase in the price of electricity—are significant.[19] In other words, the dangers and risks of climate change are not yet perceived to be so threatening that 'trivial' economic and pecuniary arguments still dominate the discussion and frustrate in principle useful methods to avert such dangers. In the end the world will have to come round and be prepared to spend astronomical amounts to keep the Earth suitable for living. If in the meantime R&D are able to reduce the required energy and monetary costs of separation in a meaningful way, this time will come earlier and would greatly reduce the barrier to CCS.

In summary we need carbon capture in the form of CCS and Direct Carbon Removal (IEA 2021c and Ringrose 2020, p. 6):

1. to provide a mechanism for decarbonizing both the existing power supply and reducing emissions from industry (by fitting or retrofitting CCS installations); power and industrial plants can be retrofitted that may otherwise still be emitting 8 billion tons of CO_2 in 2050—around one-quarter of today's annual emissions;

2. to allow the transition to renewable energy sources to be achieved faster and cheaper than by using only renewable sources, as it can tackle emissions in sectors with limited other options, such as cement, steel and chemicals manufacturing, and in the production of synthetic fuels for long-distance transport;

3. to remove CO_2 from the atmosphere by combining it with bioenergy or Direct Air Capture to balance emissions that are unavoidable or technically difficult to avoid.

In spite of these in my view reasonable and inescapable arguments in favour of carbon capture, the technology has not yet made much progress. Many want to smother the child in its infancy.

---

[19] The IPCC in its 2005 Special Report on CCS calculates that the cost is about 3 dollar cent per kWh (see Chapter 14).

But fortunately the times seem to be changing. In its 2021 report *Net Zero by 2050—A Roadmap for the Global Energy Sector* (IEA 2021a, p. 84) the IEA says: "A failure to develop CCUS (*Carbon Capture, Utilization and Storage*) for fossil fuels would substantially increase the risk of stranded assets and would require around $15 trillion of additional investment in wind, solar and electrolyser capacity to achieve the same level of emissions reductions".

Likewise, it is gratifying to see that the US government has recently come round to the realization that CCS is an important tool in mitigating climate change. In the $1.2 trillion Infrastructure Investment and Jobs Act, signed by President Biden on 15 November 2021, $10 billion were allotted over five years for demonstration projects and R&D for carbon capture programs. This includes $3.5 billion for the Regional Direct Air Capture Hubs Program, for establishing four DAC hubs across the US that have the capacity to store and/or utilize at least 1 million metric tons of CO_2 captured by DAC technology each year, per hub, located in regions with heavy concentrations of fossil fuel industry. The Department of Energy has significant latitude for implementing these hubs.[20] Another $3.4 billion is allocated to commercial-scale demonstration plants and large-scale pilots for capturing CO_2 from point sources like fossil-fuel power plants and industrial installations (Kramer 2022a). Dizzying amounts of money unheard of in the sector. In the preceding years just a few tens of millions were put aside for such projects. So, if anything, the funding for CCS and DAC is at last coming of age.

In August 2022 this was followed by the Inflation Reduction Act, already branded a miracle before anything concrete has been achieved.[21] The Act also boosts carbon capture by transforming the possibilities for obtaining tax credits and increasing these credits. An unprecedented total amount of $369 billion will be allocated to reducing greenhouse-gas emissions and investing in renewable energy sources. A drawback of the Act, if there is one, is that it is all carrots and no sticks, meaning that no carbon tax will be levied, so for those who won't bother there will still be no incentive to change their behaviour.[22]

Finally I note that the discussion in this chapter mainly concerns the capture or removal of carbon dioxide, which indeed mostly costs money, apart from a small amount of the gas for which applications have been found. The situation is however different for other greenhouse gases, especially for methane CH_4. Captured methane has a clear use and market value as natural gas. As a result, many methane reduction measures have low or even negative costs. We do not have to discuss such measures in detail here, for knowing how the world works we can be sure that, when money can be made, methane capture will happen in due course, although it is not yet happening at present. If natural gas prices remain at their current high level, even after the war in Ukraine has been brought to an end, this will happen rather soon.

---

[20] Carbon180, DAC on track in 180 days; https://carbon180.medium.com/dac-on-track-in-180-days-85c73c0f9bc5.

[21] https://yaleclimateconnections.org/2022/08/ponder-the-miracle-of-a-u-s-climate-law/.

[22] An analysis by the Rhodium Group (https://rhg.com/research/inflation-reduction-act/) claims that the Act would reduce emissions by 31% to 44% below 2005 levels in 2030, implying cumulative emissions reductions by 2030 by 6 Gt (billion tons) of CO_2, and this at about $60 per ton (by spending $369 billion), which would be incredibly cheap.

CHAPTER 5

The Case of CCS for the Coal Sector

As an intermezzo we will in this chapter consider the coal sector and argue that a lot can be done with relatively little effort and expense when focusing on the most polluting coal-fired power stations currently in operation.

Coal is the most urgent problem as regards climate change and global warming. It is the world's most affordable energy fuel, but of all fossil fuels it is also the most polluting and the largest source of energy-related CO_2 emissions. Table 5.1 shows the current situation as regards coal production and projections until 2024.

The table shows that China produces half of the world's coal, with India a distant second. Most countries are expected to increase coal production at least until 2024. Only the EU shows a sizable decrease and the US and Indonesia a rather limited one. Coal is clearly not going away any time soon, despite its large carbon footprint, with a typical 500 MW coal-fired plant emitting about 3 Mt of CO_2 per year.[1] In 2021, as economies picked up after the Covid-19 pandemic, coal was in high demand with global coal consumption increasing by 450 Mt or about 6%,[2] reaching an all-time high and reversing the declining trend over the past two years. It pushed CO_2 emissions from coal power plants to a record 9.7 Gt, more than 100 Mt above the previous peak in 2018. To get on track with net-zero by 2050, an annual average reduction of emissions from coal-fired power plants of around 8% is needed through to 2030.[3]

The use of coal in the world is massive, no wonder that it got a lot of attention at the COP26 conference in Glasgow in November 2021, where 40 countries pledged to stop issuing permits and direct government support for new coal-fired power plants. However, the US, India, China, Australia and Russia did not join in the pledge, making it, as follows from Table 5.1, largely irrelevant. At COP26 it even looked for some time as if the world might agree on 'phasing out' unabated coal for power generation, unabated meaning without measures to capture the CO_2 emissions or reduce them by other means. India and China objected and could only agree to a phase-down, which I suppose would mean that power generation from *unabated*

---

[1] US figure, http://www.energyjustice.net/files/coal/igcc/factsheet.pdf.

[2] https://www.visualcapitalist.com/future-of-global-coal-production-2021-2024f/.

[3] https://www.iea.org/reports/coal-fired-electricity.

Table 5.1. Current global coal production and projected change until 2024 (source IEA).

Country	Coal production (2021)	Coal production (2024)	Share (2024)	Change (2021–2024)
China	3,915 Mt	3,982 Mt	50%	+67 Mt
India	793 Mt	955 Mt	12%	+162 Mt
Indonesia	576 Mt	570 Mt	7%	–6 Mt
United States	528 Mt	484 Mt	6%	–44 Mt
Australia	470 Mt	477 Mt	6%	+7 Mt
Russia	429 Mt	445 Mt	5%	+16 Mt
EU	329 Mt	247 Mt	3%	–82 Mt
Other	839 Mt	855 Mt	11%	+16Mt

coal will no longer increase but actually decrease, but whether that interpretation is correct remains doubtful. Although China may be in a situation in which such a phase-down can be contemplated, for India that is unlikely to be the case, in view of the catching-up the country still has to do. And what is true for India, equally applies to the rest of the developing world.

In this respect I would like to remark that it is unhelpful and unrealistic, immoral even, to demand from developing countries to phase out coal power as it is in many instances the only viable way for them to quickly and cheaply provide the energy they desperately need to stay on the development path. Developing nations choose coal-fired power plants because the technology is mature and the costs are low enough. Ample electricity is generated and is cheap (about 5 cents/kWh). The ongoing fuel costs are affordable with the fuel often available in the country. An example is Botswana, a country of just over 2 million inhabitants, relatively prosperous with a GDP (PPP) per capita of $18,000 in 2021, on a par with China (nominal about $8,000). Formerly one of the world's poorest countries—with a GDP per capita of about $70 per year in the late 1960s—it has since transformed itself into an upper-middle-income country, with one of the world's fastest-growing economies and the highest Human Development Index of continental sub-Saharan Africa. The land-locked country is not suitable for wind energy as average wind speeds are lower than 4 m/s, the minimum for wind energy to be viable, although there may be superior wind speeds at higher altitudes. Of course, being in the tropics it has great potential for solar energy. It is estimated that using less than 1 per cent of the country's area, Botswana could meet its current electricity consumption from solar energy.[4] But that is only electricity and even in the tropics the sun does not shine at night, or even all day. As long as massive affordable storage facilities remain unavailable, it will not bring the solution for Botswana. Moreover the country will have to spend vast amounts to build a solar energy infrastructure with the associated maintenance facilities. Fortunately, Botswana sits on coal reserves that are estimated to amount to a staggering 212 billion tons.[5] To date only two coal mines are in operation in the country, jointly producing 3.26 million tons of coal per year, with plans for extension

---

[4]   UN Environment Programme, Botswana Energy profile, UNEP.org.
[5]   Bloomberg, 27 August 2021, Botswana's Minergy Coal revives LSE listing plans.

to about 5 million tons. It would be a despicable neo-colonial and paternalistic attitude to try to prevent Botswana from exploiting these coal reserves. Instead, the developed countries should make the offer to pay for CCS installations for any coal-fired power station the country may want to build. In addition, they could also provide funding for solar energy to be put on the rails in the country, as is their duty under the Paris Agreement and in view of the debt rich countries have due to legacy emissions, such that in due course in Botswana only a few coal-fired power stations may be needed for backup. Botswana's CO_2 emissions are at a very low level, just 6.5 Mt in 2020 (2.77 tons per capita in 2020, far less than the world average of 4.77 tons and about half of what the best countries in Europe have to offer) and it would be a great and affordable way to bring this down even further or at any rate prevent it from going up and make Botswana an example to Africa and the world.

India's annual per capita energy consumption is still only about 8% of the US's and 18% of China's.[6] Its per capita carbon dioxide emissions are consequently also very low, with 1.77 ton per capita in 2020, lower even than Botswana's. India's coal consumption is bound to increase rapidly until the end of the century and it is unrealistic to assume that sustainable sources can fulfil the demand to meet this growth. In 2021, India was the world's second-largest producer and consumer of coal and its production is expected to increase further in the coming years as Table 5.1 shows. State-owned Coal India is the largest coal miner in the world, producing approximately 600 million tons of coal a year, and the country has an estimated 100 billion tons of coal reserves. According to *Climate Action Tracker*,[7] based on its current coal expansion plans, India's coal capacity would increase from current levels of over 200 GW to almost 266 GW by 2029–2030, with 35 GW expected to come online in the next five years. On 30 January 2023 Reuters reported that the Indian government has asked power generation companies across the country not to retire any of the 179 coal-fired power plants operating in the country before 2030.[8] Under current policies and action, greenhouse-gas emissions (excluding *land use, land-use change and forestry* (LULUCF)) are projected to reach a level of 3.84–4.02 Gt of CO_2-equivalent in 2030. Huge financial assistance from developed countries is needed to keep India's greenhouse-gas emissions in check. CCS can play a pivotal role here. Unfortunately the developed countries are not and have not been very forthcoming with their support, in spite of much talk. I have not seen any realistic plan put forward by climate activists and NGOs in Western countries that shows how India can lift its population out of poverty and onto a development path that provides cheap and reliable electricity to its population, assuming that this can be done without climate change dragging them back into poverty a few decades later. They visit the country travelling from one air-conditioned four- or five-star hotel to the next, proclaim that this or that part of India is very suitable for solar PV or wind turbines, and then go home to drive their donor-paid Tesla EV to the next superfluous party or meeting.

---

[6] Vijaya Ramachandran, Why India Can't Wean Itself Off Coal, *Foreign Policy* Oct. 2021; https://foreignpolicy.com/2021/10/08/india-coal-energy-climate-summit-renewable-solar-wind-electricity/.

[7] https://climateactiontracker.org/.

[8] https://www.reuters.com/business/energy/india-asks-utilities-not-retire-coal-fired-power-plants-till-2030-notice-2023-01-30/.

A short time before COP26, China also announced that it will stop building new coal-fired power plants abroad, which is precisely the wrong decision. How will these mainly developing countries where Chinese coal power plant construction is taking place now get more energy? Reuters reported in October 2022[9] that Chinese companies and government-run investment banks have so far financed a total of 171.6 GW of overseas power generation capacity, representing a total of 648 plants in 92 countries, with 113.5 GW already operational. Carbon dioxide emissions from these plants total an estimated 245 million tons per year. The decision ended 15 power projects in the planning stages with a capacity of around 12.8 GW, and could stop another 37 GW of capacity currently in the pre-construction phase. At home though China doesn't let up, starting the construction of 33 GW of new coal-fired power generation capacity in 2021, the most since 2016. Its decision is wrong. Instead it should have pledged to stop building *unabated* coal power plants at home and to only fund new coal power plants abroad when likewise equipped with CCS. It could and should have challenged the West to contribute to this towards redeeming their climate debt. But, worse still, at COP26 wealthy nations including the US, UK, Canada, and France pledged[10] an end to public financing for fossil energy projects abroad, a policy which would not affect their own fossil fuel use while restricting energy options for the poorest countries, especially in sub-Saharan Africa, Botswana for instance, if it ever gets to the point that it wants to exploit its vast coal reserves. In effect, it comes down to the rich world asking some of the world's poorest countries to take the lead on decarbonization, as Zeke Hausfather pointed out, while they emit just 1% of global carbon emissions per year.[11]

Or as Arthur Baker and Vijaya Ramachandran recently wrote:[12] "Poverty is still a global scourge. (…) Reducing poverty is not feasible without access to cheap and reliable energy. Indeed, the lack of it in low- and lower-middle-income countries, home to a large majority of the world's poorest people, is one of the world's biggest inequities. Today, for example, only 14 percent of people in Sub-Saharan Africa have access to clean fuels and technologies such as bottled gas for cooking; the rest are forced to rely on wood, charcoal, and animal dung. For them and for others around the world, governments must invest significantly in energy, particularly in energy infrastructure. More electricity generation, improvements to the public grid, and spending on roads and cold storage facilities would allow businesses to create more and better jobs, increase productivity, and boost well-being and human dignity." If you want them to use solar panels and/or wind turbines, go and install them. Make sure that there is sufficient energy storage or other backup, e.g., coal-fired power plants with CCS, a reliable electricity grid to be fed with the generated energy and an accompanying maintenance network. It is only a question of money and since, as

---

[9]   https://www.reuters.com/business/energy/emissions-china-invested-overseas-coal-plants-equal-whole-spain-research-2022-10-25/.

[10]   https://www.theguardian.com/environment/2021/nov/03/twenty-countries-pledge-end-to-finance-for-overseas-fossil-fuel-projects, the Guardian, 3 November 2021.

[11]   https://thebreakthrough.org/issues/energy/who-leads-on-decarbonization, 15 November 2021.

[12]   Arthur Baker and Vijaya Ramachandran, *Let Them Eat Carbon: Entrenching Poverty by Limiting Fossil Fuel Investment Won't Solve Climate Change*, The Breakthrough Institute, 22 March 2022.

the Covid-19 pandemic has shown, there is plenty of money around in these Western countries to save pubs, festivals, restaurants and other irrelevant spheres of activity from bankruptcy, the West has no excuse not to fork out. That could have helped start the third world on the way to gaining access to affordable electricity and to avoid carbon emissions both at home and abroad. Now the third world will look for and probably find other ways to build the coal power plants they need and will then do so as cheaply as possible and most likely without CCS.

Another recent large international study (Duan et al. 2021) showed that a 1.5°C-consistent goal requires China to reduce its carbon emissions by at least 90% and its energy consumption by more than 39% by 2050, compared with the "no policy" case. It is clear that China has no intention of taking such action in the time span required and that energy consumption will continue to grow for a considerable time to come. China is following a completely different route as made clear by the Guardian's report on 17 January 2022 that China's coal production reached record levels in 2021 as the state encouraged miners to ramp up their output to safeguard the country's energy supplies through the winter gas crisis that suddenly gripped the world in late 2021. Chinese coal production climbed to an all-time high of 4.07 bn tons, up 4.7% on 2020 and already more than the projection for 2024 in Table 5.1, but also in Europe and the US coal plants produced 20% more electricity compared to the low levels in 2020, but still remained below 2019 levels. This situation has recently been worsened by the war in Ukraine. Due to reduced supplies of natural gas from Russia, coal power plants in various European countries (e.g., Germany, Netherlands) have ramped up production to save gas for the coming winter.

In 2020 coal combustion worldwide was responsible for 14 Gt of CO_2 emissions, of which half originated in China, 40% of total global emissions (34.8 Gt) in that year.[13] Coal accounts for 72% of total greenhouse-gas emissions from the electricity sector. It is the electricity sector (together with transportation) that can and should be decarbonized first. Individual countries can do this without waiting for a global agreement and it does not distort competition (Jaccard 2020, p. 250ff.). By improving the coal sector a lot of progress can be made, not cheaply but relatively cheaply, compared to sectors that are more difficult to abate, as it is called, such as cement making and home heating. As we will see, such progress is especially possible if we focus on the top-polluting plants.

As of 2021 there are in total more than 4,000 coal power plants[14] in operation in the world, of which 1082 in China and 281 in India.[15] None of these have CCS installations. In 2020 coal power stations generated 9,421 TWh in electricity, down from a peak of 10,101 TWh in 2018; a decrease of 7% in 2 years, greatly being

---

[13] *Our World in Data* (https://ourworldindata.org/emissions-by-fuel); oil accounted for 11 Gt, gas for 7.4 Gt, cement production for 1.6 Gt, flaring for 435 Mt and other industries for 300 Mt.

[14] 6,601 coal-burning units, encompassing 4,000-plus power plants, many with multiple units (https://www.fastcompany.com/90610616/this-map-shows-every-one-of-the-worlds-remaining-coal-power-plants).

[15] As of 2021 there were 252 coal power plants operating in the US, 87 in Japan, 85 in Russia, 77 in Indonesia, 74 in Germany and 50 in Poland (Statista.com). Germany intends to phase out its coal plants by 2030.

helped by the Covid-19 pandemic. It will be no surprise when in 2021 or a year later power generation from coal will hit a new peak, but, even if it doesn't, the rate of decline shown since 2018 will not rid us of coal until all the coal has been burned, making the case for CCS all the more urgent.

China alone operates a little over 1,000 GW in power generation from coal with about 200 GW still in the pipeline. According to a 2020 analysis by *Energy Foundation China*,[16] to keep global warming to 1.5°C all of China's coal plants without carbon capture must be phased out by 2045. That will of course not happen; plants that are built now will have a lifetime of some 40–50 years. Carbon capture is the only road to rescue here. The IEA study mentioned in the preceding chapter (IEA 2016) concluded already some years ago that for China the pathway to providing an 80% chance of achieving the 2°C target requires 185 GW of coal-fired power capacity to be retrofitted with CCS technologies in 2035. Although CCS technology is essential for achieving long-term climate-change mitigation targets, China's subsidy policies tend to support photovoltaic power (solar panels), with few incentives for CCS. Focussing on retrofitting (a sizeable number of) the coal power stations in China and in the rest of the developing world with CCS installations will be the most sensible and most effective way of curbing carbon-dioxide emissions. It should have priority over net-zero projects in affluent regions of the world like the EU. If only the Chinese coal stations were equipped with CCS installations, global emissions could be curbed by about 20%, about three times total annual emissions in the EU. For some coal plants such a measure will come too late as they are at the end of their lifetime, but for newer plants that still have decades to go it is the obvious thing to do, and it should be an absolute requirement for new power plants. It will vastly improve the environment. Although it may be possible for most coal plants to be fitted with carbon capture installations, transport and storage may cause insurmountable difficulties for some of them, but such difficulties are no reason not to install CCS where it is possible.

A medium sized 500 MW coal-fired power plant emits per second about 400 m^3 of flue gas containing about 12% of CO_2 (by volume). Depending on the type of coal used, such a plant emits per year about 3 million tons of CO_2 (or 8,000 tons per day) into the air. If we assume a lifetime of 50 years, we have to sequester 150 million tons of CO_2 coming from 6.3×10^{11} m^3 of flue gas at each power plant. That is enormous, and such large volumes make some CCS critics sceptical, but it does not show or imply that we are facing an impossible task and that we should not even try to take it on. It only shows how huge the problem of continuing to emit CO_2 into the atmosphere actually is and how urgent it is that we should do something about it. On the other hand it is nothing new. These flue gases are already cleaned of other noxious substances such as sulphur and nitrous dioxide, carbon dioxide has to be added to them. If we want to do something about climate change, we simply have no other choice than to tackle this problem head on. The sooner we start the easier it will be.

---

[16] Energy Foundation China, *China's New Growth Pathway From the 14th Five-Year Plan to Carbon Neutrality* (December 2020), p. 24.

Let's look at the coal sector a little more in detail and especially at the most polluting plants. In a 2021 paper (Grant et al. 2021) it was found that 17%–49% of the world's CO_2 emissions from electricity generation (which is responsible for about 25% of total emissions) could be eliminated, depending on the intensity standards, fuels, or carbon-capture technologies adopted by hyper-emitting plants. The authors of the study conclude that this suggests that policies aimed at improving the environmental performance of 'super polluters' are effective strategies for transitioning to decarbonized energy systems.

Table 5.2 lists the ten most polluting coal power plants in the world in 2018. Together they were responsible for a little over 300 Mt of CO_2 emissions, twice total emissions of a country like the Netherlands. These top ten polluters' higher emission levels were not the result of their greater output in electricity. Rather they burned their fuel less efficiently and/or used more carbon-intensive fossil fuels, as can be seen in the last column of Table 5.2. For instance the emissions intensity (kg of CO_2 per MWh of power produced) of the Niederaussem plant in Germany, whose mere existence is an embarrassment to Germany, making a mockery of its environmental credentials, is estimated to be 45% higher than the average for all fossil-fuelled plants in Germany. It also burns lignite, the most polluting and health-harming form of coal, only good for generating electricity in highly inefficient thermal power plants.

The most polluting plant in the world is the Bełchatow plant in Poland, which emits more carbon dioxide than Switzerland, a country of 8.5 million inhabitants, as a whole. In order to reduce its CO_2 emissions, the operator of the plant sought to introduce CCS technology in the past. In 2008, it signed a memorandum of understanding for the design and construction of a pilot carbon capture plant at Unit 12 by mid-2011. A larger carbon capture plant was to be integrated with the new 858 MW unit by 2015, whereby the plant aimed at capturing 1.8 Mt of CO_2 per annum. The project was awarded €180 million by the European Commission from the

Table 5.2. Top ten polluting power plants in 2018. Relative intensity refers to relative emissions per unit of generated electricity, i.e., for the Bełchatow plant 1.756 times more than for the average fossil-fuel plant in Poland (from Grant et al. (2021)).

2018							
Rank	**Plant name**	**Country**	**Tons of CO_2**	**Primary Fuel**	**Age**	**Capacity**	**Relative Intensity**
1	Belchatow	Poland	37 600 000	Coal	27	5298	1.756
2	Vindhyachal	India	33 877 953	Coal	14	4760	1.485
3	Dangjin	South Korea	33 500 000	Coal	10	6115	1.473
4	Taean	South Korea	31 400 000	Coal	12	6100	1.481
5	Taichung	Taiwan	29 900 000	Coal	22	5834	1.282
6	Tuoketuo	China	29 460 000	Coal	10	6720	1.450
7	Niederaussem	Germany	27 200 000	Coal	38	3826	1.451
8	Sasan Umpp	India	27 198 628	Coal	3	3960	1.401
9	Yonghungdo	South Korea	27 000 000	Coal	9	5080	1.481
10	Hekinan	Japan	26 640 000	Coal	21	4100	1.394

European Energy Programme for Recovery in 2009, but was cancelled in 2013 after failing to meet the criteria for receiving a grant within the EU NER300 program.[17]

The above-mentioned paper by Grant et al. investigated the 5% most polluting power stations for each country and concluded that "in China, its extreme polluters accounted for 24.5% of total emissions or nearly five times more than would be expected if emissions were distributed evenly across plants. In other countries like the US, Japan, South Korea, Germany, and Australia, where the distribution of pollution was more extremely skewed, the top 5% of polluters were responsible for 75%–89.6% of all emissions or roughly 15–18 times more than would be expected if emissions were evenly distributed. For the world as a whole, its top 5% percent of polluters contributed 73% of all electricity-based CO_2 discharges or 14.6 times more than if pollutants were evenly dispersed."[18] Table 5.3 shows how much carbon-dioxide pollution from electricity generation could be avoided in a specific country and in the World as a whole if the 5% most polluting plants were targeted in each country.

From column 4 of Table 5.3 it can be concluded that almost half of all emissions from coal-based electricity generation can be avoided if just the top 5% polluting plants in the world are provided with CCS installations! Germany's CO_2 emissions from electricity generation can be lowered by an astonishing 75% if its top 5% polluting plants are fitted with CCS installations.[19] In spite of this seemingly easy way of making progress, absolutely nothing is being done in this respect. If just shutting them down is not an option, switching to natural gas or improving the plant's intensity can also already make a big difference, as columns 2 and 3 in Table 5.3 show. For each plant it must be investigated what is the best way to abate emissions and what is technically possible, but it is obvious that such global-oriented approaches directed at certain 'hotspots' make perfect sense, much more than desperately pursuing net-zero scenarios at all cost that have little or no effect on a global scale.

Let's look at power plant number 6 in Table 5.2: the Chinese plant Datang Tuoketuo, which with a nameplate capacity of 6.7 GW is the biggest coal-fired power plant in the world. It is located in Inner Mongolia Autonomous Region, China. Initially, the power station comprised eight 600-megawatt (MW) units commissioned between 2003 and 2006 before two additional 300 MW units were installed in 2011 to increase production. In 2017, the plant's capacity was again increased with the addition of two further units with a combined capacity of 1,320 MW. The fuel is sourced from the Junggar Coalfield located 50 km away from the plant and it meets its water requirements by pumping it from the Yellow River. All generated power is delivered to Beijing. The annual net electricity output of this station is 33.3 TWh (capacity factor 56.6%).

---

[17] NER300 is a funding programme pooling together about €2 billion for innovative low-carbon technology, focusing on the demonstration of environmentally safe Carbon Capture and Storage (CCS) and innovative renewable energy technologies on a commercial scale within the EU.

[18] Please note that this paragraph concerns emissions from electricity generation only.

[19] The new German government has promised to phase them all out by 2030, so there is no longer any need for installing CCS installations.

Table 5.3. Percentage changes in electricity-based CO_2 emissions if the top five percent of polluters adopted certain intensity standards.

Rank	Country	(1) If top 5% lowered their intensities to their sector's average for fossil fuel plants	(2) If top 5% lowered their intensities to the World's average for fossil fuel plants	(3) If coal and oil plants in top 5% switched to gas	(4) If top 5% incorporated carbon capture and storage
1	China	−3.80%	−9.60%	−11.13%	−20.70%
2	United States	−29.30%	−29.50%	−41.71%	−63.70%
3	India	−8.40%	−20.40%	−23.57%	−44.20%
4	Japan	−39.20%	−35.20%	−43.06%	−76.10%
5	Russia	−20.90%	−17.40%	−34.13%	−37.60%
6	South Korea	−40.00%	−32.70%	−39.26%	−76.00%
7	Germany	−55.90%	−35.10%	−43.71%	−75.90%
8	South Africa	−8.60%	−11.60%	−13.30%	−25.20%
9	Australia	−28.30%	−34.90%	−42.88%	−75.50%
10	Indonesia	−9.00%	−28.00%	−33.38%	−60.50%
	World	−16.81%	−25.40%	−29.50%	−48.90%

The plant consists of twelve units. If each were fitted with a CCS installation, the costs would be about $6 billion in initial capital costs.[20] Spread out over 40 years and assuming that the plant will generate 33.3 TWh of electricity per year as it does today over this whole period, it would imply a cost of $0.0045 per kWh, less than half a dollar cent. There are other costs of course, such as operating costs, compression and transport costs, interest costs, etc., but it seems in any case that the initial capital costs do not have to be an impediment. Over these 40 years it would avoid 40 × 30 Mt = 1.2 Gt of CO_2 emissions, and this is just one plant.

CCS is mostly rejected both by environmental organizations and by the power generating sector. As said, the former consider it a false solution to the climate change problem, while for the latter CCS will considerably add to their costs and in some cases may even cause them to go out of business. It is remarkable to see how two sworn enemies, climate-change activists and carbon emitters, are working hand in glove, albeit unwittingly, at least I hope so, to ensure that nothing happens. The number of power plants fired by fossil fuels is increasing rapidly and thus the emission of greenhouse gases. Developing cost-effective CCS is the only practical way to mitigate the carbon emissions of these plants. In particular, for the many existing plants built without CCS, cost-effective post-combustion carbon capture

---

[20] It is hard to find values quoted in the literature or media for such installations. I have assumed here $500 million per unit. The talk is mostly of the cost per ton of CO_2 which includes all costs and is in the region of $50 per ton (*International CCS Knowledge Centre*).

will be essential to allow their continued operation as part of a low-carbon economy. However, CCS is still only taken up at a pitifully small scale, although the need to apply CCS becomes clearer by the day. As said, China has pledged to move away from coal, but the current (mid 2021) energy crisis and shortage of coal after the resurgence from the Covid-19 pandemic has shown that this may be very hard. China's energy structure is dominated by coal. That is an objective reality, which cannot be ignored. CCS can help resolve the problem. In this respect the Global CCS Institute[21] wrote in 2017 that "regardless of costs, the business case for applying CCS on all industrial applications is strong, in that it is one of the few, complementary and necessary methods available to achieve large emission reductions from these facilities. The perception of CCS being a high cost emission reduction technology largely stems from estimates for the power sector. These estimates are inappropriately compared to renewable technologies that do not provide equivalent grid services or dispatchable power and hence have lower economic value" (Global CCS Institute 2017). As their most recent 2021 report shows the tide is slowly turning, but unfortunately far too slowly.

---

[21] An independent international think tank whose self-proclaimed mission is to accelerate the deployment of carbon capture and storage. It is founded and funded by the Australian government, and headquartered in Melbourne, Australia, with offices in Washington DC, London, Brussels, Beijing and Tokyo. It has a very diverse membership including (oil) industries, banks, (local) governments and carbon capture companies.

Chapter 6

The Need and Status of Carbon Pricing

Introduction

All the reports that have recently been published tell essentially the same story. The promises of emission reductions are most likely too late and too little. Achieving a limitation of global warming in line with the Paris Agreement around the middle of the century requires both rapid and deep reductions in global greenhouse-gas emissions and the scaling-up of removals. Since the world is obviously reluctant and/or incapable of seriously cutting emissions in the short term by scaling down the burning of fossil fuels, or politicians are incapable or unwilling to convince their voters[1] that this *must* be done, large-scale carbon capture, in the form of both carbon capture and storage (CCS) and Direct Carbon Removal, will become increasingly necessary and within a couple of decades unavoidable, the only sensible and relatively risk-free way available to us to avert disaster.

The tools available to governments to reduce greenhouse-gas emissions include voluntary and information policies (mostly ineffective), financial incentives such as subsidies and grants (always helpful), and regulations (setting efficiency standards and suchlike). They can all play a useful role in enticing people to use more efficient and more environmentally friendly equipment and in promoting renewables, but will no more than cause a small dent in emissions. It will not be enough! Moreover, the concentration of carbon dioxide in the atmosphere is already much too high. Apart from preventing carbon from getting in, it must be taken out! We are beyond the stage where policies to set efficiency standards or financial incentives, like subsidies, can do the job. These will only cut some emissions and, although emissions must be cut, for sure, the stubborn emphasis on just cutting emissions will never get us where we want to be, will never limit global warming to below 2°C. Direct Carbon Removal must be deployed, e.g., by Direct Air Capture to scrub carbon from the air, and CCS installations must be (retro)fitted to fossil-fuel fired power plants and other installations that are suitable for this. For this money is needed, lots of money and to collect this money carbon emissions must be priced.

---

[1] Even those who do not depend on voters for their political life do not seem to see the urgency of action in this area.

Any sensible policy towards combating climate change will include some sort of carbon pricing or carbon tax. The idea that billions of tons of carbon dioxide and other greenhouse gases can be spewed into the atmosphere without having to pay for the damage this causes does not make sense from any point of view, in particular also from a *free market* point of view. It creates an unfair advantage for the polluters, similar to the one you gain by illegally and secretly discharging waste water or toxins into open waters, with the only difference that it is not (yet) illegal. It is unfair competition, plain and simple. The economist Nicholas Stern, lead author of the Stern Review, a report on the economics of climate change prepared for the UK government in 2006, called the "policy" of allowing fossil-fuel users to pollute the atmosphere and cause climate change for free "the greatest market failure the world has ever seen".[2] Anybody who tries to devise a product or service (at enhanced cost) that does not involve such emissions is at a disadvantage and will have difficulty gaining a foothold in the market.

It is obvious that the current situation is untenable. So long as polluting the atmosphere with CO_2 (and other pollutants) is free, can be done without having to pay for the current and future harm this causes, there is little chance that any (costly) means of mitigation will be implemented. Competition with something that is free tends to be a useless exercise and is bound to fail. In view of the fact that the world has been aware for quite some time now that emitting carbon dioxide and other greenhouse gases into the atmosphere is harmful, it is actually beyond belief that this can still be done for a great part with impunity, with hardly any cost being attached to it and without having to pay for the damage it causes. The absence of some carbon price is also the main drawback of the current approach of the Biden administration. Political reasons prevented the inclusion of such a price in its latest legislation, the Inflation Reduction Act. The absence of a 'stick' will make it harder, if not impossible, for the 'carrot' to be effective, and a further supply of 'carrots' could well be needed when the current stock has been depleted.

In my opinion, and many agree with me (see, e.g., Jaccard 2020, Chapter 6), there is no alternative to charging those who are emitting carbon or have emitted it for the harm this causes or has caused. A sufficiently high global[3] carbon price can account for all costs, create a more level playing field and the hardest part of the struggle for climate-change mitigation might be over. It will give a boost to renewable energy sources and carbon-capture technologies, as well as to innovation as all corporations whose carbon emissions are taxed will have a powerful incentive to implement ways to avoid emitting the stuff and, by doing so, avoid paying the tax. For a properly working free market in any commodity, a levelling of the playing field is essential, so also in the energy market. A level playing field demands that *all* costs are factored in properly, and that nobody enjoys an unjustified advantage. There is no

---

[2] https://www.theguardian.com/environment/2007/nov/29/climatechange.carbonemissions.

[3] This is quite likely impossible to achieve, knowing what the world is like. But global does not necessarily mean enforced all over the world. Dirty products produced with a lot of unabated carbon emissions can also be taxed in other ways, e.g., when imported into another country or bloc of countries. The EU Border Adjustment Mechanism is one such tool. If applied by a significant number of countries with economic clout, such mechanisms will force exporting countries to adjust their production methods.

way around this. How the revenues from such a tax must be used and who must be compensated how for the extra costs that such a price imposes is a different matter.

Carbon pricing is an essential part of a policy portfolio for tackling carbon emissions. It influences energy use and investment decisions and, when designed well, can encourage a cost-effective set of emission reductions. The revenues can be used for other policies or technologies. In the first place, for the large amounts of money needed to take out the surplus of CO_2 (and possibly other greenhouse gases) in the atmosphere, and for capturing as much of it as possible at power stations and other facilities that use fossil fuels and emit greenhouse gases into the atmosphere. In view of the enormous quantities of carbon that are involved a lot of money will be needed, not only for capturing the carbon, but also for safely transporting and storing it. Others, in particular also James Hansen (Hansen 2009, p. 209ff), argue that the proceeds of such a tax should be distributed uniformly to the public. He calls it a fee and dividend system whereby a fee is charged for carbon-dioxide emissions and paid back to the people in the form of carbon dividend cheques sent every few months to every adult in the country. Such a carbon fee would raise the price of fuel and goods, but by dividing the proceeds equally among all residents of the country people have the choice to reduce their carbon footprint in which case they will receive more than they will pay extra for the products they buy. It sounds sympathetic and the prospect of free money will always attract supporters but I do not consider this a good approach, unless it is the only one feasible in a certain political climate (as is perhaps the case in the US), for although not optimal it is certainly better than nothing and could lead to something more substantial later. It is not optimal as there is no guarantee that carbon emissions will actually go down in such a system, for if a company can just pass on the costs of the fee to its customers by raising the price why take (costly) measures to do something about carbon emissions? In that case the fee and dividend system would just pump some money around whereby some would benefit and others lose out. It would be all right if curbing carbon emissions could be done without incurring any extra costs, but large amounts of money are needed to develop alternative energy sources, storage and carbon-capture technology. Revenues from a carbon tax can be used for this and it is only fair that those who are causing the problems pay for this, both producers and consumers. There is no need to reward people who have no or only a small carbon footprint, just as people are not rewarded either for refraining from crime.

Social cost of carbon

Political barriers tend to result in prices on emissions that are far lower than the true harm they cause to society, resulting in too few emission reductions. The core goal of pricing carbon is to reduce emissions by giving emitters an incentive to stop the practice, but there are often ancillary goals—such as equity, reducing conventional pollutants, stimulating economic or technological development, and cutting other taxes. How then should the price be determined? The *social cost of carbon*, representing the damage caused or the total welfare lost across the globe by carbon emissions, can be thought of as a reasonable target price. According to economic theory the carbon tax should be set equal to this social cost.

Determining this social cost is answering the questions: How much damage will a ton of carbon dioxide emissions released today cause in the future? And how can this damage be weighed against the costs and benefits of actions taken today? Can such action not better be postponed? There is no straightforward and accurate answer to these questions. It requires estimating the impacts of climate change, including impacts on human health, as measured by the amount of damage done and the cost to fix it. There is consequently no obvious and accepted way to calculate this social cost. Various models are used for the calculation, depending among other things critically on the discount rate[4] used. While under the Trump presidency the US Environmental Protection Agency had lowered the social cost to $1–7 per ton of CO_2, early in 2021 the new Biden administration raised it to $51 per ton,[5] using a discount rate of 3.1%. The figure is important as federal agencies use it when evaluating the costs and benefits of regulations. Some critiques argue that this figure is still far too low,[6] with a recent calculation quoting values from almost zero to as high as $1,500 depending on the climate change scenario and the model used, underlining the practical difficulty of obtaining a reliable value for this quantity.[7] In practice, very few carbon pricing efforts have even come close, much less surpassed, the $51 rate, although a turning-point has recently been reached, it seems. A majority of carbon prices still remains far below the range of $40–80 per ton of CO_2-equivalent needed in 2020 to meet the 2°C temperature goal of the Paris Agreement—with only 3.76% of global emissions being covered by a carbon price at or above this range. Even higher prices will be needed over the next decade if we still have the ambition to stay below the 1.5°C target.[8]

That a price must be attached to carbon has been reluctantly realized already quite some time ago and the number of carbon pricing mechanisms is steadily increasing, but most of them cover only a small part of actual emissions. The reasons for introducing such schemes have however nothing to do (yet) with the need to remove carbon dioxide and other greenhouse gases from the atmosphere or from flue gases of installations that emit such gases. In all cases carbon pricing of one form or another is and has been used as a tool to achieve certain emissions reductions. As it is clear by now that all the efforts made today will be insufficient to avoid dangerous climate change, current and future carbon pricing will increasingly have to aim at providing funds for carbon removal schemes. As soon as the price for carbon exceeds the amount needed to capture and safely store the stuff, industry will automatically seriously consider the capture option by doing what they always do, namely following the money by saving costs.

---

[4] See Glossary for a definition.

[5] The $51 figure was challenged in court by energy-producing Republican-led states, but in late May 2022 by declining to consider the challenge the US Supreme Court allowed the Biden administration to use this higher number for how much carbon pollution costs society.

[6] The German Umwelt Bundesamt recommended $202 in 2019 and Moore and Diaz 2015 estimate the true value as $220 per ton.

[7] Kikstra et al. 2021. The value of $1,500 applies to the most optimistic climate change scenario SSP1-1.9 (see Chapter 4) of the IPCC.

[8] World Bank Carbon Pricing 2021.

To preserve a level playing field everybody must participate in a taxation system and charge a similar amount for emitting carbon into the air, or must be made to pay in another way (e.g., via carbon-based import duties), else it will not work. To make progress the international community must work out a global system of one form or another for the improvement of air quality. After all, climate change is a global problem, but nevertheless a global pricing system does not seem feasible. There is no obvious reason for this, other than that so far hardly any efforts towards such a goal have been made, which is astonishing as it has been recognized that from an economic perspective pricing is the most effective means of reducing carbon emissions.[9] I guess that it will be much easier to come to international agreement on such a price than on curbing emissions. The OECD, followed recently by the G20, agreed surprisingly smoothly on a minimum global corporate tax rate. This suggests that a global price for carbon might be easier to agree upon than curbing emissions by a cumbersome system like the NDCs demanded by the Paris Agreement, which make it advantageous for parties to take a position of wait-and-see before committing themselves. It is not necessary to immediately aim for an agreement that covers the entire world. One possibility would be for the largest economies (US, China, EU, Japan and India, just five entities, covering more than 60% of carbon-dioxide emissions in 2021 and representing about 70% of global GDP) to start negotiations on a minimum price of carbon. Within such a small group of parties it will be easier to come to an agreement, certainly if the developmental status of countries is taken into account, which in this group especially applies to India, and to a lesser extent to China. The OECD and IMF have already called for such an agreement, but none of the countries mentioned seems keen to take the initiative. If agreeing on a minimum price for all emissions is a bridge too far for the time being, a start could be made by setting a price for emissions from certain sectors, such as cement and steel production that are hard to abate, or emissions that are international in nature like from aviation and maritime transport. For aviation this could fairly easily be established by levying a global carbon tax on aviation fuels. These are now not taxed at all and in fact, due to its cross-border nature, aviation is currently evading most taxes and is lagging behind on climate targets. One drawback is that, although not insignificant, the carbon-dioxide emissions from aviation only amount to 2.5% of total global emissions and from maritime transport to 1.7%. Apart from carbon dioxide planes also emit nitrogen oxides, water vapour, sulphate and soot particles at high altitudes, in total representing about 3.5 % of global warming.[10] The proceeds could be used for developing carbon-free synthetic fuels or for scrubbing carbon out of the air. A third possibility is that existing emissions trading systems of different countries or groups of countries are further coordinated. The EU, China, Canada and South-Korea all have emissions trading systems and talks are under way between these parties to better align their systems from a technical perspective, for instance

---

[9]   Pricing can also take the form of price rises. The exploding energy prices due to the Ukraine war have resulted in substantial (up to 25%) reductions in natural-gas consumption in European countries, which will also result in emissions reductions (if the burning of gas is not replaced by the burning of coal or wood).

[10]   *Our World in Data;* https://ourworldindata.org/co2-emissions-from-aviation.

in the way emissions are measured. When standards can be agreed upon it will also be easier for other countries to join. These systems still cover only a little over 20% of total global emissions.[11]

Carbon tax or cap-and-trade system?

The world community is still far from putting a uniform and stringent price on greenhouse-gas emissions as the first best strategy to reduce emissions and combat global warming, but a number of regional schemes are now in place. Carbon pricing can be achieved either by imposing a cap on emissions, using a permit trading scheme to limit emissions and achieve the most cost-effective set of reductions, or by imposing a straight carbon tax on emissions. In the latter case a carbon tax is levied on the carbon content of goods and services, predominantly in the transport and energy sectors, but could be levied on any product. It sets the price of carbon-dioxide emissions and allows the market to determine the quantity of emissions reduction, while a cap-and-trade system sets the quantity of emissions reductions and lets the market determine the price. It is choosing between price certainty and quantity certainty. Which is better? Both can successfully and cost-effectively reduce carbon emissions. Proper design is more important than the type of policy selected; the two approaches are more similar than different. Whether or not they produce a good result depends largely on design and execution (Harvey et al. 2018, p. 255). A hybrid approach whereby a carbon tax is blended with a carbon cap can also be useful in balancing environmental and economic goals. Higher-than-expected emissions or higher-than-expected carbon prices are both undesirable. Combining a cap with a tax can balance these risks and reduce uncertainty.

Both methods intend to reduce carbon-dioxide emissions by increasing prices, thereby decreasing demand for goods and services that involve such emissions, and making more environmentally friendly alternatives more competitive.

The cap-and-trade method creates a carbon market. At first glance this seems a rather paradoxical result as it has been the institution of the (free) market as a competition based mechanism with the purpose to make money that created the problem in the first place. This new market essentially trades in a public good, like clean air, which earlier came for free, and for this reason is understandably somewhat repugnant to an environmentalist. Moreover, markets generally favour developed industrialized nations and it is not a priori clear that they can achieve an equitable outcome for developing countries (See, e.g., Chichilnisky and Bal 2019, Chapter 3).

In the case of a carbon tax the emission reduction outcome is not pre-defined but the carbon price is. That implies that a carbon tax may result in no emissions reduction at all if the emitter succeeds in passing on the extra costs incurred through the carbon tax to his customers or somewhere else, or if the tax is just too low to make an impact, or if the tax burden is the same for all competitors on the market, or if there are no or very few competitors. Such an unfavourable outcome can be

---

[11]   DNB, Towards a Global Price for Carbon Emissions, 3 November 2021; https://www.dnb.nl/en/actueel/dnb/dnbulletins-2021/towards-a-global-price-for-carbon-emissions/.

avoided by starting with a tax at some level and gradually raise it until you see the desired result, a substantial lowering of emissions, emerging.

In Norway a carbon tax has been levied since 1991. It covers about 64% of Norwegian CO_2 emissions and 52% of total greenhouse-gas emissions. It is not so that this carbon tax has resulted in a decrease of Norwegian emissions[12] which in 2020 were still about 25% higher than in 1991, while other countries, e.g., Germany have managed to substantially reduce their emissions compared to 1991 without such a tax (even by more than 40%).[13]

The Norwegian taxation has however resulted in the first offshore carbon capture and storage plant at the Sleipner gas field (see Chapter 12). Before being suitable for the market, part of the CO_2 in the gas from the Sleipner field had to be taken out to bring the concentration down from 9% to 2.5%. The surplus of CO_2 could have been released into the air and a tax would have to be paid for this. The exploiter of the field, the Norwegian state-owned oil and gas company Statoil (now called Equinor), chose not to vent this surplus into the air, but to capture and store it and thus to avoid paying the Norwegian carbon tax. One million tons of CO_2, which by the way amounts to about 2.5% of total Norwegian emissions, have annually been captured and stored 1,000 metres under the seabed since 1996. Operating costs are $17/ton of CO_2 stored and the project required an initial investment of $80 million. This is the one (and so far only) clear success story of CCS. It was the tax that provided the incentive for Statoil to act as it did. Without this tax the CO_2 burden of the atmosphere would have been 25 Mt higher. The costs for Statoil have been 25 × $17 million (operating costs) + $80 million = $505 million, while without CCS the tax would have amounted to $1.275 billion (at the 1991 tax rate). Clearly a profitable business, and at the same time environmentally sound. Although it does not matter from a taxation or environmental point of view, it should be noted that it does not concern here CO_2 emissions from burning gas or another fossil fuel, but a surplus of CO_2 in the mined product that has to be taken out to make it suitable for the market.

In the cap-and-trade variant of emissions trading, a limit on access (the cap) to a resource (the atmosphere) is defined and then allocated among users in the form of permits or allowances. The system, also known as an emissions trading system (ETS), caps the total level of emissions and allows industries with low emissions to sell their surplus allowances to larger emitters. Such a trading scheme for greenhouse-gas emissions essentially works by establishing property rights to the atmosphere. The atmosphere is a global public good, and greenhouse-gas emissions are a so-called international externality.[14] An obvious and immediately apparent drawback

---

[12] It has been argued that the relatively small effect of the tax is due to extensive tax exemptions and relatively inelastic demand in the sectors in which the tax is actually implemented (Bruvoll and Larsen 2004). Still Norway seeks to increase its tax on carbon dioxide to €200 by 2030.

[13] Per capita German emissions went down from 12.8 tons per capita in 1991 to 7.7 tons in 2020, while for Norway per capita emissions in 1991 were 8 tons and 7.6 tons in 2020, hardly any decrease in 30 years. Comparisons between countries are never easy as there are always special circumstances. German emissions were high in the past with a maximum of almost 15 tons per capita in 1980, while Norway has never exceeded 10 tons per capita and its 1991 level is virtually equal to Germany's 2020 level.

[14] See Glossary.

of such a system is that the number of permits awarded is calculated according to existing levels of pollution (called grandfathering), which means that those who are polluting the most are initially rewarded with the greatest number of permits. This can be seen as a free gift of pollution rights to the worst industrial polluters who out of the blue are allocated tradeable property rights. Instead of cleaning up their own act, a polluter can trade these permits with another polluter who is able to make 'equivalent' changes more cheaply.[15] This is a fair criticism but the idea is that the availability of carbon permits will gradually be reduced, by lowering the cap. The market will then retain its value while at the same time forcing a reduction in the overall level of pollution. That is indeed how the EU ETS has turned out, as we will see below.

By creating supply and demand for emission allowances, an ETS establishes a market price for greenhouse-gas emissions. It is argued (Chichilnisky and Bal 2019, p. 240) that from trading carbon credits, or rights to emit, a carbon price emerges that represents the monetary value of the damage caused by each ton of CO_2. The carbon market introduces a carbon price that corrects the negative impact the CO_2 emissions have on the climate. Is that actually true? Nobody knows the actual damage a ton of CO_2 does to the climate and the price established in the market has nothing to do with it. It is just a cost factor for industry created by an artificial market. The fact that the price is very volatile disproves the assertion that it has anything to do with the actual damage. For if in January the price is X, how then can the price in February be Y, with Y considerably lower or higher than X? The damage done to the climate is about constant and does hardly vary over such a short period of time, and certainly not as violently with time as the carbon price does. The cap is useful of course as it helps ensure that the required emissions reduction will be achieved by keeping the emitters (in aggregate) within their pre-allocated carbon budget. Such caps on the emissions of industrialized nations can be used to guarantee that global emissions will not exceed levels that could cause catastrophic climate change. But, in that case you have to know what those levels actually are, for you can also be sure that emissions will not drop below the cap set, so the success of such a system very much depends on how and at what level such a cap is set. It is claimed that this allows the system to find the most cost-effective ways of reducing emissions without significant government intervention. That may indeed be the case, but not necessarily so, as it actually only determines a (time-dependent) bottom price for carbon emissions. If it reduces emissions that is only so because a cap has been set. It is also true though that any price on carbon, however small and volatile, will help clean energy become more profitable and fossil-fuel energy less, but with a carbon market there is no certainty that the price will be high enough for clean energy to gain the edge over fossil-fuel energy. So, there is no guarantee that it fosters green development. A straight (sufficiently high) carbon tax would actually be an easier means to achieve this.

It is generally thought that both approaches entail risks, but by building in mechanisms which allow the tax, or the cap and allowances to be lowered or raised a required emissions reduction can be obtained by both methods. A perverse additional

---

[15] Gilbertson and Reyes 2009 (https://www.daghammarskjold.se/publication/page/12/).

feature of a cap-and trade system is that it allows an emitter with surplus allowances to make money on something that in an ideal world should not be done at all, and on tradeable goods (allowances) that did not exist earlier and were created by the cap-and-trade mechanism itself. This has indeed given rise to some unsavoury practices with large polluters raking in large amounts of money in this artificially created market. A carbon tax has the further and in the current world possibly overriding advantage that it is less prone to corruption and fake deals, which will inevitably mar an ETS.

The carbon market's supporters endorse the latter as "the most environmentally and economically sensible approach to controlling greenhouse-gas emissions" and cite evidence of its success, while its opponents call it "ineffective" and claim it suffers from an inherent contradiction in pursuing its dual aims of "trying to save the world and trying to make money".[16]

So far, this discussion has not mentioned carbon capture at all, it has no relation to the cost of carbon capture. For carbon capture to become a viable option, it is essential that the carbon tax is set at a level that is equal or very close to the cost of removing the emitted carbon from flue gases or from the air, in the same way as sewer charges are set equal or slightly above the cost of treating waste water. If that costs \$100 per ton, so be it, set the tax at that amount and give it to a contractor to remove the carbon. That is not an ideal situation either as in such a system the contractor will have little incentive to try to improve the system and lower the cost of carbon capture. In this new field innovation will be essential as at \$100 per ton capturing 1,000 Gt will cost \$100,000 billion, comparable to total current global GDP, and there is obviously a lot to gain by bringing the cost down, but getting below \$100 per ton will be very, very hard indeed.

Present situation as regards carbon pricing

As of 2021 64 carbon pricing instruments are in operation and three are scheduled for implementation, an increase of six instruments compared to 2020. Carbon pricing instruments in operation now cover 21.5% of global greenhouse-gas emissions, where in 2020 this was only 15.1%. The increase is mainly due to the launch of China's national ETS, in February 2021, which will be the world's largest carbon market in volume. It aims to "contribute to the effective control and gradual reduction of carbon emissions in China and to the achievement of green and low carbon development." A laudable aim, which it has in common with most aims. A first crucial test will be set by the current energy crisis following the Covid-19 pandemic, which has forced the Chinese authorities to increase the use of coal. Not a propitious sign. Initially covering around 2,162 entities in the power generation industry, the plan regulates annual emissions of around 4.5 Gt of CO_2 (a little over half of emissions by China's coal sector). For comparison, the EU ETS emissions cap in 2021 was 1.6 Gt of CO_2. The regulated entities in the Chinese system will need to surrender allowances in 2021 to cover their 2019 and 2020 emissions. Trading started in July 2021.

---

[16] Rebecca Dobson, Carbon market corruption risks and mitigation strategies, U4 Expert Answer, Transparency International, 2015.

The biggest difference with the EU ETS, to be discussed below, is that as yet there is no absolute cap in China's ETS. All covered companies are allocated their emission allowances free of charge and each emitter is allocated allowances equal to their verified emissions. Given this approach, China's national ETS is actually not yet a cap-and-trade system. Nonetheless, those companies that are able to reduce the carbon intensity of their production can generate a surplus of allowances to sell.[17] However, without an absolute cap there will at first be little demand for them.

The national carbon market will be a tool to promote China's commitment to peak carbon before 2030 and to achieve carbon neutrality before 2060 (World Bank 2021). It did not get off to a good start with reports soon speaking of "China's new carbon trading market isn't working".[18] The biggest problem apparently is that far too many allowances were handed out, a problem the EU ETS also had to grapple with in its early days. Most companies have plenty of credits to spare and there is hardly any need for trading. It started off with a carbon price at just under $8 per ton of CO_2e, and hovered between $8-$9 per ton of CO_2e in 2022. At this price it will not do much to reduce Chinese emissions. For comparison, on August 20, 2021 when the price in China hit $7.72 it was $63.80 in the EU ETS. Activity in 2021 was limited with the Chinese ETS trading a total of 412 Mt of allowances. The EU ETS, however, also had a slow start with only 321 million allowances transacted in its first year 2005, topping 12 billion by 2021. In the EU ETS prices were also too low from 2005 for its entire first period, until one to two years ago. We should also grant the Chinese system some time to get settled, although they could have drawn some appropriate lessons from the European experience. There is not always a need to repeat mistakes.

Before the Chinese ETS came online, the EU ETS was the largest carbon trading system in the world. It has been operating since 2005 and we will discuss it and other European initiatives separately below.

European Union Emissions Trading System

The EU has been operating a cap-and-trade system, the European Union Emissions Trading System (EU ETS), since 2005. Until the Chinese ETS came on line, it was the most comprehensive approach for emissions pricing in the world, setting a cap on carbon emissions from energy-intensive sectors in the EU. A cap (maximum) is set on the total amount of greenhouse gases that can be emitted by all participating installations (energy-generating installations and energy-intensive industries, as well as aviation), which are collectively responsible for close to half of the EU's anthropogenic CO_2 emissions and 40% of its total greenhouse-gas emissions. Allowances for emissions are then auctioned off or allocated for free, and can subsequently be traded. Installations must monitor and report their CO_2 emissions, ensuring they hand in enough allowances to the authorities to cover their emissions. If emissions exceed what is permitted by allowances, an installation must purchase allowances from others. Conversely, if an installation has performed well at reducing

---

[17] https://www.energymonitor.ai/policy/carbon-markets/carbon-trading-the-chinese-way/.

[18] https://qz.com/2050774/chinas-new-carbon-trading-market-isnt-working/.

its emissions, it can sell its leftover credits. The primary goal of this policy is to achieve a given reduction target for aggregate CO_2 emissions at minimal cost. A longer-term objective, to be achieved by gradually reducing the cap, is to stimulate innovation that will help with the transition to a low-carbon economy.

The scheme started with a trial phase in 2005, with almost all emission credits being given to companies for free. The available evidence suggests that, after a shaky start, the EU ETS has had a robust negative impact on carbon emissions (Martin, Muûls and Wagner 2016; Bayer and Aklin 2020) and that emissions in the EU have been reduced at costs that are significantly lower than projected. The system succeeded in decreasing the equivalent of over 20% of the EU's carbon emissions. Overall, the estimated cost was a fraction of 1% of GDP. On the other hand, in the early days it was criticized for several failings, including: oversupply of emissions allowances (too many credits allocated as in China), windfall profits from trading in allowances,[19] price volatility, and in general for failing to meet its goals.[20]

The oversupply of emissions allowances had the consequence that, when the economic crisis hit a few years later, production and consumption contracted and emissions dropped on their own. The emissions market was drowning in excess permits causing the price of carbon to drop dramatically to as low as a few euro per ton of carbon, removing the incentive to shift away from dirty energy (Klein 2014, p. 224).

It is probably fair to say that these vicissitudes were due to "learning problems". It was a completely new mechanism that had to establish baselines and create the infrastructure for a carbon market, not to achieve significant reductions right from the beginning. There are still complains though, and rightly so, that also in the revised EU ETS free allowances should be better targeted and assigned to installations that are investing in clean production processes and have a clear decarbonization plan to contribute to the EU's short and long-term climate targets.[21]

Some critics are also blaming the EU ETS, in combination with the closure of nuclear power stations, for the current (2021) global energy crisis, because of record carbon price increases (the price in the EU ETS now (February 2023) stands at the sizeable amount of more than €100 per ton, where in earlier years it was much lower with a maximum of €30 in 2006 and the negligible amount of just €0.10 per ton in September 2007). Much of this criticism is rather disingenuous (in the early days the complaint was that the price was too low and now as it finally starts to bite it is blamed for other woes) and is made by those who are opposed to any carbon pricing or to carbon pricing that affects their industries. The EU Commission must be praised

---

[19] A report of May 2021 by CE Delft, entitled "Additional profits of sectors and firms from the EU ETS 2008–2019," shows that energy-intensive industries across Europe have profited up to €50 billion from 2008 to 2019 as a result of the free allocation of pollution permits under the EU ETS. These profits arise from overallocation of free emission allowances, from using cheaper international offsets for compliance, from passing through (part of) the opportunity costs of freely obtained allowances into product prices.

[20] EU ETS: failing at the third attempt; https://corporateeurope.org/sites/default/files/sites/default/files/files/article/eu-ets_briefing_april2011_0.pdf.

[21] November 2022 Open letter from a number of organisations to ETS negotiators, https://carbonmarketwatch.org/wp-content/uploads/2022/11/Letter-on-ETS-and-CBAM-for-trilogue.pdf.

for mostly sticking to its policies. When in 2012 aviation emissions were included, it encountered opposition from both the US and China, with the US enacting legislation to forbid its airlines to participate in the EU ETS and China threatening to cancel $60 billion in orders for Airbus. In the end, after a delay of a year, only flights within the EEA are covered, while transcontinental flights are not. All airlines operating in Europe, European and non-European alike, are required to monitor, report and verify their emissions, and to surrender allowances against those emissions. They receive tradeable allowances covering a certain level of emissions from their flights per year. According to the European Commission,[22] the system has so far contributed (e.g., by improvements in fuel efficiency) to reducing the carbon footprint of the aviation sector by more than 17 Mt per year.

One of the major changes introduced in phase 3 of the EU ETS (the years from 2013–2020) is that power plants will no longer receive free allocations and have to pay the EU ETS carbon price for their pollution.[23] To address further deficiencies in the system the European Commission proposed some structural reforms in 2014, including a linear reduction factor for the overall emissions cap from 1.74% (2013–2020) to 2.2% each year from 2021 to 2030, thus reducing EU CO_2 emissions in the sectors covered by the ETS by 43% compared to 2005.

The EU ETS has contributed positively to the impressive emissions reduction in the EU from 3.74 Gt CO_2 in 2005 to 2.60 Gt in 2020 (a reduction of 30%). Per capita the reduction is even more impressive: from 8.60 ton per capita in 2005 to 5.84 ton in 2020, just 1 ton above the World average for one of the most developed areas of the world.

With the announcement of the European Green Deal in 2020, the EU ETS will be revised to bring it in line with the 2030 Climate Target Plan to reduce net emissions by at least 55% by 2030 (compared to 1990 levels), which means to 1.74 Gt of CO_2,[24] and to achieve climate neutrality by 2050. (World Bank 2021.) The deal sets binding new targets on energy efficiency for EU member states; increases the proportion of renewables in the EU's energy mix to 40% by 2030; and expands carbon sinks by at least 15% through new regulations on agriculture, forestry and land use, as well as the planting of 3 billion trees. The most far-reaching proposal is that "all new cars registered as of 2035 will be zero-emission", thanks to a continent-wide network of charging stations. The plan will tighten the bloc's existing Emissions Trading System, while also extending it to the building and transportation sectors, including aviation and shipping.[25] The targets set are ambitious and the question is whether the war in Ukraine will not derail these plans, for instance as coal-fired power stations are kept open longer than intended. On the other hand, energy is no longer cheap, making savings in energy more attractive and for many people a necessity in order to continue to be able to pay their bills.

---

[22] https://ec.europa.eu/clima/eu-action/transport-emissions/reducing-emissions-aviation_en.

[23] LIFE ETX (2021) EU ETS 101, A beginner's guide to the EU's Emissions Trading System (2022), p. 13.

[24] This reduction in the ten years from 2020 to 2030 is slightly higher than the reduction achieved from 2010 to 2020.

[25] https://www.cfr.org/article/eu-green-deal-just-raised-bar-climate-policy.

The EU Market Stability Reserve

As a response to the large surplus of allowances that accumulated in the EU ETS market following the economic crisis of 2008–2012, a further reform of the system involved the creation of the EU Market Stability Reserve (MSR), a 12% reserve mechanism of verified annual emissions in the fourth ETS period from 2021 to 2030, which creates a quasi carbon tax or *carbon price floor* with a price range set each year by the European Commission. It works by automatically moving allowances to the MSR when there are too many allowances in the market and releasing allowances from the MSR when there are too few, as a means of providing price stability. As a further reform, the share of allowances put into the MSR was increased from 12% to 24% and from 2023 onwards all allowances in the MSR above the total number of allowances auctioned during the previous year become invalid. These measures have played a key role in reducing the surplus in the market. The market participants factored in the expected future scarcity of allowances in their business plans, which resulted in increasing prices in the EU ETS from around €5/tCO2 in 2017 to over €80/tCO2 by the end of 2021, holding on to this price level through 2022. In August 2022 it touched an all-time high of €99/tCO2 and has now in February 2023 reached more than €100, as can be seen in Fig. 6.1.

Experts also credited the MSR with the quick bounce-back in prices following an initial drop at the beginning of the Covid-19 pandemic in early 2020.

Fig. 6.1. Price development of the EU ETS carbon permit price (in € per ton CO$_2$) (*Source* LSEG (Refinitiv)).

Carbon border adjustment mechanism

In view of the absence of a global climate treaty which prescribes a legally binding carbon price or legally binding emission caps for individual countries, the EU is currently contemplating the imposition of carbon-based import tariffs on carbon-intensive goods (called the carbon border adjustment mechanism) in order to reduce carbon leakage, i.e., the relocation of emissions from countries that regulate emissions to parts of the world economy subject to no or weaker regulations, and increase the cost-effectiveness of unilateral CO_2 emission pricing in the EU ETS. The ETS only applies to entities within the EU and, if carbon-intensive goods are imported into the EU free of charge, it creates unfair advantages to manufacturers outside the EU that are not subject to a carbon-pricing system, or entices European producers to move production abroad. Since the prospects for globally coordinated stringent emission pricing are bleak, tariffs on the carbon embodied in goods imported from regions with lax or no emission regulations are seen as an important means to foster the efficacy of unilateral emission pricing. Importers of carbon-intensive goods will pay a charge linked to what they would have had to pay if they had been covered by the EU ETS. At the outset the measure will be limited to a few sectors, with power, cement, steel, aluminium, and fertilizers the likeliest candidates. Europe imports electricity from Russia, Ukraine, and the Western Balkans. The biggest sources of cement imports are Belarus, Colombia, Turkey, and Ukraine, while steel is brought in mainly from China, Russia, Turkey, the UK and Ukraine. The mechanism will be gradually extended to other industries in the coming years.[26] Such a tax will of course be vehemently opposed by others, in particular India and China. The momentum however seems to be in favour of such measures, with Canada recently announcing that it is studying the introduction of its own carbon border adjustment mechanism and the US exploring the idea of a border adjustment tax on countries that fail to meet their climate obligations, which makes the US the only country without a national carbon price that openly considers border adjustments.

Other Carbon Taxation in Europe and Elsewhere

In this section carbon taxation is illustrated with examples from various parts of the world, without aiming to be exhaustive. If anything, they show that there is in general a tentative willingness to do something about greenhouse-gas emissions, but not wholeheartedly in view of any possible negative economic effects.

In Europe there are currently 18 countries that have some sort of carbon tax (most of them in addition to or complementary to the EU ETS). In 1990, Finland was the first country in the world to introduce such a tax. Since then, many European countries have followed, implementing carbon taxes that range from less than €1 per ton of carbon emissions in Ukraine and Poland to over €100 in Sweden. The scope of greenhouse gases covered also varies from a puny 3 and 4% in Spain, respectively Poland to 71% in Ukraine. The tax rate in Ukraine is however negligible at €0.37 per ton.

Carbon taxes can be levied on different types of greenhouse gases, such as carbon dioxide, methane, nitrous oxide, and fluorinated gases. Obviously a low carbon tax on just a few per cent of emissions, like in Poland (€0.09 on 4% of emissions), Ukraine, Estonia and Spain is completely useless. One of the biggest challenges in designing carbon pricing is finding a way to lower emissions while minimizing economic impact.

---

[26] Bloomberg, 8 April 2021. No doubt the war in Ukraine will also affect these import flows.

It should also be noted though that for some countries it is harder to levy such a tax than for others. For France and Sweden, for instance, which generate most of their electricity from respectively nuclear power and hydropower, it is easier than for Germany or the Netherlands which almost by necessity have to burn a lot of fossil fuels.

The tax in Sweden is primarily levied on fossil fuels used for heating purposes and motor fuels. Sweden's total CO_2 emissions rank among the lowest in Europe; just 36 million tons in 2021 (3.4 tons per capita). For the Netherlands these figures are 141 million tons and 8.1 tons per capita, more than twice the Swedish figure; and for Germany 675 million tons and 8.1 tons per capita (2021 figures). Still even for these countries a tax at €100 per ton would not be outrageous; less than €1,000 per capita[27] and adding up to total carbon tax revenues of €14.1 billion and €67.5 billion, respectively (if levied on all carbon-dioxide emissions). These are by all means sizeable amounts but still fairly modest compared to their respective GDP, and dwarfed by the subsidies and support doled out during the recent Covid-19 pandemic, which for the Netherlands alone amounted to more than €80 billion. If such amounts can give a substantial boost to resolving the impending climate crisis, then one should not let such an opportunity slip by.

The Irish government agreed to a steeper trajectory for the carbon tax, increasing the target rate for 2030 from €80/tCO2 to €100/tCO2. Germany launched its national fuel ETS on January 1, 2021 at a fixed price of €25, covering all fuel emissions not regulated under the EU ETS (mainly heating and road transport—around 40% of national greenhouse-gas emissions). Some fuels (e.g., coal, waste) will be phased in subsequently in 2023. In the next years, the fixed price will continuously to rise from €30 in 2022 to €55 in 2025. In 2026, allowances will be auctioned within a price corridor that ranges between €55 and €65. From 2027 onward, allowance prices will be set by the market unless the government proposes a new price corridor in 2025. The cap is set based on Germany's mitigation targets for sectors not covered by the EU ETS. Revenues are used for a variety of measures, in particular to support decarbonization, to lower electricity rates for consumers, and to deduct transport costs from income taxes for commuters. The Netherlands Industry Carbon Tax Act also entered into force on January 1, 2021, with a rate of €30/tCO2e. The policy applies to industrial installations in the Netherlands that are subject to the EU ETS— acting as a top-up fee—as well as to waste incinerators and facilities emitting large amounts of nitrous oxide that are not covered under the EU ETS (World Bank 2021, p. 66).

Figure 6.2 shows the hotchpotch of carbon pricing measures taken in various countries and regions in the world.

Upon departure from the EU, the United Kingdom stopped participating in the EU ETS on January 1, 2021. On the same day, the UK ETS came into operation, closely resembling the EU ETS. Covering the power, industry, and domestic aviation

---

[27] Most to be paid by industry, not by citizens. For an average Dutch family the carbon tax to be paid for home heating and suchlike would be very modest at a couple of hundred euro per year, in addition to €0.23 per litre of petrol (one litre of petrol generating 2.3 kg of CO_2).

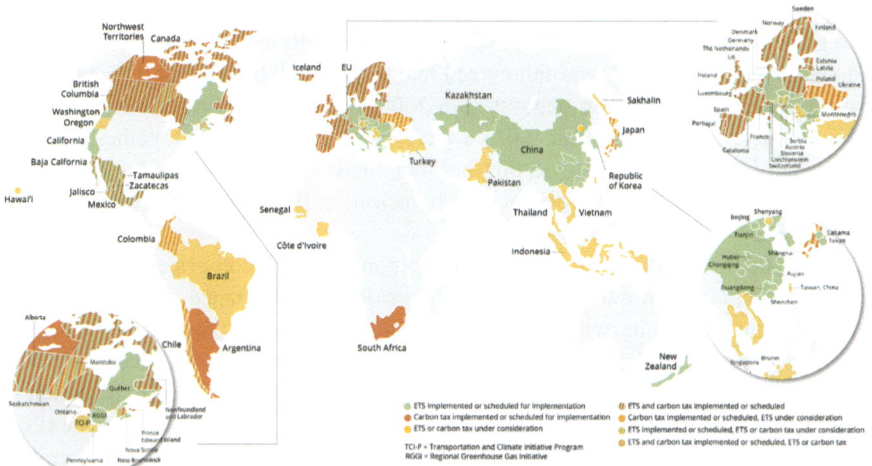

Fig. 6.2. Countries and regions in the world where a carbon pricing scheme is implemented, scheduled or under consideration (*from* World Bank 2021).

sectors, the cap will reduce emissions by 4.2 Mt annually (where total UK CO_2-emissions in 2021 were 347 Mt) and will be revised in 2024 in line with the country's 2050 net-zero trajectory.

Two regional carbon taxes came into effect in Mexico. The Baja California carbon tax entered into force on May 1, 2020, for the sale of various petroleum and natural gas products. Revenues will be redistributed to the five municipalities and allocated to the state budget. In July 2020, the Mexican state of Tamaulipas passed legislation enacting a carbon tax starting in 2021, equivalent to about \$12/tCO2e to fixed sources and facilities that emit more than 25 tCO2e monthly.

Slightly surprising is that Canada, one of the most polluting countries, has some pricing system in place since 2019. In that year most Canadian provinces and territories introduced initiatives in response to the federal government's Pan-Canadian Approach to Pricing Carbon Pollution. Furthermore, Canada's federal backstop system—which includes an ETS and a fuel charge similar to a carbon tax—has been imposed on provinces and territories that do not opt into the system, or that do not put in place a sufficiently ambitious carbon pricing mechanism. This federal backstop consists of two components: (i) a regulatory charge on fossil fuels set at \$14/tCO2e in 2019 that rises by \$7/tCO2e per year to \$35/tCO2e in 2022; and (ii) an Output-Based Pricing System (OBPS) that sets emission intensity standards for power generation and a wide range of activities, and applies to facilities in backstop jurisdictions emitting more than 50 kilotons of CO2e per year or any eligible facility that voluntarily chooses to participate. The two parts of the federal system can be implemented together or separately. Revenue from this system amounted to \$204 million in 2020.

Finally we mention New Zealand that has an ETS in place since 2008, covering 51% of emissions. In 2019, the New Zealand government announced measures to strengthen its ETS and help New Zealand meet its climate change and NDC targets, including a phaseout of free allocations for the industrial sector. From 2021 to 2050 free allocation will be reduced by at first 1% per year in the first decade from 2020 and growing to 3% in the last decade to 2050, which still would not get rid of all free allocation. In late 2019 it was announced that a price will be put on GHG emissions from the agricultural sector, beginning in 2025. For livestock emissions, a separate alternative pricing mechanism will be developed at farm level, together with the agricultural sector. However, if this has not made enough progress by 2022 for implementation in 2025, emissions will be priced at the processor level and likely through the New Zealand ETS.

To date, no single country imposes any mandatory carbon price on agricultural emissions and current evidence suggests considerable reluctance to the application of other climate policies with comparable stringency to agriculture. New Zealand is the first country to at least develop plans in this direction. It is clear that more must follow as agricultural practices contribute significantly to atmospheric pollution of greenhouse gases, and all global emission mitigation pathways, detailed by the IPCC in its Special report on Global Warming of 1.5°C, that would achieve the temperature goal of the Paris Agreement include substantial reductions in agricultural greenhouse gases (methane and nitrous oxide). These mitigation pathways achieve the necessary emission reductions by applying prices of typically several hundred \$/tCO2e by 2050 to agriculture as well as the energy sector. From what we have seen in this chapter it is clear that there is a large gap between such model scenarios and reality (Specifically on agricultural emissions, see Leahy, Clark and Reisinger 2020).

Further information on carbon pricing in place in other countries can be obtained from Fig. 6.2 and from the annual *State and Trends of Carbon Pricing* published by the World Bank.

Concluding remarks

A carbon tax levied on products from developed countries will increase the import cost of such products for developing countries with, in any case in the short run, potentially negative welfare impacts. This is actually true and unavoidable for any mitigation policy that drives up the price of carbon (Ellis et al. 2010).

Apart from the price in the EU ETS in recent years, most carbon prices still remain far below the \$40–80/tCO2e range, recommended for 2020 in the World Bank's report of the High-Level Commission on Carbon Prices. A carbon price at these levels will not be sufficient to generate the kind of changes at the speed and scale required to achieve the Paris targets. Other complementary policies, and higher prices, will be needed to limit warming to 1.5°C.

Reduced economic activity as a result of Covid-19 saw allowance prices briefly dip before quickly recovering in most ETSs. This provides a stark contrast to the 2007–2008 financial crisis and subsequent economic downturn, which led to sustained price depressions across multiple systems.

As the World Bank states in its State and Trends of Carbon Pricing 2021, carbon pricing schemes can play a role in incentivizing low-carbon action by internalizing the cost of greenhouse-gas emissions. However, in order to work they must be sufficiently ambitious with prices in the above-mentioned \$40–80 range, must be well-designed and adapted to the specific context of a jurisdiction, must form part of a supportive policy package—policies to drive research and development, unlock non-economic barriers to mitigation and to target emissions reductions with very high abatement costs. In the EU, allowance prices have hit all-time highs as the bloc steps up both long and short term climate ambitions and the market foresees caps tightening following the announcement of the Green Deal. Prices are increasing in countries like Canada, Germany and Ireland. New Zealand's Climate Change Act sets out changes to its ETS and outlines a national mitigation framework in line with a 2050 net-zero target. This greater ambition is also leading more governments to consider carbon border adjustments, which will force non-complying parties to join on the penalty of being excluded from international trade (World Bank 2021, p. 9).

In 2020, carbon pricing instruments generated $53 billion in revenue globally, with $22.5 billion in the EU at a price of $49.8 levied on 39% of emissions. This is an increase of around $8 billion compared to 2019, largely due to the increase in the EU allowance price, but still laughably little in view of the size of the problem and hardly a burden for any industry (which anyway will be able to pass it on to its customers).

Finally, we note that the war started by Russia in Ukraine will put all carbon pricing, indeed all mitigation measures started or contemplated, in jeopardy, certainly if the war drags on for a couple of years. Energy prices have gone through the roof since the invasion started. Russian fossil fuels are now shunned by a substantial part of the world, making drastic efficiency and saving measures necessary, especially in Europe, to keep the economy going. It is hard to predict what the eventual outcome will be, but a positive impact on the carbon pollution of the atmosphere is not to be expected.

CHAPTER 7

Materials and Processes for Carbon Capture

Introduction

The most tried approach to capture CO_2, be it from flue gases of power stations or from ambient air, is currently by chemical means. This requires special types of materials, sorbents or solvents, that have a high affinity for CO_2, i.e., easily chemically bind with CO_2 with little or no energy input. A sorbent is a liquid or solid material used to absorb or adsorb liquids or gases, while a solvent dissolves a substance, resulting in a solution. Adsorption and absorption mean quite different things. In the case of absorption a liquid is soaked up, e.g., into something like a sponge, cloth or filter paper, absorbing it completely into the absorbent material. Adsorb means hold as a thin film on the outside surface or on internal surfaces within the material. Individual molecules, atoms or ions gathering on surfaces, where they interact with the surface of a material via intermolecular interactions, making them 'stick', or adsorb, to the surface. If a material has a very high surface area, lots of molecules can stick to its surface.

Since the concentration of CO_2 in the various emission sources varies greatly, it is most appropriate to use different capture techniques for different sources. In air the CO_2 concentration is 'only' 0.04%, while for flue gases of iron and steel plants it is about 20–30%, for refineries 30–40% (rising to 80% for natural gas and ethanol refineries) and for power plants 12–15%.

To capture 90% of the CO_2 emitted by a 500 MW coal power plant 2–3 Mt of CO_2 will annually have to be separated.[1] It can be done but requires an effort, both as regards energy and as regards disposing of the CO_2, once it has been captured. In spite of the extreme dilution there is nothing revolutionary either about capturing CO_2 from ambient air. Everybody knows that it can be done. Green plants do it all the time by photosynthesis, and so-called bio-sequestration, storing the carbon dioxide in biomass by planting more trees, is part of most climate mitigation strategies, so much so that there is not enough land on Earth for everybody's tree-planting schemes

---

[1] Burning one kg of bituminous coal produces 2.42 kg of CO_2; and about 1 ton of CO_2 is released for every MWh of electricity generated.

(Dooley et al. 2022). Based on extrapolating known capture technologies to the extreme dilution of CO_2 in the air, it has generally been concluded that conventional technologies for capturing CO_2 from air are most likely too expensive. This may be true, but does not mean that CO_2 capture from air is impossible. More likely it will imply that the technologies for such capture will dramatically depart from conventional technologies (Shi et al. 2020).

The first DAC technologies were developed in the 1930s and applied later, e.g., in the space shuttle, in every submarine in the world, and in various industrial settings. For climate-change mitigation the concept was first introduced in 1999 by Klaus Lackner, a German-American physicist and climate scientist working at Arizona State University as director of the Center for Negative Carbon Emissions (Lackner et al. 2012; Lackner, Ziock and Grimes 1999; Lackner, Grimes and Ziock 1999). A few years later, in 2004, Lackner and his colleague Frank Zeman were the first to describe a specific process, called alkali scrubbing, to scrub CO_2 out of the air with a solution of alkali hydroxide, i.e., sodium hydroxide (NaOH) or potassium hydroxide (KOH) (Zeman and Lackner 2004). Sodium hydroxide, also known as caustic soda, is the most widely used alkaline neutralizing chemical in use in industry today. It is easy to handle and very effective for the neutralization of strong or weak acids. The hydroxide reacts with the CO_2 in the air to produce a carbonate, Na_2CO_3, dissolved in water. This is the easy bit and does not require any energy input, but also means that your sodium hydroxide has gone. Since sodium hydroxide is fairly expensive, around \$350 per ton (April 2022), you would like to use it again and again for other CO_2-scrubbing rounds. Therefore the significantly cheaper calcium hydroxide $Ca(OH)_2$ is added to convert the sodium carbonate to calcium carbonate (CaCO3). In this step, which does not require any energy either, the calcium carbonate precipitates and the NaOH is regenerated. The final step of releasing the CO_2 from the calcium carbonate is the most energy-demanding step as it requires heating to above 900°C and has a major impact on the cost of carbon capture. The final result is a concentrated CO_2 stream which can be compressed, transported and stored. The entire scheme is illustrated in Fig. 7.1.

From the above it can be imagined that the choice of sorbent or solvent material is an extremely important part of a CCS or DAC system, since it determines most other aspects of the overall system. Successful carbon capture agents must (1) have fast reaction kinetics (i.e., the absorption and desorption reactions must be fast), (2) be low in cost and easily available in large quantities, (3) be able to regenerate with a low energy barrier to complete the whole CO_2 capture-release cycle, (4) be non-toxic and non-polluting, (5) have high CO_2 selectivity and high capacity (e.g., not capture inert gases like nitrogen), (6) have high CO_2 capture capacity (i.e., a large amount of CO_2 captured per gram of solvent or sorbent), and (7) have high cyclic ability (i.e., not lose its effectivity even after many cycles). Most CO_2 sorbents fail as regards the third point as they have to overcome a large energy barrier to regenerate (Shi et al. 2020). The development of improved sorbents and solvents is a key research priority and a vast global effort is being made to develop new materials for carbon capture, with literally thousands of new materials having been proposed since the beginning of the millennium. We will discuss some aspects of them in the following.

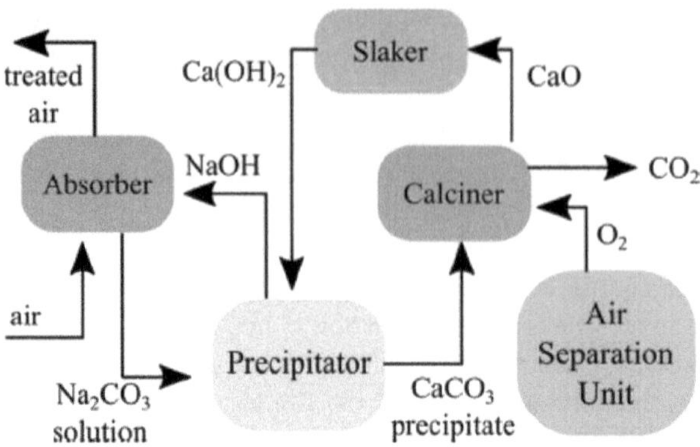

Fig. 7.1. Scheme of a plant for CO$_2$ capture from air that uses NaOH as the absorber. In the precipitator calcium carbonate is precipitated and sodium hydroxide is regenerated (Leonzio et al. 2022).

If however the last step in the process of Fig. 7.1 is omitted, we are left with calcium carbonate to dispose of. The bottom price for capturing one ton of CO$_2$ would then be the price of calcium hydroxide with about half a ton of calcium hydroxide needed to capture one ton of CO$_2$. Calcium carbonate has many uses, e.g., as a raw primary substance for building material, but can also be buried deep in the ground to permanently sequester the carbon it contains.

Liquid solvents

Solvents involve the chemical or physical absorption of CO$_2$ into a liquid carrier. The idea is rather simple: if we have a solvent with a different solubility for CO$_2$ and N$_2$, we can design a separation process such that the exhaust gasses have the composition we desire. However, depending on the type of solvent, the size of the separation equipment can vary and energy requirements can differ. When applied to flue gases, it works by bubbling the gas through an absorber column packed with liquid solvents which separate the CO$_2$ from the flue gas.

Let us consider the situation a little more in detail for a coal-fired power plant, which is the most logical point source for carbon capture systems. A generic 500 MW power plant emits about 400 m^3 of flue gas per second that have to pass through the CO$_2$ absorber. This corresponds to about 8,000 tons of CO$_2$ per day! The absorber must be designed such that with reasonable size this amount of CO$_2$ can be handled and absorbed per day. On top of that, there are restrictions for the technology at a capture site, such as the requirement that the equipment fits on the site of the power plant. Improving absorption design will clearly be a huge factor in successfully translating this technology to practical carbon capture systems (Smit et al. 2014, p. 164–165).

If we use water as an absorber, which is a theoretical possibility as CO$_2$ dissolves quite well in water, the volume of water needed to deal with the above-mentioned volume of gas is 750,000 litres per second. That is a lot of water. While flue gas is coming in at the bottom of the absorption column, all that water must be pumped to the top of the column before it comes down and is subsequently passed over to the stripper which separates the CO_2. Making matters worse, those 750,000 litres of water per second must be heated in the stripper so that the CO_2 can be separated for storage. From this it will be clear that water is not ideal as an absorber. Adding CO_2 to water makes it slightly acid (as it forms carbonic acid (H_2CO_3) in a rather slow chemical process), so the addition of a base will improve the solvability of CO_2. Using a very strong base is problematic, as the absorption of CO_2 as bicarbonate (also called hydrogen carbonate (HCO_3^-)) is a very exothermic reaction (i.e., releasing a lot of heat to the surroundings). As a result, the process will require a significant amount of heat to subsequently separate the CO_2 from the solvent in the regeneration step. That means more energy and more expense, so a strong base probably wouldn't be a good idea for a traditional absorber. However, it is very efficient at capturing CO_2 even at very low concentrations, so it might be good to use strong bases to capture CO_2 directly from air. For the problem at hand here, a weaker base will be more suitable. That is where the amines come in.

Amines (also called alkylamines) are the benchmark for CO_2 capture from flue gases, either a single amine or a blend of several amines in aqueous solutions. They are compounds that contain a basic nitrogen atom and a pair of valence electrons not shared with another atom; derivatives of ammonia (NH_3) with one or more of the hydrogen atoms replaced by a group of atoms, such as a methyl ($-CH_3$) group. In general, such a group of atoms can be any collection of atoms bound together, but not forming an entire molecule or ion. The number, maximum three, and size of the groups that an amine has will influence the molecule's properties and ability to react with other compounds. This will be important for figuring out what type of amines we can "tune" to make better solvents for carbon absorption.

Amines are used in amine gas treating (Smit et al. 2014, Chapter 5), also known as amine scrubbing, gas sweetening or acid gas removal, which refers to a group of processes that use aqueous solutions of various amines to remove hydrogen sulphide (H_2S) and carbon dioxide (CO_2) from gases. Because amines are weak bases, they tend to absorb CO_2 by forming carbonate ions. With current technology it is the most effective method of CO_2 capture from the flue gas of a pulverized coal plant and for other applications that require CO_2 removal. For instance, the Sleipner project in Norway uses amine scrubbing technology to remove CO_2 from the high pressure natural gas stream.

Amine scrubbing and alternatives

In this process CO_2 is absorbed from a fuel gas or combustion gas by an aqueous solution of amine. The scrubbing process consists of two columns: the absorber and the stripper. A typical process setup is shown in Fig. 7.2. The flue gas enters at the bottom of the absorber and rises through the column, which is segmented into a number of tray decks. Each tray consists of a flat plate with holes and a weir, or dam-like structure, that maintains a thin layer of liquid on each tray. The idea is to increase the interface between the solvent and the flue gas as much as possible to ensure optimal contact. Liquid solvent (lean amine) enters at the top of the absorber and flows down the plates in the opposite direction to the flue gas. The tray "holds up" just enough solvent to ensure that the flue gas vapour bubbling up through the

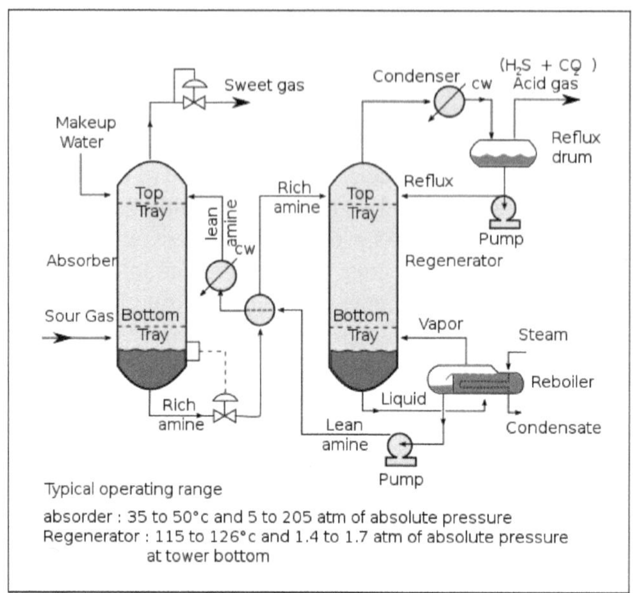

Fig. 7.2. Process flow diagram of a typical amine treating process (*Wikipedia*).

holes in each plate allows the solvent to absorb CO_2. As a result of the falling solvent and rising flue gas passing through many such plates, the exhaust at the top of the column will be significantly CO_2 depleted (sweet gas), with typically 85 to 90% of the CO_2 having been captured.

The capacity of the solvent to absorb CO_2 is of course limited. Once it has become completely saturated (rich amine), it is moved to another device (the stripper) where it is stripped of its CO_2.

The stripper, also called regenerator, looks and acts as the opposite of the absorber. The CO_2-rich solvent from the bottom of the absorber is injected at the top of the stripper where it cascades through a similar construction of plates with holes and weirs. Heat is added (in the form of water vapour at 100–120°C) at the bottom of the stripper to release CO_2 from the liquid solvent, forming a gas phase that rises through the stripper. At the top of the stripper a condenser returns the solvent, and nearly pure CO_2 is ready for subsequent cooling, compression, and transport for storage. The hot liquid solvent "stripped" of its CO_2, is recycled back into the top of the absorber. The regenerated solvent is much warmer and a heat exchanger is used to recover some of the heat used for the regeneration process.

The large amount of energy required for regeneration remains a serious drawback for amine solvents. Amines have the further disadvantages that most are volatile, toxic and nonbiodegradable. This calls for the development of alternative carbon-capture materials (Ozkan and Custelcean 2022).

The technology was first evaluated in 1991 and at that time deemed to have unacceptable energy use and costs, but is nevertheless currently widely applied in hundreds of plants to remove CO_2 from natural gas, hydrogen and other gases. Lime or limestone slurry scrubbing for flue gas desulphurisation, which is essentially the same technology, was also at first deemed unacceptable for the same reasons but has nevertheless become the dominant technology, once it was made compulsory to scrub sulphur dioxide out of flue gases (Rochelle 2009). Ideal low-cost processes do not always exist and as the need arises we have to settle for what is available, while continuing to search for better solutions.

In nature and in human tissue, the enzyme carbonic anhydrase catalyzes the reaction of carbon dioxide and water to bicarbonate (HCO_3^-) and protons, and is very adept at helping with CO_2 uptake. Without carbonic anhydrase as catalyst the reaction is very slow, and with the catalyst 10 million times faster! So can we use an enzyme to improve CO_2 capture? Unfortunately carbonic anhydrase is not very stable at high temperature (i.e., in a flue gas), at pH greater than 7, or at high salt concentration during absorption, while its operation can also be inhibited by flue gas impurities. To make enzymes work for carbon capture, researchers are studying members of the family of carbonic anhydrases found in organisms that live in very harsh conditions, and are also looking at protein engineering techniques to create thermo-tolerant enzymes. Several methods have been developed to improve the stability of the enzyme and contribute to the development of highly stable and reusable carbonic anhydrase for faster and more efficient CO_2 conversion and utilization (Alvizo et al. 2014; Talekar et al. 2022).

We can also use a completely different solvent. In this context, there is a great deal of interest in ionic liquids, salts in a liquid state, largely made up of ions. In an ionic liquid the cations (positive ions) and anions (negative ions) are both bulky, asymmetric molecules, preventing them from forming the crystalline lattice of a typical salt. By changing both these ions, it is possible to form millions of different ionic liquids. They can be chemically manipulated to enhance the CO_2 solubility, making ionic liquids promising candidates for CO_2 capture.

Of the available commercial solvents Selexol and Rectisol have to be mentioned. In the Selexol process the solvent dissolves (absorbs) CO_2 from the feed gas at relatively high pressure, usually 2.07 to 13.8 megapascal. The rich solvent containing the CO_2 is then let down in pressure and/or steam stripped to release and recover the CO_2. The Selexol process can operate selectively to recover hydrogen sulphide and carbon dioxide as separate streams. Selexol is a physical solvent, i.e., based on the solubility of the gas in the solvent, unlike amine based acid gas removal solvents that rely on a chemical reaction with the acid gases. Since no chemical reactions are involved, Selexol usually requires less energy than amine based processes. Physical solvents tend to be favoured over chemical solvents when the concentration of acid gases or other impurities is very high. Unlike chemical solvents, physical solvents are non-corrosive, requiring only carbon steel construction. One of the other main advantages is that they do not require the use of water. The disadvantage, however, is that the solvents used have higher viscosities than water, and thus require more energy to be pumped around.

Rectisol is another commercial process that uses methanol (CH_3OH) as solvent. Methanol binds CO_2 sufficiently well, but the entire system must be run at very low temperatures (between –40°C and –62°C), which necessitates expensive stainless steel refrigerated vessels. Most of these processes were developed in the context of cleaning natural gas, and were later adopted for flue gasses. Much of the current research is focussed on developing novel solvents or processes to reduce the costs associated with carbon capture from flue gasses.

An example is the process researched at the Pacific Northwest National Laboratory in Richland, Washington (Kothandaraman et al. 2022), that avoids the last energy-demanding step by using the carbon dioxide to produce methanol. Methanol is normally produced by reacting carbon monoxide (CO) with molecular hydrogen (H_2), usually in the presence of a catalyst, whereby pairs of hydrogen atoms are added to the molecule. The idea here is to use CO_2 instead of CO.

This process starts with the capture of CO_2 from a flue gas using a liquid solvent as shown in Fig. 7.1. Instead of sending the CO_2-rich solvent to a solvent regenerator, the entire rich solvent is pumped to the desired pressure and then heated and fed to the main reactor packed with a catalyst along with H_2, produced off-site from fossil or renewable sources, preferably renewable ones of course. Out of the reactor comes CO_2-lean solvent, unconverted H_2, methanol (CH_3OH), and various other alcohols and hydrocarbons. All these products can be used, while the lean solvent is recycled back to the absorber for further CO_2 capture, and there is no further need to dispose of CO_2. Technoeconomic analyses suggest that methanol can be produced with a minimum selling price of \$4.4/gallon (\$1,460/ton) when using CO_2 captured from a 650 MW natural gas combined cycle plant, bringing the cost of CO_2 capture down to less than \$50 per ton of CO_2 in 2021.

Liquid solvents for direct Air capture

The liquid solvent-based approach for DAC is based on a similar process of two chemical loops. In the first loop, taking place in an air contactor unit, gaseous CO_2 is absorbed by a liquid solvent, resulting in a CO_2-depleted gaseous exiting stream (the original air without the carbon dioxide) and a CO_2-rich liquid exiting stream (the solvent plus the absorbed CO_2). The second loop releases the captured CO_2 from the solvent solution in a series of units operating at high temperature (between 300 and 900°C) by an exchange of negative ions (such as chloride or hydroxide ions for carbonate ions) that ultimately results in precipitated calcium carbonate ($CaCO_3$) pellets. Heating to high temperatures is subsequently needed to recover the CO_2 from the precipitated calcium carbonate. This process is no different from the one schematically shown in Fig. 7.1.

Both alkali carbonates, like sodium or potassium carbonate, and alkaline hydroxides (composed of a positively charged ion of an alkali metal and the hydroxide anion (OH^-)) are used for such CO_2 scrubbing, in which process they form carbonates and/or bicarbonates. For instance, potassium carbonate with water captures CO_2 as potassium bicarbonate:

$$K_2CO_3 + H_2O + CO_2 \leftrightarrow 2KHCO_3.$$

Let us illustrate the alkaline hydroxide process with the procedure used by the Canadian company Carbon Engineering (see Chapter 13), as detailed in Fig. 7.3. Most air capture processes use solid sorbents, Carbon Engineering is an exception in this respect.

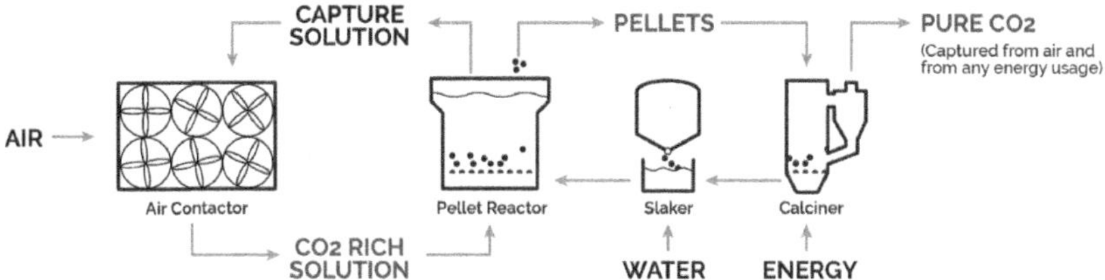

Fig. 7.3. Carbon Engineering's Direct Air Capture technology (*Credit* Carbon Engineering Ltd).

The process starts with an **air contactor**—a large structure modelled after industrial cooling towers. Fans pull air into this structure, where it passes over thin plastic surfaces that have the liquid sorbent (potassium hydroxide[2] (KOH, also called caustic potash)) flowing over them. Efficient air contact with the sorbent is extremely important as large volumes of air have to be processed to capture a relevant amount of CO_2.

The solution absorbs atmospheric carbon dioxide via the reaction:

$$2KOH + CO_2 \rightarrow H_2O + K_2CO_3,$$

removing it from the air and trapping it in the liquid solution in the form of the carbonate salt K_2CO_3. The remaining air, minus the carbon, is released.

After exiting the air contactor the solution is pumped to a regeneration facility (the **pellet reactor**) in which the KOH is regenerated by reacting with calcium hydroxide ($Ca(OH)_2$) to form pellets of calcium carbonate ($CaCO_3$), exactly in the same way as the sodium hydroxide was recycled in the process for flue gases described earlier,

$$K_2CO_3 + Ca(OH)_2 \rightarrow CaCO_3 + 2KOH.$$

The pellets are separated out of the solution and heated to 900°C in the **calciner** to form calcium oxide (CaO) and water, releasing the CO_2 in pure gaseous form. The calciner is a steel cylinder that rotates inside a heated furnace in which indirect high-temperature processing (550–1150°C) takes place within a controlled atmosphere, similar to equipment used at very large scale in mining for ore processing. The CaO is subsequently fed into a **slaker** to be mixed with water to form $Ca(OH)_2$, which can then again be used in the KOH regeneration facility. See Fig. 7.1.

To capture one ton of CO_2 this process requires energy, either from 240 m³ of natural gas, equivalent to about 2400 kWh, or from 150 m³ of natural gas, equivalent to 1500 kWh, coupled with 366 kWh of electricity.

The trade-off between having a strong capturing agent and the energy required for regenerating the solvent is similar to the solid sorbent DAC case. However, the extreme dilution of CO_2 in air requires a strong base for adequate separation, which further drives up the energy requirement of this separation. Both the solid sorbent and liquid solvent DAC approaches require roughly 80% thermal energy and 20% electricity for operation. The electricity in the solid sorbent approach is required for the contactor fans and the vacuum pumps, which remove residual air from the

---

2 Commonly used for the manufacture of soft soaps and also used in certain types of batteries.

contactor during regeneration. The liquid solvent system requires electricity for the contactor fans as well as for the pellet reactors, steam slaker and filtration units.

Solid sorbents

Solid sorbents use physical adsorption (adhesion of atoms, ions or molecules from a gas, liquid or dissolved solid to a surface) to capture the CO_2 from the flue gas, while membranes use permeable or semi-permeable materials that allow for the selective transport and separation of the CO_2. Solid sorbents are the main alternative to amine solvents and current research efforts explore a variety of different structures to use as sorbents for carbon capture, including amines that are supported on (i.e., attached to) metal-organic frameworks (so-called MOFs),[3] zeolites,[4] alkaline sorbents such as sodium, lithium and potassium hydroxide (NaOH, LiOH, KOH; the first materials to be suggested for DAC), activated carbon (carbon processed to have small, low-volume pores that increase the surface area), silica materials, carbon nanotubes, porous organic polymers, and carbon molecular sieves. Carbon molecular sieves, also used for instance for separating nitrogen from air, are carbonaceous materials with a very sharp pore size distribution while activated carbon materials have a broad range of pore sizes.

In the solid sorbent approach, the sorbent being a solid material to absorb the CO_2, CO_2 molecules interact with porous materials that can remove CO_2 from the incoming gas mixture. Sorbents generally come in three categories depending on the sorption mechanism: physisorption (physical adsorption), chemisorption (chemical adsorption), and moisture-swing sorption. Chemisorption involves the formation of a chemical bond with the sorbent, whereas physisorption relies on typically weaker physical interactions such as Van der Waals or other weak intramolecular forces; it typically occurs on the surface of a sorbent. Moisture-swing sorption seeks a compromise between these two mechanisms. It has been developed by Klaus Lackner and is a chemisorption process, in which a special resin is used that acts as a medium for ion exchange. It absorbs CO_2 from the air when dry, and releases it when exposed to moisture. Rather than binding and releasing CO_2 on a substrate whose sorption characteristics for CO_2 are fixed, moisture-swing sorbents change their affinity to CO_2 through interaction with water. When dry they bind CO_2, when wet they release CO_2. The energy required for concentrating the CO_2 is provided by the evaporation of water.

The ad- or absorption takes place at ambient temperature and pressure. After all the sorbent has been used up in sucking up CO_2 (there is no need to suck up all the CO_2 in the air, as there is a lot of air around, but it is important that all the sorbent

---

[3]   A class of compounds consisting of metal ions or clusters to form one-, two-, or three-dimensional structures with small, adjustable pore sizes and high void fractions that make them promising as an adsorbent to capture CO_2.

[4]   Zeolites are naturally occurring minerals that form when volcanic rocks and ash layers react with alkaline groundwater. The name is derived from the Greek words ζέω (zéō), meaning "to boil" and λίθος (líthos), meaning "stone". They are aluminosilicate minerals with a porous structure (molecular sieve) with pores of diameter less than 2 nm and have the potential of providing precise and specific separation of gases, including the removal of H_2O, CO_2, and SO_2 from low-grade natural gas streams.

is used) it has to be stripped of the CO_2 (desorption) and recycled into the scrubber. This recycling phase accounts for most of the energy and operating costs. It happens at low pressure and medium temperature (80–100°C). That is why it is important that all the sorbent is used; recycling unused sorbent without having reacted with CO_2 just wastes energy in the recycling process. Collecting CO_2 is the easy part, as sorbents bind CO_2 from the air without energy input.

Both methods are characterised by an adsorption/desorption cycle, and one of the challenges in scaling up is to swing fast enough from the adsorption to the desorption stage and back, to have a short swing time. Each mechanism has its advantages and disadvantages, and one of the main issues is the energy needed to release the CO_2 from the liquid sorbent. For example, various chemical compounds such as hydroxides, oxides, or alkaline salts can easily scrub CO_2 from air and convert it into carbonates (salts of carbonic acid (H_2CO_3)) through chemisorption. This works well because of a high chemical binding energy. The CO_2 is subsequently released in a calciner, whereby the compound is raised to high temperature without melting under restricted supply of ambient oxygen. This high-energy regeneration step makes capture with liquid sorbents technically and economically difficult. In the table below we summarise the key features of DAC with solid sorbents and liquid sorbents.

Overall, the energy demands of solid sorbent and liquid sorbent systems do not differ greatly from one another. Both need electricity for (heat) pumps, fans and suchlike. However, the quality of thermal energy required differs greatly. The solid sorbent system requires thermal energy on the order of 80°C–130°C, which may

Table 7.1. Key features of solid sorbent-DAC and liquid-sorbent DAC (adapted from IEA 2022, p. 23–24).

	Solid sorbent	Liquid sorbent
Energy consumption (kWh/tCO$_2$)	2,000–2,700	1,500–2,500
Share as heat consumption	75–80%	80–100%
Share as electricity consumption	20–25%	0–20%
Regeneration temperature	80–130°C	Around 900°C
Capture capacity	Modular (e.g., 50 t/year)	Large scale (e.g., 0.5–1 Mt/year)
Net water requirement (tH$_2$O/tCO$_2$)	–2 to none	0–50
Advantages	1. Possible net water production 2. Less capital intensive 3. Modular 4. Operation only with low-carbon energy 5. Novel hence better prospects for cost reduction	1. Less energy intensive 2. Large-scale capture 3. Operation with commercial solvents
Main trade-offs	1. More energy-intensive 2. Manual maintenance required for adsorbent replacement	1. More capital intensive 2. Relies on natural gas for solvent regeneration

be met via industrial waste heat or other sources of lower quality thermal energy, while the liquid sorbent system as mentioned above requires heat near 900°C for the decomposition of $CaCO_3$ into CaO and CO_2. Thus, solid and liquid systems are most efficiently paired with different thermal energy sources.

The distribution of electricity and thermal energy requirements for DAC are similar to those of petroleum refineries, which also exhibit an energy breakdown of 20% electricity and 80% thermal for their operations. Electricity includes energy for pumping, motors, and instrumentation where thermal energy primarily involves steam production and process heating. Both the breakdown of thermal energy and electricity, as well as the operations requiring this energy closely mirror the DAC process. In this sense, DAC can be viewed as a 'refinery for the sky.' Across the US refineries use an energy amount which is roughly equivalent to 880 TWh[5] per year. If the same amount of energy was allocated to DAC with each DAC facility requiring approximately 300 MW of consistent energy to capture 1 MtCO2 per year, the same amount of energy could be used to capture 370 Mt CO_2 per year (McQueen et al. 2021). This illustrates the feasibility of DAC, at considerable energy costs of 2400 kWh per ton. The energy consumption of DAC must be reduced through process improvements and new designs to meet the challenge of deploying DAC at an industrial scale from an energy standpoint. In addition to efficiency, the relative cost and greenhouse-gas emissions of those energy resources are also an important factor. A further drawback of both methods is that they are at present still generally too slow for the vast scale we are after.

Summary and concluding remarks

In summary, the capture processes described here are regenerative carbon removal systems. In these processes capture is only the first step, after which the capturing agent, either liquid or solid, must be able to release the captured CO_2 at conditions (temperature and pressure) that do not require too much energy, so that the capturing agent can be used repetitively and the CO_2 can be prepared for some form of secure sequestration. The CO_2 is extracted from the saturated sorbent or solvent through either a temperature swing or a pressure swing. In a temperature swing regeneration, the solvent is heated to often rather high temperatures (around 200°C) until the CO_2 desorbs. In a pressure swing, the pressure is decreased until the CO_2 desorbs. This desorption is the most energy demanding step in the process and hence the most costly. The capture itself with use of these chemical agents generally happens spontaneously. It is an exothermic process, releasing energy to the surroundings, while desorption, the subsequent process of getting the captured CO_2 out of the capturing agent, is an endothermic process. The most significant energy costs are therefore incurred in the step for recovering and concentrating the captured CO_2. One of the main problems is to realize these CO_2 capture and release processes fast enough in order to achieve the required scale.

---

[5] This is a non-negligible amount when compared with total electricity consumption in the US of 3,930 TWh (in 2021).

The liquid sorbent/solid sorbent methods are pure chemistry, both the adsorption and desorption processes. The question is whether they are not too slow to ever achieve the scale we need for climate mitigation. They function a bit like sponges. The CO_2 comes in contact with the sorbent, is soaked up and trapped inside the structure. Subsequently, the CO_2 is squeezed out by applying large amounts of heat, breaking the bonds between the CO_2 and the sorbent. This process uses considerable amounts of energy and is difficult to scale. The CO_2 is released in a concentrated, high-purity stream in solution or gaseous form and the regenerated sorbent is re-used for further CO_2 capture.

The sorbent goes through an adsorption/desorption cycle, the duration of one such cycle being the swing time. In general, the energy penalty for regeneration, high costs, and loss of active sorbent are major barriers for wider implementation of these carbon-capture technologies. Electrochemically-mediated CO_2 capture and concentration may offer solutions to overcome some or all of these barriers. This concerns the utilization of electrochemically active sorbents, meaning that they can release or accept electrons. In this so-called electro-swing adsorption the soaking process is more efficient by allowing gases to flow through with less resistance and instead of squeezing out the CO_2 with heat, a specific voltage is applied to the capture material to release the CO_2. A solid electrode adsorbs CO_2 when negatively charged and releases it when a positive charge is applied. Before we talked about temperature or pressure swing, while here the electric charge swings. It is a nearly isothermal separation process that consumes electrical energy to facilitate effective CO_2 capture and regeneration processes under more benign conditions of sorption and desorption than in traditional continuous wet-scrubber operations, and thus may limit thermal degradation. It is claimed that this approach allows for far more efficient capture and release of CO_2 using only electricity, and without the need for heat or water. It is also relevant to carbon capture from point sources as it has the potential to significantly reduce the difficulty of retrofitting CO_2 capture units to existing fossil-fuel fired installations. The process was developed first at the Massachusetts Institute of Technology (Renfrew et al. 2020; IEA 2022, p. 24) and is now being tested at lab scale by a spin-off company, called Verdox.[6] According to their website the process passes through a number of steps. The first step is that air containing CO_2 enters the system, after which a specific voltage is applied that activates the electrodes. These activated electrodes attract CO_2 and bind it. While the remaining gas that is not captured exits the system, a different voltage is applied that releases the CO_2. Finally the CO_2 exits the system in high concentration. Verdox's system essentially functions as a battery, absorbing CO_2 when electrically charged and releasing it when discharged. The battery's electrodes have an electrochemical affinity for carbon dioxide. When gas and electricity are fed into the system, CO_2 molecules at any concentration are captured. Turn the power off and the electrodes release those same molecules as a pure stream of carbon dioxide. Such electrochemical procedures may well be able to provide the solution for CCS and DAC to become viable and cost effective in the near future.

---

[6]   https://verdox.com/.

CHAPTER 8

Post- and Pre-combustion Carbon Capture

Introduction

Now that we have left it too late by ignoring the massive problem of climate change (and air pollution) for the past fifty years, deploying CCS technology on a commercial scale will be a vast undertaking. It has always been a bad idea to just burn coal and dump all the exhaust gases into the atmosphere. Its capacity for air pollution (with sulphur dioxide (SO_2) contributing to acid rain and respiratory illnesses, nitrogen oxides (NOx) contributing to smog and respiratory illnesses, and particulates contributing to smog, haze, respiratory illnesses and lung disease) has been known for a long time, even at the time when CO_2 was not yet seen as a problem. Recalling the London smog years of the 1950s will be a sufficient reminder of this. According to the WHO, air pollution is still the cause of 7 million premature death per year, of which 4.4 million due to outdoor air pollution and about 2.6 million due to indoor air pollution from the indoor burning of solid fuels (coal, wood, dung, crop waste) and kerosene for cooking and heating. This is not only a problem for the third world since 99% of the global population live in places where air pollution exceeds WHO guidelines. Nonetheless, to date in the 21th century coal still supplies over one third of global electricity, and is used in huge quantities in heavy industry, such as steel making. For electric power generation natural gas has now taken the lead and will probably keep this position for a few more decades, although the prospects have become very uncertain due to the current war in Ukraine.

When a fossil fuel is burned in air, as happens in an ordinary power plant, it reacts with the oxygen in the air: fuel + O_2 → heat + CO_2 + H_2O. Although this is true, in reality the situation is not that simple. The main component of air is nitrogen, while fuel, e.g., coal, does not only contain carbon and hydrogen, but also other elements (nitrogen (N) and sulphur (S) that form the nitrogen and sulphur oxides mentioned above) and trace metals (e.g., mercury (Hg)). Already for some time it has been realized that (some of) these impurities have to be removed before the flue gas can be vented into the atmosphere, and such clean-up processes require money and effort. Now we also know that CO_2 has to be added as a component for removal and for this even more effort, money and energy are required. The main thing is that whatever capture method is chosen or irrespective of how it is done, a lot of separation is necessary, as our aim is to end up with an (almost) pure CO_2 stream (as in the ideal reaction above). The final step of getting rid of the water, always produced in addition in these processes, is the easiest as water can just be removed by condensation.

In Chapter 5 we already made the case of CCS for the coal sector and in the previous chapter we discussed some materials and processes that can help capturing carbon from flue gases or from air. In this chapter some of the capture technologies to be used in fossil-fuel burning power stations will be set out.

An integrated Carbon Capture and Storage (CCS) system, including possibly intermediate or final utilization (CCUS), would include three main steps:

(1) capturing CO_2 and separating it from the other gases present;

(2) purifying and compressing the captured CO_2;

(3) and transporting it for use in industry or somewhere else or for injecting it in geological reservoirs for permanent storage or storing it anywhere else where it can safely be tucked away, never to be seen again.

The first step, carbon capture, is the most costly and technologically challenging step in the process (accounting for 75% of the costs). Carbon-capture equipment is capital intensive to build and energy intensive to operate. Power plants can supply their own energy to operate CCS equipment, but at the expense of the electricity the plant can sell to its customers. This difference, sometimes referred to as the energy penalty or the parasitic load, can be as high as around 20% of the plant's capacity. That is a considerable fraction, but it still leaves 80%. In this respect it applies that the idea that CCS would or could be free is preposterous. Nothing comes for free and a power plant with CCS that does not pollute the atmospheric commons is of course less 'profitable' than the same power plant without CCS that uses the atmospheric commons to dump its noxious fumes and other harmful waste. Nothing can beat the 'parasitic' way power plants have been and are operating. If it were free, CCS would have been utilized ages ago. So long as the energy penalty is fairly modest like the 20% mentioned and lacking a realistic alternative to burning fossil fuels, fossil-fuel power plants should only be allowed with CCS. To complain about the energy and effort required is similar to complaining that treating sick people in hospitals comes with a 'penalty or parasitic load', i.e., costs money.

However, most of the flue gases belched out of smokestacks are hot (typically 150°C) and this and other residual heat of a power plant can also be used for capturing the carbon. According to some (Chichilnisky and Bal 2019, p. 281)[1] it is enough to capture twice as much CO_2 as the plant emits.

As far as such capturing of CO_2 is concerned a number of general ways of capture can be distinguished:

(1) Post-combustion: the most obvious and simplest method in which the fuel is burned and the resulting carbon dioxide is captured from the exhaust gas of the plant by fitting a carbon capture installation; a special application is BECCS, the burning of biomass for generating energy with carbon capture and storage;

---

[1] The authors are associated with the firm Global Thermostat and claim that carbon capture, including direct air capture, can be powered by free or low-cost, widely available residual low-temperature (85°C) heat.

(2) Pre-combustion: capturing the carbon dioxide from the fuel before it is combusted, for instance via gasification or reforming (i.e., producing a syngas (a mixture of hydrogen and carbon monoxide) by letting the hydrocarbons react with water). Various refinements of this general method exist:

   (a) Oxyfuel combustion: combusting the fuel in a power station in pure oxygen, resulting in a CO_2 stream that is more pure and easier to capture;

   (b) Integrated gasification combined cycle (IGCC);

   (c) Chemical looping;

   (d) Allam-Fetvedt Cycle.

They all involve pre-combustion, but differ in the way the technology is integrated into the power plant. Oxyfuel combustion is often treated as separate from both post- and pre-combustion, but I include it here as a pre-combustion technique.

Post- and pre-combustion methods of carbon capture utilize a solvent or sorbent as discussed in the previous chapter to separate CO_2 from the other gases. They can be implemented at coal-fired power plants, natural gas fired plants, cement plants, steel mills, petrochemical plants, in short at any carbon-emitting point source.

(3) Direct Air Capture: capture from the atmosphere after having been emitted; a method of last resort but one that we will be forced to utilize, now that the CO_2-concentration has already become dangerously high, and will continue to increase further. This will be dealt with in the next chapter.

Post-combustion capture

This is essentially the only method that can be utilized for existing power stations. After the complete combustion of the fuel, CO_2 is separated from the flue gas that consists mainly of nitrogen, water and CO_2. It can be applied to gas-fired plants and to conventional (pulverized) coal-fired power plants where fuel is burned to generate electric power. In the direct combustion step of a coal-fired power plant, pulverized coal is mixed with air and burned. The heat of this combustion is recovered to produce high-pressure steam, which in turn drives an electric turbine to produce electricity. The flue gas contains various impurities that have to be removed before venting the remainder into the air. A further complication and cost factor is now added by also demanding that the CO_2 be removed from the flue gas.

The principal challenge is separating the CO_2 generated during combustion (12–15%) from the large amounts of nitrogen (from air) found in the flue gas. The main advantage is that the CCS installation can be added on as an accessory to an existing power plant without requiring the construction of an entirely new plant. Main drawbacks are the reduction of energy production for the power plant, and the fact that the size of such a separation unit can be significant—too big, even, to easily fit on the power plant site. It must probably be accepted that not all existing coal-fired power plants can be retrofitted with such installations but the world will anyway be busy for quite a while before all suitable plants have been fitted with separation units. If pre-combustion capture is not deemed feasible, CCS must in any case be fitted to all new plants.

As mentioned in the previous chapter, the benchmark technology for post-combustion carbon capture is amine scrubbing. It has been around since 1930, has been applied in hundreds of gas-separation processes and is fully ready to be implemented. With current technology it is the most effective method of CO_2 capture from the flue gas of a pulverized coal plant and the most mature and scalable technology for post-combustion carbon capture, but subject to considerable energy and capital investment costs.

Bioenergy with carbon capture and storage

Bioenergy with carbon capture and storage (BECCS) is an application of post-combustion capture, in this case in a biomass burning power plant. It has already been discussed in Chapter 4. It is argued that BECCS amounts to 'negative' emissions when the carbon dioxide released in the air when the biomass is burned is captured and stored. When this is not done, the process can be considered carbon neutral: the carbon dioxide captured by the biomass in the photosynthesis process is again released into the atmosphere when the biomass is burned, but does not increase the net carbon dioxide concentration in the air. That is why energy from biomass counts as green energy. We have already noted that this is only true when the biomass would go to waste anyway, releasing its carbon via natural decomposition processes. If, as is often the case, whole forests are cut down to produce biomass for transportation from, e.g., the US to be burned in biomass power plants in Europe (like the Drax plant in the UK discussed in Chapter 4), a few question marks on the carbon neutral character of this procedure are in order. Burning biomass without CCS, which is actually done in most small biomass power plants, has the added negative feature that apart from carbon dioxide it also emits other harmful pollutants into the atmosphere.

Many biomass combustion facilities are relatively small and inefficient, compared to the typically much larger coal plants. Furthermore, raw biomass normally has a higher moisture content compared to common coal types. When this is the case, more of the wood's inherent energy must be spent solely on evaporating moisture, compared to the drier coal, which means that the amount of CO_2 emitted per unit of produced heat will be higher. Some research groups (e.g., Chatham House) therefore argue that the use of woody biomass for energy will release higher levels of emissions than coal (*Wikipedia* page on biomass). Others don't agree, and the topic is hotly debated but the IPCC estimates the extra CO_2 for biomass at roughly 16% more than for coal in general. Although such aspects make classifying energy from burning biomass as renewable/sustainable questionable, the IPCC is actually extremely biased towards BECCS, much more than towards CCS applied to ordinary coal- or gas-fired power stations (Kolbert 2021, p. 160). Whether or not biomass is truly carbon neutral depends on the time frame being studied, what type of biomass is used, the combustion technology, which fossil fuel is being replaced (since the combustion of both fossil fuels and biomass produces carbon dioxide), and what forest management techniques are employed in the areas where the biomass is harvested.[2] This is the more pertinent as biomass makes up a large part of what

---

[2]   See, e.g., https://news.climate.columbia.edu/2011/08/18/is-biomass-really-renewable/.

counts as renewable energy. At 54% biomass was the largest source of sustainable energy in the Netherlands in 2020, followed by wind (23%) and solar energy (14%).[3] Globally too bioenergy is the largest of the energy sources that are called sustainable, accounting in 2017 for 70% of sustainable energy consumption. It is important to keep this in mind when hearing exalted stories about sustainable energy generation, for biomass for sure has a serious drawback. In view of this it is decidedly odd that CCS for coal-fired power plants is frowned upon while biomass installations *without* CCS are called sustainable.

Pre-combustion capture

Pre-combustion capture is based upon a gasification process in which the biomass or fossil fuel is converted into a gas (gasified) before burning it. In such process the fuel is partially oxidized in a controlled amount of oxygen and/or steam under high temperature and pressure to form a synthesis gas, which mainly consists of carbon monoxide (CO), methane (CH_4), hydrogen (H_2) and CO_2. An important feature of the process is that there is insufficient oxygen for a complete combustion. The next step is to convert these gases into a mixture of CO_2 and H_2 via the water-gas shift reaction (see below) which converts CO and water into CO_2 and H_2. After separation from the CO_2, the H_2-rich fuel gas can be used to fire a gas turbine or run a fuel cell. Such gasification can convert any carbon-based raw material into fuel gas. The process, its use and versatility (almost any feedstock is suitable) are illustrated in Fig. 8.1.[4]

The name synthesis gas (or syngas) comes from its use as intermediate in creating synthetic natural gas (also called substitute natural gas) that can be produced from fossil fuels, and for producing ammonia and methanol.

Synthesis gas can also be produced by dry reforming of methane (also known as carbon-dioxide reforming, contrary to the above-mentioned method which also goes under the name of steam reforming). In dry reforming the synthesis gas is formed by reacting carbon dioxide with hydrocarbons such as methane with the aid of noble metal catalysts (typically Ni or Ni alloys): $CO_2 + CH_4 \rightarrow 2H_2 + 2CO$. The result is that two greenhouse gases, carbon dioxide and methane, are consumed and useful chemical products, hydrogen and carbon monoxide, are formed. A challenge for the commercialization of this process is that the produced hydrogen tends to react with carbon dioxide. Another issue with dry reforming lies in the fact that it operates under conditions that produce water. As a result, this water can result in an unwanted back-reaction to CO_2 via the water-gas shift reaction (the CO reacting with water to form CO_2 and H_2; see below). It has been argued that from a practical standpoint this approach will, at its best, make negligible contributions to CO_2 emission relief. "Broad deployment of this method for such a purpose will be a politically, financially, and energetically costly distraction. It remains a valuable reaction that may serve the business need for value-added manufacturing, but the goal of CO_2 emission relief has to be realized with low-carbon energy" (Parsapur et al. 2020). It must be noted though that this is the standard argument used to oppose any CCS

---

[3]   https://www.cbs.nl/en-gb/news/2021/39/more-and-more-renewable-energy-from-biomass.

[4]   https://netl.doe.gov/research/Coal/energy-systems/gasification/gasifipedia/intro-to-gasification.

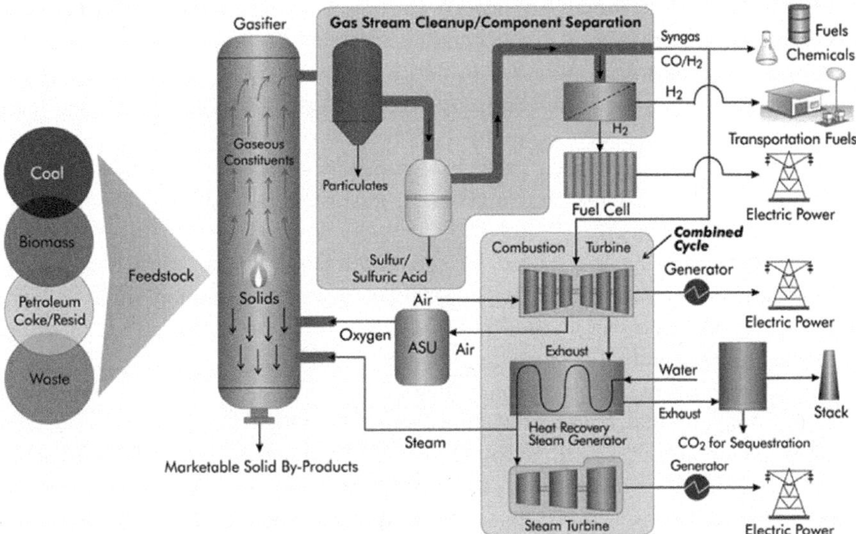

Fig. 8.1. A representation of the gasification process, depicting both the feedstock flexibility inherent in gasification, as well as the wide range of products and usefulness of gasification technology (*from* NETL).

(or DAC) technology, especially by those who believe that renewables can solve all our problems.

In steam reforming the syngas is a mixture consisting primarily of hydrogen (H_2), carbon monoxide (CO), carbon dioxide (CO_2), water and natural gas (methane (CH_4)). This syngas will subsequently be subjected to the water-gas shift reaction in which the carbon monoxide and water are converted into hydrogen and CO_2: $CO + H_2O \rightleftharpoons CO_2 + H_2$. The result is a gas rich in H_2 and CO_2. The concentration of CO_2 in this mixture can range from 15–50%, where typical flue gases from coal-fired boilers contain 12–14% CO_2 and from natural gas-fired power plants 8–10% CO_2. The CO_2 can be captured and separated, compressed and transported, and ultimately sequestered, and the H_2-rich fuel used for combustion. The main cost factor in this procedure is that a separate air separation plant is needed to separate oxygen from air before it is burned with the coal in the gasification process.

The water-gas shift reaction, mentioned a few times above, was discovered as far back as 1780 by the Italian physicist Felice Fontana (1730–1805), but its industrial value as a means of producing hydrogen was only realized much later. From the early 1900s until well into the 1930s, hydrogen was obtained in the steam-iron process[5] by reacting steam under high pressure with iron to produce iron oxide and hydrogen. With the development of industrial processes that required hydrogen, such as the Haber-Bosch process for the production of ammonia (especially for use in the fertilizer industry), a less expensive and more efficient method of hydrogen production was needed. This problem was solved by combining the water-gas shift reaction with the

---

[5] The interest in this process has grown in recent times, due to its simplicity, the high purity of hydrogen obtained, which is especially important for the use of hydrogen in fuel cells, the feedstock flexibility and the possibility to use renewable energy sources in the process (see, e.g., Peña et al. 2010).

gasification of coal to produce hydrogen. Large-scale coal gasification installations are currently primarily used for generating electricity or for producing chemical feedstocks. The hydrogen obtained in this way can be used for various purposes such as making ammonia, powering a hydrogen economy, or upgrading fossil fuels. As the idea of a hydrogen economy is presently gaining popularity, there is an increasing focus on hydrogen as a replacement fuel for hydrocarbons, and the gasification of the currently almost generally despised coal may become one of the sources for this hydrogen.

Coal-derived syngas can also be converted into transportation fuel such as gasoline and diesel through additional treatment, or into methanol or methyl alcohol (CH_3OH) which itself can be used as transportation fuel or fuel additive, or can be converted into gasoline.

As far as carbon capture is concerned, gasification has a significant advantage over the conventional burning of mined coal, in which the CO_2 resulting from combustion is considerably diluted by nitrogen and residual oxygen in the combustion exhaust, making it relatively difficult, energy-intensive, and expensive to capture the CO_2 (the post-combustion capture option, discussed above). In gasification some oxygen is supplied to the gasifiers and just enough fuel is combusted to provide the heat to gasify the rest. Moreover, gasification is often performed at elevated pressure. The resulting syngas is at higher pressure and not diluted by nitrogen, allowing for much easier, efficient, and less costly removal of CO_2. Gasification and the cycle's ability to easily remove CO_2 from the syngas prior to its subsequent use is one of the significant advantages over conventional coal utilization systems. Due to the higher CO_2 concentration, pre-combustion capture is typically more efficient but the capital costs of the base gasification process are often, not surprisingly, higher than for traditional pulverized coal power plants, which just emit CO_2 and any other unwanted combustion products into the atmosphere.

It should also be noted that water use is a problem for coal gasification, even when the CO_2 is captured, as it is one of the more water-intensive forms of energy production.

The gasification procedure can also be applied to unmined coal seams using injection of oxidants and steam (so-called underground coal gasification or UCG), whereby the product gases (predominantly methane, carbon dioxide, hydrogen and carbon monoxide) are brought to the surface. Oxygen and steam are pumped into the coal seam through a small borehole to produce a small and controlled combustion, whereby the actual coal is converted from a solid into gas. The product gases are then siphoned off through a second borehole. The technique can be applied to coal resources that are otherwise unprofitable or technically complicated to extract by traditional mining. It has been around since the 19th century but has yet to become commercially viable on a grand scale. The technique avoids some of the air pollution that is inherent to coal power plants, and produces less CO_2. The real advantage lies in the possibility to capture this CO_2. The process is different from fracking, which involves pumping fluid into coal seams to cause fractures, while in UCG the entire process takes place underground within the coal body. UCG eliminates the need for mining, hence the dangers to miners (no more mining disasters) and the associated environmental degradation are greatly reduced. Compared to surface gasification,

UCG requires much smaller gas clean-up equipment, because tar and ash content of UCG-based syngas is substantially lower than that obtained from a surface gasifier. Challenges with underground coal gasification stem from the potential leaching of unwanted substances into groundwater, while subsidence, where the surface actually sinks as the deep seam is gasified, can also be an issue.[6]

Oxyfuel combustion

Oxyfuel combustion is essentially a refinement of the pre-combustion discussed above. The fuel is burned in pure oxygen (or actually, in a mixture of oxygen and CO_2 recycled from the power plant, to moderate the otherwise excessively high flame temperature—too high for the currently available boiler materials—that would result from burning in pure oxygen). In this process virtually all the waste gas will be CO_2 and water vapour. Since no nitrogen is heated, fuel consumption is reduced, and higher flame temperatures can be achieved. Condensing out the water, the CO_2 can be put to a productive use or stored. In this oxyfuel system the challenge is to separate large volumes of (liquid) oxygen out of air, which can use up to 15% of the power produced by the power station. To produce a stream of pure oxygen an air separation unit must be built.

Oxyfuel combustion can be applied to several fuels, including coal (oxy-coal combustion), natural gas or blends of biomass and coal. Interest has mainly focused on oxy-coal combustion due to the abundance, reliability and high carbon content of the fuel.

In summary, oxyfuel combustion has significant advantages over traditional air-fired plants, including a reduction of the flue gas mass and volume by approximately 75%. This implies that less heat is lost in the flue gas and the size of the flue gas treatment equipment can be reduced by 75%; the flue gas is primarily CO_2, suitable for sequestration; the concentration of pollutants in the flue gas is higher, making separation easier; most of the flue gases can be condensed, making compression separation possible, while the condensation heat can be captured and reused rather than it getting lost in the flue gas, and finally because there is no nitrogen from air present, nitrogen oxide production is greatly reduced. As said, the challenge is separating large volumes of oxygen out of air, which makes this method more costly than a traditional air-fired plant, but the benefits enumerated above will increasingly gain weight.

Many fossil fuels also produce ash as a result of combustion. This ash must also be disposed of and will have a possible impact on the environment. Studies indicate that, in general, oxyfuel combustion does not significantly affect the composition of the ash produced. One notable exception is that oxyfuel ashes often have lower concentrations of unburned free lime (i.e., calcium oxide (CaO) that survives processing without reacting) (*Wikipedia*).

---

[6] National Energy Technology Laboratory, https://netl.doe.gov/research/Coal/energy-systems/gasification/gasifipedia/underground.

Several pilot plants are undergoing initial proof-of-concept testing to evaluate the technologies for scaling up to commercial plants, including the Callide A Power Station in Queensland, Australia; the Schwarze Pumpe Power Station in Spremberg, Germany; CIUDEN in Cubillos del Sil, Spain; and the NET Power Demonstration Facility in Texas, discussed below as it employs the Allam Cycle. These and others will be discussed in a future chapter.

Integrated gasification combined cycle

An integrated gasification combined cycle (IGCC) is another similar technology which turns coal or other carbon based fuels into synthesis gas in a closed pressurized reactor with a shortage of oxygen.[7] This oxygen shortage ensures that the coal is broken down by the heat and pressure as opposed to burning it completely, and the chemical reaction between coal and oxygen produces a syngas, which is a mixture of carbon and hydrogen. The exhaust heat from the production of the syngas is used to produce steam, which is subsequently used for generating electricity. Impurities from the syngas can be removed prior to using the gas for power generation, whereby some of these pollutants, such as sulphur, can be turned into re-usable by-products. This results in lower emissions of sulphur dioxide, particulates, mercury, and in some cases carbon dioxide. With additional process equipment, a water-gas shift reaction can increase the gasification efficiency and reduce carbon monoxide emissions by converting them to carbon dioxide. Large-type IGCC systems can improve power generation efficiency by approximately 15% and reduce CO_2 compared with conventional coal-fired power systems. The resulting carbon dioxide from the shift reaction can be separated, compressed, and stored through sequestration. Such CO_2 removal is much easier than CO_2 removal from flue gas in post-combustion capture due to the high concentration of CO_2 after the water-gas shift reaction and the high pressure of the syngas. During pre-combustion in an IGCC, the partial pressure of CO_2 is nearly 1,000 times higher than in post-combustion flue gas (*Wikipedia* (Integrated gasification combined cycle); Davidson 2011).

Such a plant is called *integrated* because (1) the syngas produced in the gasification section is used as fuel for the gas turbine to generate electricity and (2) the steam produced by the syngas coolers in the gasification section is used by a steam turbine to do the same. Regular coal-fired power plants burn coal and use the heat to produce steam. The steam powers steam turbines to generate electricity. In the case of IGCC coal is gasified and a combined cycle (a combination of gas and steam turbines) is used to generate electricity. A schematic representation of the cycle is shown in Fig. 8.2.

During pre-combustion in an IGCC, the partial pressure[8] of CO_2 is nearly 1,000 times higher than in post-combustion flue gas. Due to the high concentration of CO_2,

---

[7]   See Glossary for a short description.

[8]   Each gas in a mixture exerts a pressure. The partial pressure of any one gas in such a mixture is equal to the pressure that would be exerted by this gas if it occupied the same volume on its own.

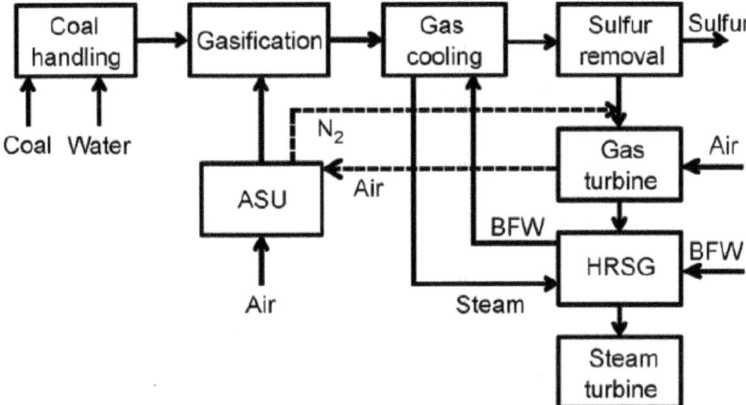

Fig. 8.2. Schematic representation of the IGCC cycle (ASU stands for air separation unit; HRSG for heat recovery steam generator and BFW for boiler feed water) (*from* Emun et al. 2020).

physical solvents[9] are preferred for the removal of CO_2. Physical solvents work by absorbing the acid gases without the need of a chemical reaction as in traditional amine-based solvents. The solvent can then be regenerated, and the CO_2 desorbed, by reducing the pressure. The biggest drawback of physical solvents is that the syngas must be cooled before separation and reheated afterwards for combustion. This requires energy and decreases overall plant efficiency.

An IGCC consumes less water than a traditional pulverized coal plant. In a pulverized coal plant, coal is burned to produce steam, which is then used to create electricity through a steam turbine. Steam exhaust must then be condensed with cooling water, and water is lost by evaporation. In an IGCC water consumption is reduced by combustion in a gas turbine, which uses the generated heat to expand air and drive the turbine. Steam is only used to capture the heat from the combustion turbine exhaust for use in a secondary steam turbine.

Currently, the major drawback of implementing IGCC is its high capital cost compared to other forms of power production, which prevents it from competing with other power plant technologies. Ordinary pulverized coal plants are still the lowest cost power plant option. The advantage of IGCC comes from the ease of retrofitting existing power plants that could offset the high capital cost.

There have been various IGCC demonstration projects, some even on a commercial scale. IGCC plants, two in the U.S. and two in Europe, which achieved operational success during their lifetimes, were the following. The Wabash River Coal Gasification Repowering Project was the first full-size commercial IGCC plant in the US and successfully repowered a 1950s era, pulverized coal plant. Its impending retirement has apparently been reversed. The Tampa Electric Polk Power Station, which is a state-of-the-art IGCC power plant, is not currently operating, but has not been retired either. The ELCOGAS IGCC Plant in Puertollano, Spain, producing 330

---

[9] See previous chapter. Physical solvent systems are similar to chemical solvent systems but based on the gas solubility within a solvent instead of a chemical reaction. Well-known physical solvents are Selexol and Rectisol.

MW (net), operated with syngas from 1998, but was recently closed. The Willem Alexander IGCC Plant in Buggenum, Netherlands, commissioned in 1994, was one of the first commercial IGCC plants in the world, using Shell gasification technology. Both coal and biomass (wood waste) were gasified in the plant. It began delivering electricity in 1998 (253 MW (net)) and operated until 2013. Emissions were minimal with NOx levels typically below 10 ppm and the elimination of sulphur compounds exceeding 99%. The CO_2 was not captured, but the plant, owned by the Swedish firm Vattenfall (formerly NUON), experimented with CO_2 capture in 2010, but two years later it was decided to close the facility.[10] The intention had been to use the experience at another much bigger IGCC plant, the 1410 MWe multi-fuel Magnum plant, to be built in Groningen in the north of the Netherlands. In the end this became a gas-fired power plant without CCS.

Each of the plants mentioned above was operational for well over a decade operating on coal, experiencing similar availabilities (around 80%) and similar operational issues, including the fouling of heat exchangers. The most recent major IGCC project to begin commercial operation in the United States is the Edwardsport IGCC Project.

In general it is fair to say that the experience with the IGCC cycle has been mixed and few modern coal power stations are designed to run on this cycle. In China, for instance, the GreenGen IGCC plant (265 MW) is the first and, to date, the only IGCC facility in that country. Since IGCC plants cannot currently compete with pulverized coal plants in China, it has been controversial whether to launch more IGCC pilot projects in China. As the IGCC plant has struggled to achieve profitability, policy support has become necessary to promote the development of the technology (Xia et al. 2020).

Capture of carbon dioxide from coal gasification is achieved at low marginal cost in some plants. One (where the high capital cost has been largely written off) is the Great Plains Synfuels Plant in North Dakota, where 6 million tons of lignite is gasified each year to produce clean synthetic natural gas.

A success with IGCC can also be reported from South Korea where the Korea Western Power's Taean IGCC Power Plant marked 4,000 hours of consecutive accident-free operations in 2021, setting a new world record for IGCC power plants. In 2013, Mitsubishi Power wrapped up a successful demonstration of the technology at the 250-MW IGCC unit at Joban Joint Power Co.'s Nakoso Power Plant Unit 10 in Iwaki, Fukushima. In April 2021, a Mitsubishi consortium put online the 543 MW Nakoso IGCC facility in Fukushima. Another unit using the same technology is planned to come online later in 2021 at Tokyo Electric Power Co.'s Hirono Power Station in Futaba-gun, Japan.[11]

The technology is still developing, but the costs of these plants are substantially higher than conventional coal plants and more expensive still compared to gas-fired plants. The US Department of Energy estimates that the addition of carbon capture to the plant would not only increase the cost by approximately one-third, it would also result in roughly 15% less power output.

---

[10] For more details on this plant see Promes et al. 2015.

[11] https://www.powermag.com/taean-igcc-continued-operation-continued-achievement/

In view of the cost picture of IGCC, it does not really come as a surprise that IGCC hasn't come off the ground yet. The overriding factor in this lack of success is that coal is past its prime. The gas fracking boom in the US has resulted in many coal-fired power plants being replaced by gas-fired plants, which emit less, but still significant amounts of carbon dioxide. Gas-fired plants too must be equipped with CCS, if CO_2 emissions are to be sufficiently curbed. However, the fracking boom is probably temporary and it is not excluded that coal will then make a comeback, certainly when it manages to 'come clean' by deploying a pre-combustion technology with carbon capture.[12]

Chemical looping

Chemical looping combustion (CLC), also known as oxygen looping cycles, was originally proposed as a means of increasing efficiency. In the context of reducing energy loss in fossil fuel combustion it was first put forward in 1983. For more than two decades it has been investigated as a promising approach for power generation with integrated CO_2 capture (Zhu et al. 2020). The technology can be used to reduce the costs involved in oxyfuel combustion for separating oxygen out of air. Key in the method is a metal oxide employed as a bed material providing the oxygen required to burn the coal. Metal oxides that act as oxygen transporters (instead of air and hence without the nitrogen) for oxidizing coal into natural gas are injected into the boiler. The reduced metal (i.e., the metal stripped of the oxygen) is then transferred to the second bed and re-oxidized before being reintroduced back to the fuel reactor and completing the loop.

The idea is to separate the combustion process into two separate reactors. In the first reactor air (i.e., nitrogen plus oxygen) reacts with a metal forming a metal oxide ($2Me + O_2 \rightarrow 2MeO$), which is then transported to the other reactor, while the nitrogen plus some remaining oxygen is vented to the air. So, the metal is used to extract the oxygen from the air and transport it from one reactor to another. In the second reactor, the fuel reactor, combustion takes place, the metal is freed from the oxygen and fed back into the first reactor, while the oxygen is used to oxidize the fuel, producing CO_2 and water. The metal circulates between the two reactors and is employed as a bed material providing the oxygen for combustion in the fuel reactor. In this scheme, the flue gas does not contain any nitrogen, therefore the separation only involves CO_2 and H_2O.

Chemical looping avoids the need to separate oxygen from air, so that no expensive air separation plant is needed. Carbon capture is made easier because the two reactions generate two intrinsically separated flue gas streams: one stream from the air reactor, consisting of atmospheric N_2 and residual O_2, but free of CO_2; and a second stream from the fuel reactor containing mainly CO_2 and H_2O with very little nitrogen. The flue gas from the air reactor, mainly ordinary air without most of the oxygen, can be discharged to the atmosphere causing minimal CO_2 pollution, while

---

[12] More information on 'clean coal' can be found at the website of the World Nuclear Organisation (https://www.world-nuclear.org/information-library/energy-and-the-environment/clean-coal-technologies.aspx#ECSArticleLink3).

exit gas from the fuel reactor contains almost all the CO_2 generated by the system. Chemical looping can therefore said to involve 'inherent carbon capture', as water vapour can easily be removed from the second flue gas via condensation, leading to a stream of almost pure CO_2. This gives chemical looping clear benefits compared with competing carbon-capture technologies, as the latter generally involve a significant energy penalty associated with either post-combustion scrubbing systems or energy required for air separation plants (Smit et al. 2014, p. 147; Wilcox 2012, p. 21).

A 2019 paper (Lyngfelt et al. 2019) reports on the progress with chemical looping combustion so far. The CLC process has been operated in 46 smaller chemical-looping combustors for more than 11,000 hours. The experience shows a large variation in performance depending on pilot design, operational conditions, solids inventory, oxygen carrier and fuel. There is at present no experience with the process at commercial or semi-commercial scale.

Operational experiences have advanced greatly in the last years, especially with more complex operation involving solid fuels and oxygen carriers of more complex composition. Experience has been gained with a number of different oxygen carriers, creating a portfolio of materials varying in properties and costs and forming a solid basis for the commercialization and optimization of the process. The possibility to have a choice of materials will facilitate upscaling and investment decisions.

At least with solid fuels, the added costs of chemical looping are small, much smaller than for competing technologies. While most CLC research has focused on fossil fuels, an important future use could be with biomass, in order to accomplish negative CO_2 emissions, i.e., remove atmospheric carbon. There is at present no market for CLC, because of lacking or insufficient incentives, but this could change rapidly. The only way to meet the Paris Agreement targets is introducing strong measures to curb CO_2 emissions. The paper concludes on the positive note that the future prospects for CLC should be excellent. It is clear that the concept works and can be scaled up.

A second related form of looping concerns lime-based CO_2 looping cycles, where calcined limestone is used for in situ CO_2 capture. Limestone is calcium carbonate ($CaCO_3$) and in the calcination process decomposes into CaO and CO_2:

$$CaCO_3 \rightarrow CaO + CO_2.$$

Temperatures needed for this are from 900 to 1050°C. The essential idea of the cycle is to feed the solid CaO, cooled to approximately 650°C, into a *carbonator* (a device that turns CO_2 into a carbonate) where it is brought into contact with a flue gas containing a low to medium concentration of CO_2. The CaO and CO_2 react to form $CaCO_3$, thus reducing the CO_2 concentration in the flue gas to a level suitable for emission to the atmosphere.

The solid calcium carbonate is then again fed into the *calciner*, heated and caused to thermally decompose into gaseous carbon dioxide and solid calcium oxide (CaO), as above. The almost-pure stream of CO_2 is then removed and purified so that it is suitable for storage or use (Anthony 2008).

Allam Cycle or Allam-Fetvedt Cycle

A novel and possibly game-changing approach is a power plant design based on an application of the oxyfuel-combustion discussed above, called the Allam or Allam-Fetvedt Cycle,[13] which uses supercritical[14] CO_2 (sCO_2) to drive an electricity-generating turbine. sCO_2 is CO_2 held at or above its critical temperature and pressure, where the carbon dioxide can adopt properties midway between a gas and a liquid. It does not change phases, but undergoes drastic density changes over small ranges of temperature and pressure. The cycle solves the CO_2 problem by exploiting the special thermodynamic properties of carbon dioxide as a working fluid.

In power plants that use this cycle a gaseous fuel (natural gas or a gasified solid fuel) is burned with oxygen and a hot, high-pressure, recycled supercritical CO_2 working fluid. The recycled CO_2 stream serves the dual purpose of lowering the combustion flame temperature to a manageable level and diluting the combustion products such that the working fluid is predominantly CO_2. Combustion of this oxy-fuel mixture in the carbon dioxide environment creates high-temperature products that subsequently enter the carbon dioxide turbine. The high-pressure CO_2 moves along into the turbine, where it expands and goes into the heat exchanger. Any water is removed, and the remaining CO_2 is compressed and pumped back into high-pressure. Most of the high-pressure CO_2 is reheated in the heat exchanger and returned to the combustor, where the whole cycle begins again. The cycle is illustrated in Fig. 8.3.

In the process, a high-quality CO_2 by-product is generated at no additional cost, which is ready for removal by pipeline.

The CO_2 produced by the cycle is sufficiently pure to be directly transported or used without requiring an additional capture or purification step. For power plant operations, sCO_2 may be more efficient than steam. Initial estimates indicate that power plants using the Allam Cycle could have comparable efficiencies to natural gas combined cycle power plants without CCS. It dramatically reduces energy losses compared to steam- and air-based cycles. For a small reduction in performance the cycle can run substantially water free. The system employs only a single turbine, utilizes a small plant footprint, and requires smaller and fewer components than conventional hydrocarbon fuelled systems (Allam et al. 2017). The cycle recycles its exhaust heat and eliminates all air emissions, including traditional pollutants and CO_2. As fuel and oxygen are continuously added to the system for combustion, an excess of CO_2 is created as a by-product (alongside water, which is continuously removed). This excess CO_2 is removed from the high pressure recycle flow at high purity and pressure, enabling delivery to a CO_2 pipeline without negatively impacting the performance of the overall system.

A key feature of the Allam Cycle is that heat goes in at two temperature levels. High temperature heat via fuel combustion, and low temperature heat, less than 400°C, which is vital in attaining high efficiency. The latter can be derived from waste

---

[13] Invented by the English engineer Rodney John Allam (b. 1940) and the American engineer Jeremy Eron Fetvedt (b. 1974).

[14] See Glossary under supercritical fluid.

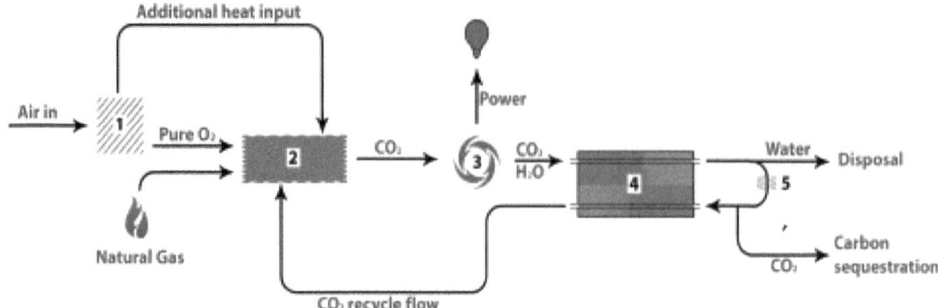

Fig. 8.3. Illustration of simplified Allam-Fetvedt Cycle (1 = air separation unit; 2 = combustor; 3 = turbine; 4 = heat exchanger; and 5 = water separator) (*source* Wikimedia commons).

heat from the air separation unit, waste heat from a gasifier unit, LNG regasification, solar power, or an existing power station.

The NET Power demonstration facility in La Porte, Texas, is the first power plant to use the Allam Cycle. It is a zero-emissions 50 MW natural gas power plant owned and operated by NET Power. It started operations in 2018.[15] The facility uses natural gas as fuel, but coal may also be used. It generates non-intermittent, competitively-priced power with full carbon capture. The company strives to push the cost of CO_2 capture so low that it will be more economic to re-use and sequester carbon than to emit it into the atmosphere. Its successful start-up, connecting with the Texas electricity grid in November 2021, proving that the Allam-Fetvedt Cycle is capable of generating power at 60Hz, has resulted in a flurry of new activity as reported on the Net Power website.[16] NET Power plans to bring a full-scale 300 MW plant online in 2022. Plans for two commercial-scale Allam Cycle power plants—the Coyote Clean Power Project in southwest Colorado and the Broadwing Clean Energy Complex in Illinois—were announced in April 2021. In the UK the 300 MW Whitetail Clean Energy NET Power project based on the Allam Cycle is being developed on Teesside, Yorkshire, with government support.

Conclusion

This chapter has shown that other means of using fossil fuels to generate energy have existed for a long time, dating back to the 19th century, and that fossil fuels can be combusted while simultaneously capturing the concomitant CO_2 emissions without any drastic penalties.

The methods discussed here only involve the first step in the process, CO_2 compression, transport, and subsequent geologic sequestration will be dealt with in future chapters. Engineered geological storage of CO_2 is a relatively straightforward and established technology (Ringrose 2020, p. 1).

---

[15] In 2018 the demonstration plant won the prize for ADIPEC Breakthrough Technological Project of the Year.

[16] https://netpower.com/news/.

Let us close this chapter with a few remarks on costs. The capture costs are directly related to the concentration of CO_2 in the gas. Flue gasses from coal-fired power plants have a higher concentration of CO_2 (10–15%) than those from natural gas-fired power plants (5–8%). Because of this difference, carbon capture from gas-fired power plants tends to be relatively more expensive than capture from coal-fired power plants. Figure 8.4 below shows the work needed for the various methods described above, showing that the Integrated gasification combined cycle is by far the most favourable. It also shows that Direct Air Capture is by far the most costly procedure, up to five times the energy input compared to capturing CO_2 directly at a point source, such as a power plant. This explains why most CCS research to date has focused on coal-fired power plants, and should continue to do so; they produce more CO_2 and are therefore better entities to first consider for capturing CO_2 more efficiently (Smit et al. 2014, p. 153–154). The number of coal power plants fitted with CCS installations is however still pitifully few, just a handful of units at the 4,000 plants in operation in the world (Global CCS Institute 2021a, p. 18–19).

Another way of looking at the minimum energy requirement is to express it as a fraction of the total energy produced. The average US coal-fired power plant generates 950 kWh net electricity per ton of CO_2 emitted. For capturing 100% of the CO_2 with a flue gas of 15% CO_2, the minimum energy needed is 5.12% of the electrical energy generated by the power plant. If we capture 90% of the CO_2, this number reduces to 4.22% of the electrical energy generated by the plant. This calculation shows that it costs 20% more energy to separate the last 10% of CO_2 from the flue gas—a disproportionate amount of energy for a small outcome. For this reason most regulations do not require 100% capture, but a much more sensible 90%. For a typical coal-fired power plant (with 12% CO_2 in its flue gas), the minimum energy for 90% separation is about 44 kWh/ton CO_2 (Smit et al. 2014, p. 154), a

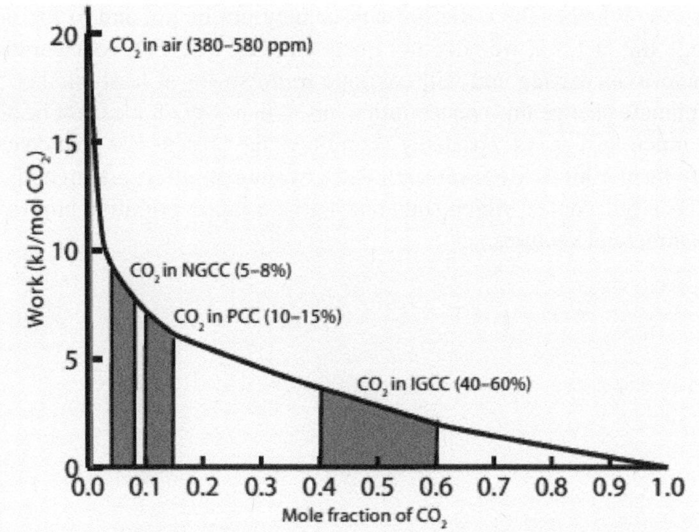

Fig. 8.4. Minimum work required to capture CO_2 as a function of the initial concentration of CO_2 in the flue gas. NGCC is carbon capture from natural gas, and PCC from coal (*from* Smit et al. 2014).

very small amount, just a few dollars, while for Direct Air Capture this will be about $100/ton at a minimum! In a recent publication (Brandl et al. 2021) it is argued that capture rates higher than 90% are both technically feasible and economically reasonable with negligible increase of the overall system costs. Residual emissions will likely be a minimum of 2% for concentrated streams, like power plant flue gases, up to 10% for low concentration streams.

All this does of course not include the capital costs of the carbon capture unit, estimated at $400–550 million, as well as compression, transport and storage costs. The IEA estimates the levelized cost of CO_2 capture for the power generation sector at $50–100 per ton (IEA 2021c, p. 101). It is peculiar that, while often the argument is heard that CC(U)S cannot compete with wind and solar energy because of the spectacular fall in costs of the latter, the increase in coal power, especially in developing countries, is just continuing. Apparently for those countries it is still cheaper to build coal plants than wind and solar energy parks. The comparison is partly false as the intermittent nature of wind and solar is not mentioned when talking about these costs. A fair comparison should also include the costs of storing the surplus energy generated by solar and wind, which is still very expensive and will in all probability remain so for a considerable time. Moreover, by the dogged fixation on costs the bigger picture is lost sight of, namely that we have a problem here that urgently needs to be solved.

Low-cost, highly efficient CO_2-capture technologies as discussed in this chapter must be developed further so that emissions reductions can become a reality, provided this is paired with policies and technologies that address other pollutants from fossil fuel production and combustion. Carbon-capture technologies have faced fierce criticism from environmental justice and other groups concerned with local air quality and land use impacts. Policy support for these technologies should include consideration of all air quality and economic impacts (Breakthrough Energy 2021, p. 13). Whether the criticism may or may not be apt and to the point does not change the fact that we have not many choices. The CO_2 concentration in the atmosphere is increasing and will continue to do so for at least another 50 years. To stop climate change this concentration has to be lowered, it cannot be allowed to increase much further. As repeatedly mentioned, the present 420 ppm concentration is already far too high. More research and government effort are urgently required to turn these procedures, which sometimes give a rather primitive impression, into viable commercial ventures.

CHAPTER 9

Direct Carbon Removal

Introduction

The too high and still increasing concentration of carbon dioxide in the air makes it absolutely essential that, in addition to phasing out as soon as possible a considerable part of current emissions, a large part of the carbon emitted into the atmosphere in the last 150 years is filtered or scrubbed out of the air, subsequently stored underground or used in some other way. As noted before, for life as we know it to be sustained, the maximum carbon-dioxide concentration in the atmosphere that can be tolerated is about 350 ppm. That value was exceeded around 1988 and it has now risen to 420 ppm. To avoid overshooting the goal of 1.5°C warming, the CO_2 concentration should be held at no more than 430 ppm, a value we will exceed within a decade. CO_2 emissions are still rising at over 2 ppm per annum and it is virtually certain that 450 ppm will be reached before mid-century with 500 ppm being very well possible. Once emitted, CO_2 remains in the air for a few centuries, implying that direct removal of carbon dioxide from the air is the only way to reduce the global atmospheric concentration if combined with long-term permanent storage or usage. In the end it will be the only way to stop global warming. Just reducing emissions will not be enough!

Before continuing, let us dwell for a moment on what is meant by permanent removal. There is currently no consensus definition of permanent removal but it seems to equate to some tens of millennia (Mac Dowell et al. 2022). This does however not imply that storage of large volumes of CO_2 for shorter periods, even much shorter periods of just a few centuries or even decades could not be useful, e.g., in the form of biochar (see below) which draws carbon from the atmosphere, providing a carbon sink on agricultural lands and sequestering it in the soil for centuries. The latter however remains at best a leaky, non-permanent storage method. Actually, afforestation, which in most schemes is proposed as an important climate mitigation method, similarly amounts to non-permanent storage and is always in danger of catastrophic failure due to pests, diseases or forest fires, as witnessed by the extensive wildfires in the summer of 2023 in Canada. The latter have resulted in a massive surge of global carbon emissions in addition to other kinds of air pollution. Utilizing carbon in 'permanent' products is also seen as a valuable means of removal although these products will rarely last longer than a few decades. We should however not be too dogmatic about the 'permanence' of storage and be careful not to demand too much of carbon capture in this respect, making it impossible in practice and killing it off as being unfeasible, when it cannot immediately attain the gold standards that some may require of a technology. Any opportunity for carbon removal that can in reason contribute to lowering the CO_2 concentration in the air for a reasonably long term and hence mitigate global warming should be seized upon with both hands. But it will be clear that products which re-release carbon within a year or so cannot be considered permanent removal. Monitoring, reporting and verification should be used to quantify the useful contribution of any method employed.

In Chapter 1 it has already been noted that carbon capture (and storage) in whatever form, apart from afforestation, has no large fan club. There seems to be an overall reluctance, a general dislike and lack of enthusiasm when trying to get new techniques on the rails. Such attitude can be lethal where enthusiasm and perseverance are required. One of the worries that has been vented (Cox et al. 2018, also cited in Beerling et al. 2020) is that deployment of any carbon capture strategy may erode society's perception of the climate threat and the urgency of mitigation measures, i.e., the emergence of such a technological solution might blur the necessity to actually avoid emissions. People who adopt that position have made the necessity of measures to *avoid* emissions to gospel and do not want to hear of any remedies to combat climate change other than by halting the burning of fossil fuels. I find that an unreasonable position as the culprit is not the burning of the fossil fuels as such but the practice that by doing so greenhouse gases are released into the atmosphere, without paying the costs attached to that practice.[1] In the previous chapters I have made it sufficiently clear, I hope, that avoiding and reducing emissions is indeed an urgent task but that it is not enough as it will not *lower* the concentration of greenhouse gases in the air. Moreover, it is an unrealistic position as the practice of the last decades has shown that we will not be able to get rid of fossil fuels anytime soon, even if we would want to. Unhelpful and harmful statements like "[m]ore work is needed to ensure that removals and their impact are fully understood before investment in scaling them up is made"[2] have been heard for years, resulting in the ultimately fateful postponement of useful action.

For those who mainly think in economic terms, the main hurdle to carbon capture seems to be that CCS and Direct Air Capture are just "investment and running costs, but no revenue." So, whatever it does it will drive up energy prices. Mark Jacobson, for instance, says in his book (Jacobson 2023, p. 186): DAC[3] "is basically a cost, or tax, added to the cost of fossil fuel generation, so it raises the cost of using fossil fuels while increasing air pollution due to its energy requirements and providing no energy security. To the contrary, it permits the fossil fuel industry to expand its devastation of the environment and human health by allowing mining and air pollution to continue at an even higher cost to consumers than with no carbon capture." The first part of

---

[1]   An extreme example of this reasoning can be found at https://rethinkresearch.biz/articles/time-to-call-bullshit-on-carbon-capture/.

[2]   Carbon Market Watch, https://carbonmarketwatch.org/2022/12/09/carbon-removals-are-no-magic-climate-or-development-bullet/.

[3]   He calls it SDACCS/U which stands for synthetic direct air carbon capture and storage/use.

the argument is correct, although DAC is not a tax. The word 'tax' is completely misplaced in the context of DAC. But indeed, as said, carbon capture is all cost and no revenue. This is also true for sewage systems and waste water treatment, for instance, which nobody dismisses as casually as done here with DAC. Neither does DAC provide energy security as it is not a source of energy, in the same way as sewage and waste water treatment do not ensure the availability of clean drinking water. However, his argument ignores the two overriding facts that make carbon capture inevitable, namely that the CO_2 concentration in the air is just too high and *must* be lowered, and that fossil fuels will remain with us for a considerable time to come, implying that the concentration will become even higher still. Climate change cannot be halted without reducing this concentration. There is no escape from these facts. Jacobson's own scheme of drastically reducing energy consumption and get the little we are still allowed to consume from renewable sources simply does not work in the world we live in, as set out in Chapter 3, and, what is more, does nothing towards reducing the already too high CO_2 concentration in the air. In his scheme the concentration will keep on rising until the end of the century or thereabouts. The second part of his argument has nothing to do with carbon capture. It is not a hard and fast rule that we must permit the fossil fuel industry, or anybody else for that matter, to pollute the air and devastate the environment. We allow them (and others) to do so as we want energy *cheaply*. The argument can equally be used against the construction of wind turbines, solar farms, hydroelectric dams, etc. They also use large quantities of materials that are mined and processed to the detriment of the environment to ensure that the resulting commodity remains affordable. Mankind's exploitation of Nature is by no means the exclusive prerogative of the fossil-fuel industry.

The negative attitude of the environmental movement to carbon-capture technology does not help either. With this they actually unwittingly and foolishly play into the hands of the fossil-fuel industry, which is all too happy not to be forced to deploy a costly technique that will probably indeed reduce profits and in any case will involve a lot of hassle. The industry knows, better than anyone else, that the world cannot do without fossil fuels for at least half a century. The frantic and at times clownish scramble after Russia's invasion of Ukraine in early 2022 made this abundantly clear. There is no reason for them to contradict Saul Griffith who in his book (Griffith 2021, Appendix A) dismisses out of hand all forms of carbon capture as being uneconomical. About CCS in general he states that "the economic argument against sequestration is that renewables are already competitive with coal and natural gas in most energy markets, and the added expense of carbon sequestration is not going to help fossil fuels compete. It is not unreasonable to say that the expense of carbon sequestration would be the death knell of fossil fuels." If all that is true and if you are bent on pushing the fossil-fuel industry out of business, why not demanding that fossil-fuel plants and other fossil-burning point-source installations capture and safely store the carbon they otherwise would emit? Griffith apparently does not realize that, if his reasoning is true, demanding the deployment of CCS actually kills two birds with one stone as it prevents carbon from being emitted and kills off the fossil-fuel industry. If you want them to die, what is better than letting them dig their own graves? If by some miracle they might survive, they have in any case become harmless. Of course, as everybody by now knows, his argument about the competitiveness of renewables with fossil fuels is only partly true as it disregards the energy storage problem for that type of energy, a problem that becomes more pressing every day, and makes a considerable part of the energy generated by wind and solar worthless, energy that could in any case be used for DAC or CCS.

The main problem we are faced with as regards Direct Carbon Removal in whatever form is to capture carbon at scale and the scale must indeed be gigantic (gigatons must be taken out) for it to have any effect in combating global warming. According to the 2022 IEA report on DAC (IEA 2022), DAC is today removing just 10,000 tons of CO_2 per year, and to reach the gigaton scale by 2050 will require annual growth rates of nearly 50% for thirty odd years (McQueen et al. 2021).

The need to remove carbon from the atmosphere has belatedly also been recognized by the IPCC, the IEA and others (See, e.g., Rogelj et al. 2018; Rogelj et al. 2015 and Luderer et al. 2013). It is claimed that to limit global temperature rise to 2°C, 10 $GtCO_2$ per year must globally be directly removed from the atmosphere by 2050, going up to 20 $GtCO_2$ per year after 2050.[4] This is of course a good start, but should be accompanied by sizeable emission reductions, as at current emission rates of more than 30 Gt the concentration of carbon dioxide in the air will still increase to unsustainable levels. In the IEA's Net Zero Scenario by 2050 (IEA 2022) DAC technologies must capture more than 85 Mt in 2030 and around 1 Gt in 2050. This will not make a great impact. A much larger and accelerated scale-up is needed. The scenarios assessed by the IPCC are better in that respect. They have a median value of around 15 Gt of CO_2 to be captured per annum by carbon capture, utilization and storage by 2050. The amounts to be captured and stored by DACCS and BECCS vary between 3.5 and 16 Gt of CO_2 by 2050. The difference between the IEA's Net Zero Scenario by 2050 and the pathways set out by the IPCC is that the IEA demands a much faster phasing out of fossil fuels, with, e.g., no new oil and gas fields or unabated coal plants (i.e., without CCS) to be approved for development from 2021, with the last unabated coal plant being completed in 2025, 60% of new car sales electric in 2030, coal use phased out by 50% in 2030 compared to the 2020 level etc. A very beautiful set of wishes but of questionable realism in the current world, as for instance pointed out recently in the *Guardian Weekly*,[5] which revealed that the world's biggest fossil-fuel firms, in collusion with governments, a number of which have been elected in democratic processes, are planning to exploit new fossil-fuel projects that would destroy (if carried out without carbon capture) any chance of keeping within the agreed temperature limits of global warming. But, even if the IEA's wishes were to come true, the CO_2 concentration would still be far too high and rise to over 450 ppm by 2050 or earlier. This rise would only stop after net zero had been achieved, with global warming still continuing for a considerable time until the global energy balance has been restored, on the condition that no tipping points are reached that cause the process to get out of control.

Studies of model calculations of carbon capture by DAC alongside other carbon-capture technologies and approaches find that gigaton scale DAC is necessary in

---

[4]   This also shows, by the way, that limiting temperature rise to 1.5°C is no longer on the cards.
[5]   Carbon Bombs, *The Guardian Weekly*, 20 May 2022, p. 10–14.

1.5°C pathways, and should be utilized heavily in 2°C pathways. One study (Marcucci et al. 2017) finds that limiting warming to 1.5°C is not feasible *without* DAC and that annual DAC of 21–40 GtCO$_2$ by 2100 is (globally) needed to limit warming to 2°C or lower, while another (Realmonte et al. 2019) foresees that DAC must remove between 16 and 30 GtCO$_2$ per year over the period 2070–2100. The latter study also mentions that a significant fraction (from 10 to 19%) of the carbon removed would be released back to the atmosphere from the oceans, requiring an additional removal of 1.7 to 9.5 GtCO$_2$/year to meet the same carbon budget. In total 600–700 GtCO$_2$ must be cumulatively removed by DAC from 2070 to 2100 to limit warming to 2°C or lower (an average of 20–23 GtCO$_2$ per year), while they too state that limiting warming to 1.5°C is not feasible without DAC.

Since the world does not show any great urgency in curbing the use of fossil fuels, DAC and other Direct Carbon Removal technologies will soon be the only feasible way to do something about global warming that has a *real* effect by physically removing the culprit from the air. The other methods of carbon capture involving fitting or retrofitting CCS units to carbon-emitting installations as discussed in the previous chapter do not reduce the carbon concentration already in the air but only let this concentration rise less fast. Already now the situation is urgent but could become desperate towards the end of the century, for if the current practice continues we will end up with a 4°C rise in temperature, making Earth's environment intolerable for humans and many other life forms. In that connection it is incomprehensible that governments and environmental NGOs pay so little attention to carbon capture. It has always been decried as being too expensive, and expensive it certainly is, but so were wind turbines, solar panels and almost any technology you can think off in its early days. Moreover, expensive is a relative matter. Any climate cure will always be expensive if compared with emitting carbon dioxide free of charge into the air, but it may still be a bargain if it prevents the world from becoming uninhabitable.

A promising set of technologies, mostly using similar chemical approaches to capture CO$_2$ from the air as discussed in Chapter 7, is available, but most are not well-developed yet, left 'criminally' underdeveloped as someone recently wrote (Book 2022). They still have to prove themselves and up till now too little has been invested in these technologies and in subsequent geological storage. The tide is changing though and in recent years carbon-capture technologies have been growing rapidly, with an increasing number of researchers shifting their attention to developing materials and processes for these technologies, along with several start-up companies pushing the technology from lab scale to demonstration and pilot scale. As we will see in Chapter 13, a multitude of new methods is being proposed, some of which may be more promising than the chemical methods that have so far dominated the scene.

The momentum for technological carbon removal greatly accelerated in the last few years. It started in 2020 with Microsoft creating a $1 billion fund for investment in emerging climate technology solutions that need capital to scale in the market and culminated in 2022 with the Biden administration allotting $3.5 billion for pilots and R&D in DAC and $4.6 billion for CO$_2$ transport and storage infrastructure to be spent in the coming five years for bringing this technology and others forward. In the years before, the US had made available just $60 million for carbon removal research and development. In addition, the tax credit per ton of CO_2 sequestered was increased to \$180 in 2022 (from just \$35–50 set in 2018). The US, as part of its Energy Earthshots Initiatives, also announced its "Carbon Negative Shot" effort to bring down the cost of carbon removal to \$100 per ton of CO_2 over the next decade.

In April 2022 the Xprize foundation,[6] funded partly by Elon Musk, awarded a dozen aspiring carbon capture start-ups in the US, Europe, Kenya, the Philippines and Australia with \$1 million each towards the further development of their ideas. The ultimate prize of \$100 million for the breakthrough carbon removal idea is still awaiting a winner.

In the same month a new privately backed nearly \$1 billion funding mechanism was unveiled by the payments company *Stripe*, in collaboration with others. It announced the formation of *Frontier*,[7] an advance-market commitment to buy close to \$1 billion of services for permanent carbon removal over the next eight years. Buyers are asked to commit themselves to buy a product (in this case captured carbon dioxide) that does not yet exist, from (future) suppliers, which are carbon removal companies with high potential vetted by *Frontier.* It facilitates purchases from these carbon removal companies on behalf of the buyers, aiming to accelerate the development of carbon-removal technologies by guaranteeing future demand (Kramer 2022b).

A carbon removal platform similar to *Frontier* is *Puro.Earth*,[8] based in Finland. Its member suppliers develop carbon net-negative processes or products, i.e., remove carbon from the atmosphere. *Puro.Earth* certifies suppliers based on their Puro Standard and an independent verification by a third party. It issues CO_2 Removal Certificates (CORCs) that can be bought by climate conscious companies directly from suppliers or through a third-party marketplace of their choice. Its registry contains a number of companies (34 as of 20 December 2022) that are mainly involved in Biochar, Carbonated Building Elements, Geologically Stored Carbon, and Woody Biomass Burial.

Others followed with the World Economic Forum's First Movers Coalition including DAC as a key technology; the Breakthrough Energy Catalyst Fund targeting DAC for \$1.5 billion investment; LowerCarbon Capital setting up a \$350 million fund to invest in carbon removal start-up companies; and Mission Innovation—a global initiative of 23 countries, including the US, China and the EU to accelerate clean energy innovation—adding carbon removal to its agenda with the still very modest goal of achieving a net reduction of 100 $MtCO_2$/year by 2030.

Ways of direct carbon removal

The various ways of Direct Carbon Removal from the atmosphere have been neatly summarised in Fig. 9.1. Some of them will be discussed in greater detail in this chapter.

---

[6]   https://www.xprize.org/.

[7]   https://frontierclimate.com/.

[8]   https://puro.earth/.

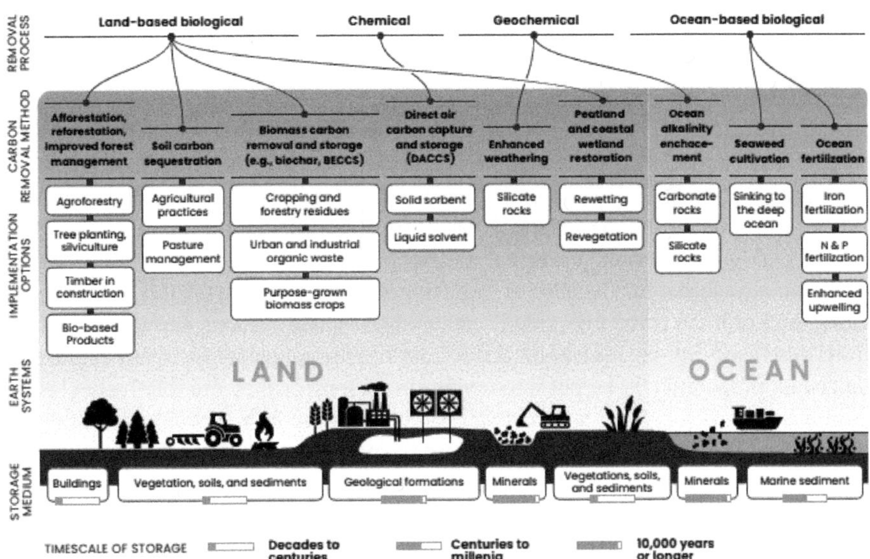

Fig. 9.1. Direct Carbon Removal approaches on land and in the ocean (*source*: Boehm et al. 2022, p. 134).

As can be seen, the methods can be divided into nature-based approaches and active carbon-removal technologies. Afforestation, reforestation and improved forest management, consisting of tree planting, using timber in construction, storing carbon in products and suchlike, are obvious, common-sense measures and will not be considered any further. Their timescale of storage is only a few decades and is insufficient to turn the tide. Timber in construction, used in furniture or trees that are only allowed to grow for 10–20 years cannot provide the scale of carbon capture needed. The same applies for most of soil carbon sequestration, as well as peatland and coastal wetland restoration. Any approaches that allow storage for only a couple of decades, although useful and necessary for all kinds of other reasons, such as Nature restoration, cannot halt global warming. We will only consider Direct Air Capture, biomass carbon removal (especially biochar), enhanced weathering technologies and some ocean-based approaches that may have a chance of being vastly scaled up and have storage timescales of centuries to millennia or even longer. I note here that biomass solutions to the carbon problem, including BECCS (bioenergy with carbon capture and storage), are often based, probably out of desperation and/or wishful thinking, on the use of unsustainably large areas of land for biomass feedstock production causing conflict with other uses. See also Chapters 4 and 8 about BECCS.

The main approach to Direct Air Capture technology involves installations using chemicals that react selectively with carbon dioxide in the air, which is then separated and stored. The various materials used in such installations have been discussed in Chapter 7. Their purpose is to produce a more concentrated stream of CO_2 from an extremely dilute source (only 1 in 2500 molecules in the air at present being CO_2 (about 0.04%)). This concentrated stream can then be utilized in a variety of ways, but to achieve positive climate intervention the real challenge is to do this at large scale with (geological) storage of huge quantities of CO_2 as the end goal.

DAC has the prominent advantage over capture from point sources that it can address emissions from distributed sources for which fitting CCS installations is no option (e.g., aviation, private cars, shipping, land-use change and individual home heating). If driven with power from renewable sources, DAC is a negative CO_2 emissions technology. Roughly half of annual CO_2 emissions are due to such distributed sources and capture solely from point-sources can never lower present atmospheric concentrations of CO_2. But, of course, as said before, just deploying DAC without cutting emissions from point sources makes no sense either.

One of DAC's advantages is that it is not location-specific; capture facilities can be set up anywhere (for instance at sea using surplus wind energy). It is immaterial where on Earth the carbon dioxide is captured and DAC systems can hugely benefit from this flexibility of placement. It does not require much land and any number of units could theoretically be put into operation, as long as they are relatively close to established geological storage sites *and* can be powered by renewable sources. This can reduce the need for expensive pipelines from the capture site to the sequestration reservoir and save considerably in costs.[9] What we do with waste water and sewage as a matter of course, we also have to do with air and dot the landscape (or the sea) with air scrubbing installations.

Furthermore, DAC does not have to deal with high concentrations of other contaminants (SOx, NOx, mercury, etc.) as are present in flue gases and cause degradation of performance of sorbents or solvents used in the capture processes.

Although DAC is generally evaluated by comparison with flue gas capture, it should be realized that it concerns here two different technologies with differing end goals. It is not a matter of either/or, the technologies should be developed in parallel. The scale of the CO_2-emissions problem is sufficiently vast that all technologies that can reduce emissions or reduce current concentrations of CO_2 in air should be pursued with vigour to identify those that scale most effectively and offer the most compelling economics, although, in view of the impending climate catastrophe, we should not be too preoccupied by economics (Sanz-Pérez et al. 2016).

DAC is still limited by the large amounts of energy it requires as well as by the location and volume of global storage capacities. This implies that only Direct Air Capture powered by renewable energy with dedicated storage would, in principle, be a proper approach to reducing the stock of atmospheric CO_2 (Sekera and Lichtenberger 2020).

Size of the problem

As already mentioned before, an inherent and substantial difficulty of DAC compared to the methods applied to flue gases discussed in the preceding chapter is that dilute streams are more difficult to separate. Large volumes of air have to be processed to capture a relevant amount of CO_2 and the separation process aiming for the same end CO_2 purity will therefore be more expensive, even thought to be too expensive, than

---

[9]   Approximate cost of CO_2 transportation via pipeline is \$2.24/ton CO_2 per 100 km of dedicated pipeline (NAS).

capture from flue gases of fossil fuel power plants. Klaus Lackner, who arguably is the inventor of negative emissions and one of the very first to emphasize the necessity to take carbon out of the air (see Kolbert 2021, pp. 149–153), had different ideas by looking at the problem from an energy standpoint, in the following way (Broecker and Kunzig 2008, p. 202–203).

One m^3 of air weighs 1,293 grams. The molecular weight of CO_2 is about 44, while the average molecular weight of air is about 29 (of N_2 about 28 and of O_2 about 32, and 99% of air is N_2 (78%) and O_2 (21%)). The current concentration of CO_2 per volume in air is 0.042%. To convert this to a mass fraction we simply have to scale this percentage by 44/29 which gives 0.064%. So the CO_2 component in air weighs 1,293 g/m^3 × 0.064 × 10^{-2} = 0.83 grams/m^3.

One litre of gasoline weighs 750 grams. Gasoline consists for 87% of carbon, or 652 grams of carbon per litre. To combust this carbon to CO_2, about 1740 grams of oxygen are needed (2 atoms of oxygen for each atom of C, hence 2 × (16/12) × 652). This results in 652 + 1,740 = 2,392 grams of CO_2 emissions per litre of gasoline. A litre of gasoline also generates 31,536 kilojoules in energy. This implies that burning just 0.35 millilitre of gasoline produces 0.83 grams of CO_2, as much as is currently contained in one m^3 of air, and generates about 11 kilojoules of energy, namely 0.35 × 31,536/1,000. That same m^3 of air blowing at ten metres per second (36 km per hour) contains only 65 joule of kinetic energy ($\frac{1}{2}$mv^2 = $\frac{1}{2}$ × 1.293 × 100 kg m^2/s^2 = 64.65 joule). Even if all of that could be harvested, which it can't, there is 170 times more energy generated in burning just the tiny amount of gasoline that produces 0.83 grams of CO_2. If some of it is used to remove the CO_2, a lot will still be left, and hence making it economically sound to try to do this. This calculation immediately shows how hard it is to harvest sizable amounts of wind energy by using wind turbines and how wonderful fossil fuels are as regards their energy content. The point Lackner makes is that you will need a lot of carbon dioxide scrubbers to make a dent in the CO_2 emissions, but not as many as the number of windmills needed to have an equivalent effect on atmospheric CO_2.

Although this perhaps sounds promising, it still means that for capturing a ton (1,000 kg) of CO_2 from the air, we would need to scrub it out of about 1.3 million m^3 of air, if we were 100% efficient at capturing the CO_2 molecules from the air (there is no need to be 100% efficient, though). To get one year's worth of global CO_2 emissions out of the atmosphere we have to filter the staggering volume of 50 million billion (or 50,000 trillion) m^3 of air. The numbers are so big that they no longer make any sense, and most people give up on contemplating a solution to such a problem. It clearly is a very, very tall order, but there is hardly any other choice. The amount of carbon dioxide in the atmosphere is just too high to leave it sitting there for another couple of centuries and let global warming continue for a very long time, even after we have managed to bring new emissions down to zero, which we may never achieve. The CO_2 concentration has to be brought down, preferably to 350 ppm, the number mentioned by James Hansen, as we have seen, and by McKibben's organization 350.org. This means that there is roughly 550 Gt of CO_2 (70 ppm) in the atmosphere that must be taken out, not counting what we are still each year putting in.

To put it in perspective let us compare this with waste water treatment. A country like the Netherlands, with about 17 million inhabitants, annually treats about 1.9 billion m^3 of waste water. Extended to the entire Earth this would imply treatment of about 900 billion m^3 of waste water, in weight almost twice the above-mentioned 550 billion tons of CO_2 that have to be filtered out of the air (1 m^3 of water weighing about 1 ton). This already sounds much more manageable. Everybody agrees that waste water treatment is necessary, feasible and affordable, in spite of it also being just "investment and running costs, but no revenue". Or should we stop with this *uneconomical* practice?

The world annually produces about 30 Gt, 30 billion tons, of CO_2 from burning fossil fuels and other activities, that is 80 million tons per day, so 80 million scrubbers with a capacity of just one ton per day are needed to suck all the CO_2 out of the air. Most of newly emitted CO_2 can however be captured by other means at point sources like smoke stacks and does not have to be scrubbed out of the air. A quarter of CO_2 comes from mobile sources and 10% from heating buildings, where there is hardly any other choice than emitting the generated carbon dioxide into the air. About 40% of CO_2 cannot be readily captured at source. So, if air scrubbers can be built with a capacity of 1 Mt of CO_2 per year, we would globally need 12,000 of them to scrub the air clean of annual emissions that cannot otherwise be avoided. Double that number would add an extra annual capture capacity of 12 Gt and the problem of legacy emissions (see Chapter 2) in the atmosphere could be tackled and solved in 60–80 years. Actually the problem is somewhat smaller if we take into account that about half of the emitted CO_2 is absorbed by the oceans and by the biomass on land. It is however difficult to assess what happens to the CO_2 already absorbed by the oceans and still contained in the top layer of ocean water when we start scrubbing CO_2 out of the air. In order to establish a new equilibrium, the oceans might start to give off CO_2 into the atmosphere when the current situation of an ever increasing CO_2 concentration is reversed. This carbon cycle feedback is known from the literature (see, e.g., Realmonte et al. 2019) and from IPCC's Fifth and Sixth Assessment Reports, as mentioned in Chapter 4. CO_2 removal from the atmosphere would be partly offset by the oceans giving off some of the carbon dioxide absorbed earlier, hence it would also reverse ocean surface acidification. In other words, looking at it from the positive side, Direct Carbon Removal would kill two birds with one stone: it cleans the air of CO_2 *and* reverses ocean acidification.

To summarise, Direct Air Capture (DAC) facilities (with subsequent carbon storage, of course) have the following key advantages:

1. capture can be carried out close to the storage location, reducing transport costs;

2. the facilities can be deployed in windy locations reducing the costs of operating fans; and

3. at the same time be located close to renewable energy sources like wind turbines.

So, ideally wind turbines could be built at sea, close to depleted gas or oil fields, or any other geological storage location, to provide the electricity for operating the DAC facility which stores the captured carbon in the depleted oil or gas field. The North Sea would be a perfect place for such facilities. In this connection I note that in the Norwegian Sleipner project (see Chapter 12) carbon dioxide is stored

800 metres below the seabed of the North Sea in a saline aquifer with a storage capacity of up to 600 billion tons. This would be sufficient for a very long time for storing all the CO_2 we want to scrub out of the air *globally*!

Methods of direct air capture

Chemically based liquid solvent or solid sorbent DAC, as discussed in Chapter 7, are at present the two procedures furthest along in development (McQueen et al. 2021). Only a small number of DAC facilities of small size have so far been built, so the development of DAC technology is far from complete. It can be expected that significant cost reductions can be achieved by upscaling, hopefully down to some $100 per ton of carbon dioxide captured by the middle of the century. More research is urgently needed here and funding provided by government agencies has been long overdue, but is now being made available. Virtually all practical efforts still come from private enterprises. The first plants are being operated in Switzerland, Italy, Iceland, the USA and Canada, with the Canadian firm Carbon Engineering having started construction of a DAC plant with a capture capacity of 1 $MtCO_2$ per year, using the alkali scrubbing process detailed in Chapter 7. The Swiss firm Climeworks, so far working on a shoe string, raised $650 million in funding in 2022, with its DAC plant in Iceland, removing 4,000 tCO_2 per year, just coming online in 2021.

By 2050, DAC facilities must play a significant role in the global effort to achieve net zero emissions and restore safe levels of CO_2 in the atmosphere. In Chapter 13 we will discuss the companies and projects that develop a great variety of technologies for DAC and other Direct Carbon Removal approaches. Most technologies developed earlier were characterized by high costs and low scalability, making them not directly suitable for climate-change mitigation purposes. Modern versions employ a combination of mechanical systems and chemical processes, include fans to move large volumes of air, and liquid or solid sorbents to capture the CO_2. The ideal properties of these sorbents have been described in Chapter 7. Overall, sorbents are critical to the success of DAC and there remains much room for improvement. Development needs to progress quickly, because global warming will not wait. Energy-efficient and low-cost CO_2 sorbents must be applied to solve the issue of global climate change (Shi et al. 2020).

A solid sorbent DAC plant is usually designed to be modular and can include as many units as needed. A single adsorption/desorption unit has a capacity of several tens of tons of CO_2 per year, with the added advantage that it can be used to extract water from the atmosphere (IEA 2022, p. 21). A large-scale liquid sorbent-DAC plant can capture around 1 Mt of CO_2 per year from the atmosphere. Depending on weather conditions, large amounts of water (up to five tons per ton of captured CO_2) would be required for such a plant.

Carbon removal methods based on natural biological processes

Here we will discuss some of the other methods of Direct Carbon Removal shown in Fig. 9.1, including carbon mineralization (referring to any process that sequesters

CO_2 as a solid carbonate) via enhanced weathering and ocean fertilization, which in essence are beefed up natural processes.

• Carbon mineralization

Carbon mineralization or mineral sequestration via enhanced weathering has so far received little attention but is an attractive and natural means of carbon capture. Weathering is a term used for the deterioration of rocks, soils, minerals, wood and artificial materials through contact with water, atmospheric gases and biological organisms, hence the process by which most things go to pieces when you leave them exposed to the elements. For our purposes it involves in particular the formation of solid carbonate minerals, like $CaCO_3$ and $MgCO_3$, through CO_2 reacting with rocks rich in calcium and magnesium.

Carbon mineralization is one of the main mechanisms the planet uses to recycle carbon dioxide across geological time scales in the carbon cycle, and natural processes of this nature already remove without our intervention about 1.1 Gt CO_2 per annum from the atmosphere (Ciais et al. 2013). Hence this scale is already in the right ballpark and a reflection of the fact that rocks form one of the largest carbon reservoirs on Earth. Suitable rocks are so-called ultramafic or mafic[10] igneous rocks, which are all abundant at or near the Earth's surface. The abundance and wide distribution of such rocks means that many nations and companies could use carbon mineralization as a CO_2 reduction and removal strategy at scale. In addition, some industrial waste, such as steel slag and fly ash, is suitable for carbon mineralization.

In the geological weathering cycle CO_2 dissolved in rainwater, in the form of carbonic acid (H_2CO_3), reacts with rocks that are rich in calcium, magnesium and silicon to form long-lived carbonate minerals (minerals containing the carbonate ion CO_3^{2-}). These find their way into the oceans, where marine organisms digest them and convert them into the stable, solid calcium carbonate ($CaCO_3$) that makes up their shells and skeletons (Temple 2020). The natural process is very slow (reaction rates depend on temperature and on the mineral in the rock that has to react with CO_2), but can be speeded up by grinding suitable rocks to a very fine reactive powder. This may however not always be suitable because of possible cost, health and environmental problems, but in general it applies that the technical risks of enhanced weathering are assumed to be rather low (Moosdorf et al. 2014).

Enhanced weathering is just that: the acceleration of such natural weathering by spreading finely ground silicate rock, such as basalt, onto land. This speeds up the chemical reactions between rocks, water and air; forms carbonate minerals in soils and/or bicarbonate ions that are transported to the ocean by rivers. It removes CO_2 from the atmosphere by trapping it permanently in solid carbonate minerals. But, as with all carbon capture methods, the challenge is to develop technologies that operate at a scale commensurate with the scale of greenhouse-gas emissions.

The technique is by no means new and has been actively discussed since the 1990s (Lackner et al. 1995). It is essentially a geoengineering technique but, as a strategy for removing CO_2 from the atmosphere, it shares many characteristics with DAC.

---

[10] See Glossary.

Both are early stage and currently expensive, have clear pathways to reduce costs, can be deployed in many locations around the world and show considerable promise in helping achieve net-zero emissions. One of the plants currently in operation (in Iceland by Climeworks, see Chapter 13) combines this method with solid sorbent-DAC and captures 4,000 tCO_2 per year.

The great advantages of carbon mineralization are that no sorbents or solvents are needed, and that the chemical reactions occurring in the process require no energy input and inherently provide a mechanism for long-term (10,000 years or longer; Fig. 9.1) and non-toxic storage of CO_2 in solid form (stable carbonates), minimizing the risks of leakage, while other DAC processes produce a concentrated stream of CO_2 that has to be piped to a storage location. A small drawback is that measurement and verification of CO_2 removal is more difficult for most carbon mineralization processes (Sandalow et al. 2021).

Carbon mineralization can be done in various ways: (1) grinding silicate rock to small particles and transport it to a site, where it is made to react with CO_2 in a high temperature and pressure reaction vessel (*ex situ*), (2) on the surface (*surficial*), where dilute or concentrated CO_2 is reacted with finely ground rock on-site at the surface (e.g., spread a slurry of water and ground mine tailings or smelter slag on land and let it rapidly draw down and mineralize carbon dioxide from the air), and (3) let CO_2-bearing fluids circulate through subsurface porous geological formations (*in situ*). For this latter form of in-situ carbonization CO_2-bearing fluids or supercritical CO_2 are injected into geological formations.

Ex situ mineralization is especially attractive for mine tailings or other industrial waste products as it could reduce hazards due to chemical contamination associated with such waste, but such *ex situ* methods, which require the waste to be transported to a place where it can be heated, are more expensive than the projected costs of other DAC procedures, e.g., about ten times more expensive than injection and sequestration of CO_2 in geological reservoirs (Kelemen et al. 2019).

When enhanced weathering is used on croplands by spreading finely ground rock, main economic cost factors concern the mining, crushing, and grinding of rocks, and the transport to and distribution on crop fields. Mine tailings can also be used for this purpose provided that they only contain (ultra)mafic rock without any other (hazardous) additions, although health hazards presented by mine tailings containing asbestos can actually be mitigated by carbon mineralization. It has been estimated (Strefler et al. 2018) that the cost of absorbing one ton of CO_2 would be just $60 when using dunite, a type of rock consisting for 90% of olivine (a magnesium iron silicate and the most abundant type of rock in the crust of the Earth), with minor amounts of other minerals, including unfortunately traces of harmful minerals such as chromium and nickel. The chemical reaction of this capture process is:

$$Mg_2SiO_4 \text{ (olivine)} + 2CO_2 \rightarrow 2MgCO_3 \text{ (magnesite)} + SiO_2 \text{ (silica)}.$$

The magnesite is a (stable) mineral form of magnesium carbonate.

The equivalent cost when basalt is used instead of dunite would be $200 per ton of CO_2. Basalt contains less harmful elements, and could even act as a fertilizer by providing elements that are in deficit in many tropical areas, but has a lower weathering efficiency. More than 90% of all volcanic rock on Earth is basalt.

It is claimed (Strefler et al. 2018) that the potential for carbon removal on cropland could be as stupendously large as 95 GtCO$_2$ per year for dunite, about three times current annual emissions, and 4.9 GtCO$_2$ per year for basalt, much less than the phantastic amounts estimated for dunite but still an extremely useful amount; after all, there is no need to scrub all CO$_2$ out of the air in a few years or a single decade. The IPCC in its Sixth Assessment report (IPCC 2021, Table 5.36, p. 766) quotes > 3 GtCO$_2$ per year as the potential for CO$_2$ absorption by enhanced weathering, while the Innovation for Cool Earth Forum (ICEF) (Sandalow et al. 2021, p. 3) states that carbon mineralization processes could remove 1 GtCO$_2$ per year from the atmosphere by 2035 and 10 GtCO$_2$ per year by 2050. There is of course a CO$_2$ penalty to be paid for the mining, grinding, and spreading of rock powder, and the need to mine and process enormous amounts of virgin rock is actually the principal challenge for large-scale use (i.e., absorption of multiple gigatons of CO$_2$ per year) of enhanced weathering. The best suited locations are warm and humid areas, particularly in India, Brazil, South-East Asia and China, where almost 75% of the global potential can be realized.

In the face of dwindling phosphate rock resources, enhanced weathering can play an important role regarding alternative fertilization methods in addition to its role as a CO$_2$ removal method. In this connection it has been argued (Moosdorf et al. 2014, p. 4814) that the effect of enhanced weathering on agricultural output will be one of the main factors determining the success of the method.

In a paper in *Nature* in 2020 by Beerling et al., including James Hansen as one of the authors, the potential for enhanced rock weathering on croplands was analyzed for a variety of countries. It was concluded that China, India, the US and Brazil have great potential to help achieve average global carbon removal goals of 0.5 to 2 Gt of CO$_2$ per year with extraction costs of approximately \$80–180 per ton of CO$_2$, hence cost levels comparable to (and generally lower than) current estimates for DAC technologies. Potential ancillary benefits include limiting coastal zone acidification and improving food and soil security.

This sounds all very positive but the current assessment of the promise of enhanced rock weathering for CO$_2$ removal may well be too optimistic as environmental and human health risks may have been overlooked. Silicate rocks and slags from industrial wastes may contain higher levels of potentially toxic elements and radionuclides than soils and conventional inorganic and organic fertilizers (Choi et al. 2021).

Enhanced weathering can also be used in the oceans. Carbon-removing sand made of the abundant and cheap mineral olivine is added to the ocean. When olivine interacts with water and CO$_2$, it creates a bicarbonate (HCO$_3^-$) that eventually sequesters the carbon as rock in the sea floor and deacidifies the ocean. So, the sand counters ocean acidification and permanently removes carbon dioxide from the ocean and indirectly from the atmosphere.

- **Ocean fertilization**

Enhanced fertilization of the oceans, also considered a form of geoengineering, refers to the strategy of enriching ocean waters with limiting nutrients such as iron (Fe), to enhance ocean biological production and CO$_2$ sequestering. Iron is a limiting nutrient that is crucial to the growth of marine algae (phytoplankton). While the other nutrients required by marine algae are abundant—or at least relatively abundant—in the ocean, iron is less readily available, so marine algae growth is limited according to iron levels. If iron, or an iron-based compound, is released into the ocean, this limitation is removed. It can result in an increase in organic productivity and marine algal growth and thus stimulate photosynthesis.

Algae are photosynthetic organisms that use sunlight to make organic matter and oxygen. They are unusually influential in the Earth's climate as they remove carbon dioxide from the air and are the source of the gas dimethyl sulphide (C_2H_6S), which oxidizes in the air to become the tiny nuclei that seed the droplets of clouds. Their photosynthesis will remove large quantities of CO_2 in the ocean's surface, which will eventually end up at the bottom of the ocean. We have more to say about this in the next chapter when discussing storage.

As any form of geoengineering, it is viewed with suspicion by environmental organizations, because of fears that tinkering with the planet could lead to catastrophe and that it lets governments and energy companies off the hook. This short-sighted attitude, which they adopt to carbon capture in general, may in the end well be the final nail in mankind's coffin. They may be waking up too late to the fact that carbon capture is no longer a choice, but a necessity for limiting global warming. Having said this, it is certainly true that critics are rightly concerned that ocean fertilization could create harmful algal blooms and other yet unknown harmful effects. Hence, it stands to reason, as Germany decided in august 2018, that ocean seeding will only be allowed for research purposes and under strict conditions.

• Biochar

Biochar is the name given to a charcoal-like substance produced by heating organic material (also called biomass) at temperatures of 300–600°C in an oxygen-limited environment. It is biomass (wood, leaves, straw, or other biosolids) heated at high temperatures without oxygen. The process is called pyrolysis and concentrates carbon in a form that is very resistant to biological decomposition. As a result biochar is a stable solid rich in carbon which can endure in soil for thousands of years, possibly making it an ideal technology for scalable carbon removal. The organic carbon-rich molecules in the plant matter undergo some rather unique chemical changes that make the biochar chemically very similar to coal. Biochar can store large amounts of carbon for very long periods, in contrast to the very rapid release of carbon from the burning or natural decomposition of biomass. The biochar can be mixed with existing soil, acting as a fertilizer and sequestering carbon. Biochar production and burial removes carbon dioxide directly from the atmosphere through uptake by plants, allowing, in principle, an actual reduction of atmospheric carbon dioxide levels. By acting as fertilizer it also boosts plant growth allowing such plants to capture more carbon (Glaser et al. 2009). In addition, biochar has multiple commercial uses at potentially industrial volumes, e.g., as greenhouse additive, in soil regeneration and in waste water treatment.

Assuming that the biomass has been sourced as sustainably as possible such that there is no carbon debt to repay, the quantity of CO_2 removed is greatest when the biochar is first applied to the soil. Thereafter, the char begins to decay such that after a period of decades to centuries, there may be very little left (Mac Dowell et al. 2022; Jeffery et al. 2016). Biochar has to be made from plant material that comes as a by-product from farming. If specific crops have to be farmed for the production of biochar to boost carbon capture, it becomes far too expensive and can cause other problems including habitat loss and food supply chain issues, as is in general the case with any carbon removal strategy based on biomass.

The Institute of Carbon Removal Law and Policy of the American University in Washington DC states that carbon sequestration in soil could be scaled up to sequester 2–5 $GtCO_2$ per year by 2050, with a cumulative potential of 104–130 $GtCO_2$ by the end of the century at a cost of between \$0 and \$100 per ton of CO_2. The potential and cost of using biochar at large scales are less clear. For Europe alone it has been estimated that based on emissions of about 4 Gt of carbon dioxide, biochar could offset around 9% of these (Glaser et al. 2009). Another estimate states that biochar could sequester 0.5–2 $GtCO_2$ per year by 2050 at a cost of \$30–120 per ton of CO_2. The broader academic literature envisions sequestration rates between 1 and 35 $GtCO_2$ per year with estimates of the cumulative potential ranging from 78–477 $GtCO_2$ this century.[11] Most of the companies on the register of the Finnish carbon removal platform *Puro.Earth*, mentioned above, are involved in biochar production.

- **Artificial photosynthesis**

Being the source of all the oxygen in the atmosphere, natural photosynthesis is one of the most important processes on Earth. In the first phase of plant photosynthesis, water molecules are photo-oxidized to release oxygen and hydrogen, while the second phase is a light-independent reaction that converts carbon dioxide into glucose. The entire process is hampered, at any rate for our purposes, by its very low efficiency in converting sunlight to biomass.

Artificial photosynthesis tries to mimic natural photosynthesis. If we could do this efficiently, it could be an important means to lower the CO_2 concentration in the atmosphere. The term is used to describe any man-mediated process that stores sunlight energy in useful, high energy chemicals (Bozal-Ginesta and Durrant 2019). The first report in which photosynthesis is mimicked by using light to split water into hydrogen and oxygen dates from 1972. Since then many efforts have been devoted to this goal, but the most promising devices and materials still suffer from fast component degradation and/or large efficiency losses. Researchers of artificial photosynthesis are developing photocatalysts (catalysts that speed up chemical reactions involving light) that are able to perform these reactions. One of its most studied applications is the production of fuels (so-called solar fuels) to replace fossil fuels and enable the carbon neutral production of energy when these fuels are combusted. In this combustion carbon dioxide is of course still released but is equal to the carbon dioxide captured and used in producing the fuel. It might offer a low carbon or carbon neutral pathway for the synthesis of sustainable fuels (e.g., green

---

[11] Institute of Carbon Removal Law and Policy, Carbon Removal Fact Sheet: Soil Carbon & Biochar; https://www.american.edu/sis/centers/carbon-removal/upload/icrlp_fact_sheet_soil_carbon_biochar_181006.pdf; Fuss et al. 2018; Smith 2016.

hydrogen from water), as well as scalable pathways for carbon dioxide reduction and utilization. With the development of catalysts able to reproduce the major parts of photosynthesis, water and sunlight would then ultimately be the only sources needed for clean energy production. The only by-product would be oxygen. Widely used catalysts for artificial photosynthesis are metal complexes (ruthenium, manganese, etc.) that mimic the natural catalyst Mn_4CaO_5.

For many years artificial photosynthesis remained an academic field. However, in early 2009, Mitsubishi Chemical Holdings was reported to be developing its own artificial photosynthesis research by using sunlight, water and carbon dioxide to "create the carbon building blocks from which resins, plastics and fibres can be synthesized."[12] During 2010, the United States Department of Energy established the Joint Center for Artificial Photosynthesis,[13] which has the mission to find a cost-effective method to produce fuels using only sunlight, water, and carbon dioxide as inputs. Also during 2010, a team at the University of Cincinnati successfully demonstrated photosynthesis in an artificial construct consisting of enzymes suspended in a foam housing. During 2011, Daniel Nocera and his research team (Reece et al. 2011) at Harvard University announced the creation of the first practical artificial leaf, in the form of an advanced solar cell the size of a playing card capable of splitting water into oxygen and hydrogen approximately ten times more efficiently than natural photosynthesis. The cell is mostly made of inexpensive materials that are widely available, works under simple conditions, and shows increased stability over previous catalysts. In May 2012, Sun Catalytix, the start-up based on Nocera's research, stated that it will not be scaling up the prototype as the device offers few savings over other ways to make hydrogen from sunlight. Leading experts in the field have supported a proposal for a Global Project on Artificial Photosynthesis as a combined energy security and climate change solution, but the idea has not come further than a few conferences, with the last one held in 2016.

Research into finding catalysts that can convert water, carbon dioxide, and sunlight into carbohydrates or hydrogen is currently still an active field, but the catalysts developed so far are too inefficient, in particular how much of the incident light can be used in a system in practice. Photosynthetic organisms are able to collect about 50% of incident solar radiation. In contrast, the highest reported efficiency of lab prototypes for artificial photosynthesis is 22.4%. Moreover, plants are efficient in using CO_2 at atmospheric concentrations, something that artificial catalysts still cannot do.

In the next decades, artificial photosynthesis could contribute to replacing large quantities of fossil fuels or raw materials. It has the potential to make a significant contribution to more environmentally friendly energy generation and climate protection. Nevertheless, a number of challenges still remain before large-scale technical and economic implementation is possible. These include, in particular, the efficient use of carbon dioxide from the atmosphere and the production of cost-effective and long-term stable photocatalysts (Beller and Beller 2019).

---

[12] http://www.digitalworldtokyo.com/index.php/digital_tokyo/articles/man-made_photosynthesis_looking_to_change_the_world.

[13] https://solarfuelshub.org/.

Conclusion

To meet the Paris Agreement targets by the middle of this century large-scale Direct Carbon Removal systems are needed, even in a world with high levels of de-fossilisation and point-source carbon capture. High-temperature aqueous solution-based and low-temperature solid sorbent-based DAC are the two main categories of currently available technologies. Although the energy demand of a low-temperature DAC system is higher, its total energy demand could be met at about the same cost, as the major share of energy demand could be supplied by relatively cheaper low-grade heat supplied by heat pumps. In addition, although the capital expenditure of both technologies is at the same level, the low-temperature technology is the more favourable option, due to its potential for extensive cost reduction by utilization of waste heat from other sources. This has been confirmed in a recent analysis (Sabatino et al. 2021) which concludes that absorption-based processes like alkali and amine scrubbing perform generally worse than the solid sorbent process, with the latter capturing up to 10 times more CO_2, although with a large error margin. The solid sorbent process also comes out on top when energy demand and costs are considered. However, many questions are yet to be answered before effective deployment at large scale is possible, with the most pressing involving the (ad)sorbent itself, in particular the sorbent's affinity for water and the adsorption/desorption kinetics which have a tremendous influence on the performance of the sorbent process. In addition, technological issues are to be resolved. The sorbent regeneration requires considerable amounts of (low temperature) heat. Especially the design of the air contactor is crucial in this respect and should allow for efficient heat transfer and recovery. More details on costs will be discussed in Chapter 14.

In addition, the solid sorbent technology shows a higher modularity (i.e., can be deployed in small units) than the liquid solvent approach, and has no demand for external water. DAC system costs could be lowered significantly with commercialization in the 2020s followed by massive implementation in the 2040s and 2050s, making them cost competitive with point-source carbon capture and an affordable climate-change mitigation solution (Fasihi et al. 2019). But since point-source carbon capture is not particularly widely applied either at present, such comparison may not be all that relevant.

Both Direct Carbon Removal by chemical means and by carbon mineralization have in principle unlimited potential for carbon abatement, and a credible, some even say high, potential for achieving gigaton-scale carbon removal before 2040. The challenge to scale up the process is undoubtedly biggest for the chemical DAC technologies discussed above, while carbon mineralization and ocean fertilization seem most advantageously placed to meet this scaling challenge.

The cost today is still above $200 per ton for both methods, but further R&D could bring this figure down quickly. According to the ICEF there is a high likelihood that costs can be brought down to $150 before 2030. Both strategies are relatively immature. Neither has mature technical pathways or contracts to deliver CO_2 removal greater than 10,000 tons per year. Chemical DAC is somewhat more mature than carbon mineralization: three companies have commissioned projects and have commercial contracts. In contrast, while a few carbon mineralization companies have announced substantial pilot projects, they are still quite early in technical development (Sandalow et al. 2021, p. 78). As noted before, there is unprecedented action on carbon removal in the US today. In its 2021 infrastructure deal the government has allocated billions of dollars for DAC and other forms of carbon capture. The DOE's Carbon Negative (Earth) Shot aims to lower the cost of carbon capture to $100 per ton or less (for capture *and* storage), resulting in durable storage for at least 100 years. It is the DOE's largest coordinated effort to catalyze gigaton level deployment of carbon removal. The total federal investment in carbon removal has shot up from a mere $83 million in 2021 to approximately $1 billion this year for several types of solutions, including DAC. Now that climate change consequences are becoming more urgent by the day, the US government has taken the lead to finally give carbon capture a chance to come of age and set about its task to capture the minimum amount of 10 Gt of CO_2 per year from the air by 2050 and roughly twice that amount by 2100, as argued by the IPCC as being necessary for keeping global warming in check. This in addition to reducing emissions and capturing CO_2 from point sources by fitting and retrofitting CCS installations on existing coal- and gas-fired power plants. Mankind can win this race if, for once, it comes to its senses and starts to behave sensibly.

The approach favoured by the IEA in its Net Zero scenario is much more modest as far as DAC is concerned. The IEA is of the opinion that the challenges DAC faces are significant, but not insurmountable, requiring on average eight 1 Mt/year DAC plants to be built in the current decade, 50 per year during 2030–2040 and about 40 plants per year between 2040 and 2050. This adds up to about 1 Gt capture capacity (1,000 plants) by 2050 (IEA 2022, p. 34).

CHAPTER 10

Compression, Transport and Storage

The discussion in this chapter mainly pertains to CCS as there is no sufficient experience yet with CO_2 capture by Direct Carbon Removal to make a fruitful discussion useful. As far as Direct Air Capture is concerned, one of its characteristics is that it can be done anywhere so that any industrial size DAC installations can hopefully be placed very close to the final storage site.

In most of the carbon capture methods discussed in the preceding chapters, the CO_2 is captured in the form of a CO_2-rich gas stream and will in the end have to be put to use or, if that is not possible, compressed and transported (by pipeline, rail, truck or ship) to the eventual storage location, in most cases a suitable geological formation. The efficiency of CO_2 storage in such formations, defined as the amount of CO_2 stored per unit volume, increases with increasing CO_2 density, so before transport the CO_2 stream is first compressed. Increasing density also increases storage safety, because buoyancy, which drives possible upward migration of a fluid, is stronger for a lighter fluid (IPCC 2005, p. 214).

Compression

To make the best use of the available void space in the eventual storage formation, CO_2 must be compressed to a dense phase (> 74 bar). For pipeline transport, which is a proven and widely deployed technology, compression also reduces transport costs compared to transport as a gas. Pipeline transport in gaseous form is technically possible, but would result in a higher pressure drop per unit of pipeline and therefore lower throughput for a given pipe diameter. Moreover, the volumes that need to be transported make compression to a denser state necessary. For instance, a typical 500 MW coal-fired power plant would need to transport on average 2–3 Mt of CO_2 per year. The volume of a ton of CO_2 highly depends on pressure and temperature, but at standard surface conditions it has a colossal volume of 534 m^3 (1 m^3 weighing 1.87 kg). Therefore, for transport of millions of tons of CO_2 to be feasible the captured CO_2 has to be compressed and cooled to a liquid. Liquid CO_2 only exists at a pressure above 5.2 bar and temperature below 31.1°C. The liquid is at full saturation at −37°C, when its density is 1,101 kg/m^3, hence the volume has then been reduced by more than a factor of 500. In that case 1 Mt of liquid CO_2 would occupy a cube with a side of approximately 100 m long; and for a single power station two or three of such cubes need to be pushed through the pipeline per year.

For the purposes of this book it is not very important how compression is actually realized. It suffices to note that there are various methods whereby one is more suitable than the other depending on the initial conditions and the desired end conditions. In the most common type of compressor an inlet volume of gas is confined in a given space and then compressed by reducing the confined space. This rather obvious process goes under the incomprehensible name of *reciprocating positive displacement*. Other compressor types include *centrifugal* and *axial flow compressors*. Centrifugal compressors are particularly suited for compressing large volumes of gas to moderate pressures. Such a compressor imparts kinetic energy into the gas stream by increasing the velocity of the gas using a rotating element and then converting this kinetic energy into potential energy in the form of pressure. In between each stage of compression, the gas is cooled, and excess moisture is removed to further increase the efficiency and gas quality.

As far as costs are concerned, the two main components of CO_2 compression cost are the capital costs for the compression equipment and the energy costs to drive the compressor. Compression energy costs scale linearly with flow—doubling the flow will double the compression energy cost, all else being equal. As such, there is no cost advantage to increasing the scale, but up to a point compression capital costs experience economies of scale. Compressor technology is a mature field and, although incremental improvements in compression technology are still being made—mainly as regards increasing efficiency and reliability, significant cost reductions are not to be expected (Global CCS Institute 2021b, p. 40).

Compression of the CO_2 captured from the flue gases of a power plant represents a potentially large auxiliary power load on the overall power plant system; the power used for capture, compression and transport can after all not be used for generating electricity. The addition of carbon capture to a power station has an impact on efficiency. For a typical coal-fired plant with post-combustion capture, the net efficiency can be expected to drop by 7 to 12% due to the heat and power requirements of the capture process. CO_2 compression is expected to consume 30 to 50% of this (Jackson and Brodal 2018). An August 2007 study conducted for NETL, the US National Energy Technology Laboratory, showed that compression using a (six-stage) centrifugal compressor with interstage cooling required an auxiliary load of approximately 7.5 percent of the gross power output of a coal-fired power plant.[1]

The energy consumption for CO_2 compression is lower in pre-combustion capture than in post-combustion capture because some of the CO_2 leaves the separation unit at elevated pressure, while in oxy-fuel processes energy consumption for compression depends on the composition of the extracted product, which is different for a coal-fired plant than for a gas-fired one. As in the entire CCS cycle, the greatest challenge in the compression process is to reduce the power load as much as possible.

Studies of the cost of capture and compression of CO_2 from power stations completed ten years ago averaged around \$95/tCO2. Comparable studies

---

[1]   https://netl.doe.gov/coal/carbon-capture/compression.

completed in 2018/2019 estimated that capture and compression costs could fall to approximately \$50/tCO2 from 2025, showing that there is a considerable effect from learning by doing. This is borne out by the practical experience gained with the Canadian Boundary Dam and US Petra Nova plants, which were (retro)fitted with CCS (Global CCS Institute 2021a, p. 34).

It is not easy to find figures in money terms for compression costs alone as most studies include compression costs with capture costs, since both occur within the boundaries of the facility, but a 2021 IEA analysis quotes \$13–25 per ton of CO_2 just for compression,[2] while others (Jackson and Brodal 2018; Wilcox 2012, p. 35–45) state that compression is expected to consume 90–120 kWh/tCO2 of electrical power, which would amount to just \$4–5.

Transport

As noted above, in view of the volumes that need to be transported when carbon dioxide is captured from the flue gases of a power station, pipelines are the preferred option for transport as they are the most economical and can ensure continuous transport from the capture point to the storage site. Most CO_2 pipelines used for enhanced oil recovery (EOR; see Chapter 11 where this will be discussed in more detail) transport CO_2 as a supercritical fluid,[3] which is actually the most cost-effective way, but there are potential corrosion issues that may need to be mitigated, e.g., presence of water (which in contact with CO_2 forms carbonic acids), hydrogen sulphide and other impurities in the CO_2 (Mosleh et al. 2019, p. 489) Another possibility is transport in the subcooled (or undercooled) liquid state, i.e., below its normal boiling point. It has been shown (Zhang et al. 2006) that CO_2 transport in this state by pumps can result in significant energy savings of at least 9% in favourable climatic conditions. By operating the pipeline at a pressure above the critical pressure of CO_2 (73.8 bar), the formation of gas and various other complications are avoided. Typical pipeline transport conditions are approximately 110 bar and 35°C, while typical conditions for transport by ship are 7 bar and –50°C.

Already since 1970 there has been extensive commercial experience with long-distance CO_2 pipelines, some of which extend over more than 300 kilometres. The US and Canada have some 6,500 km of pipelines carrying CO_2 from various sources for injection as part of EOR, which have been operating without major concerns or incidents (Jaccard 2005, p. 201). They pose no higher risk than the safely managed risk for transporting natural gas or oil. In the US about 68 million tons of CO_2 are transported each year through these pipelines. It is estimated that North America's CO_2 transport pipeline network needs to grow to 43,000 km by 2050 (Global CCS Institute 2021a, p. 12). In the US and Canada these pipelines are mainly routed through sparsely populated areas, where the consequences of an accident are likely to be relatively small. This will be different if such CO_2 pipeline networks have to be constructed in densely populated areas like the UK or the European continent. Extra safety precautions will then have to be taken.

---

[2]   https://www.iea.org/commentaries/is-carbon-capture-too-expensive.

[3]   See Glossary.

However, transporting CO_2 streams containing impurities, as opposed to pure CO_2 streams, imposes additional challenges. In the US, pure CO_2 is regularly transported via onshore pipelines over long distances, in most cases designed purposely for EOR. But, transport of CO_2 with impurities is also fairly common in the US and Canada. An example is the 325 km pipeline transporting CO_2 containing about 0.9% hydrogen sulphide (H_2S) from a gasification plant in North Dakota to Saskatchewan in Canada for EOR. Such onshore CO_2 pipeline systems have been operational for more than 30 years without any significant incidents caused by corrosion. However, extensive experience of CO_2 transport via offshore pipelines over long distances is not yet available.

As said, CO_2 is preferably transported via pipeline in the supercritical phase because of its high density (equal to a liquid) in this phase compared with other phases. For instance, when stored at a depth of 1 km (corresponding approximately to the injection point of the Sleipner project, see Chapter 12) that volume is only 1.43 m^3 (a density of 700 kg/m^3). In estimating the costs of such a transport pipeline, the pipeline diameter is a critical factor (Onyebuchi et al. 2018). After capture and storage, transport (i.e., piping and shipping) cost is expected to be the next largest cost component of CCS.

An emerging trend is the development of CCS networks which collect CO_2 from multiple sources. Such networks offer economies of scale by sharing larger infrastructure for CO_2 liquefaction and port facilities, CO_2 compression and pipelines. After capture, CO_2 can be transported as a gas (less than 74 bar of pressure) or in the dense phase (above 74 bar). Estimated pipeline costs are very high per ton of CO_2 for small flows of CO_2, but once flows exceed about 0.25 Mtpa for the gas phase or around 1.0 Mtpa for the dense phase, most economies of scale have been exploited and costs go down to a few dollar cents per km per ton, which provides a clear incentive for the central collection of CO_2 streams from multiple smaller capture plants.

Shared storage facilities also reduce costs. Drilling wells is an expensive process with significant fixed costs that vary only little, depending on the diameter of the well and the associated CO_2 injection capacity. It makes sense to spread investment over larger tonnages of CO_2, sharing storage costs across multiple sources.

Shared transport and storage infrastructure makes smaller scale CO_2 capture projects (0.2 Mtpa or smaller) viable. In the United States in 2019, approximately 298 Mt of CO_2 was produced by 4,931 individual facilities each emitting under 200,000 tons of CO_2 a year. Jointly they represented nearly 16 per cent of US industrial facility emissions and close to 6% of total US emissions, hence it is worthwhile to try to do something about them. Other parts of the world also have significant volumes of CO_2 spread across smaller facilities. Net zero targets require these facilities to be decarbonized and CCS networks can help provide economic, essential infrastructure.

Cross-border transport networks enable nations lacking good local CO_2 storage resources to undertake CCS projects. For example, industrial regions such as Dunkirk (France), Ghent (Belgium) and Gothenburg (Sweden), are planning to aggregate their industrial CO_2, then liquefy and ship it for storage in the North Sea. For that same reason the proposed onshore infrastructure for the Porthos project (see Chapter 12) in the port of Rotterdam (Netherlands) has been oversized and is capable of handling 10 Mt per annum, while the initial participants will only provide 2.3 Mt per annum. CO2TransPorts, an EU common interest project, aiming to establish the necessary infrastructure to facilitate the large-scale capture, transport and storage of CO_2 from three of the most important ports of Europe, the ports of Rotterdam and Antwerp and the North Sea Port (a fusion of the ports of Terneuzen and Vlissingen in the Netherlands and Ghent in Belgium), is working on how best to connect Porthos to the North Sea Port and Port of Antwerp where local industrial carbon capture clusters are being developed. The North Sea offers these countries high quality storage resources with big cost advantages (Global CCS Institute 2021a, p. 53).

Storage

Potential storage options for captured CO_2 are geological storage, fixation of CO_2 into inorganic carbonates (mineral sequestration), or utilization in industrial processes. As far as the latter is concerned, only a fraction of the captured CO_2 can be used as a raw material to manufacture polymers and liquid fuels, among other valuable products, but this is not considered a major factor in reducing CO_2 emissions, while geological storage is seen as the most promising practical solution. Storage is the most critical and crucial point as there must be sufficiently hard guarantees that the CO_2 will be stored safely for a very long time. In the previous chapter we have already briefly discussed what permanent storage means. The IPCC special report on CCS technology projected that more than 99% of carbon dioxide stored through geologic sequestration is likely to stay in place for more than 1,000 years. Using carbon sinks such as the ocean, trees and soil for such storage may involve the risk of adverse feedback loops at increased temperatures, i.e., the occurrence of processes whereby changing one quantity causes a change in a second quantity, whereby this second change has in turn an uncontrolled effect on the first quantity, resulting in runaway and uncontrolled behaviour. If carbon capture is scaled up to the necessary level to keep global warming below 2°C, or preferably below 1.5°C, hundreds of gigatons must be sequestered by 2100, not only from carbon capture at point sources by CCS but also from Direct Carbon Removal.

Carbon mineralization, treated as a carbon removal method in Chapter 9, can also be seen as a form of storage as the carbon is permanently removed by conversion into a stable carbonate. The same is true for algal sequestration (ocean fertilization in the preceding chapter), which like carbon sequestration by trees uses photosynthesis and is also proposed as a storage option. Algae can absorb more CO_2 than trees because they can cover more surface area, grow faster and can be more easily controlled in bioreactors. The separated CO_2 (or the CO_2-rich flue gas) is pumped into vessels, or tanks, or ponds (algae bioreactors) where the CO_2 is taken up by photosynthesizing algae. The algae can then be harvested and turned into a range of products such as transport fuels, pharmaceutical products or animal feed. This is called bio-sequestration, but is considered a niche commercial opportunity, not a major storage option (For more details see Haoyang 2018; Singh and Dhar 2019; *Wikipedia* on Algae bioreactor).

An early proposal to use algae to solve the CO_2 problem, originates from James Lovelock (1919–2022), who in 2007 in a short letter to the journal *Nature* in collaboration with Chris Rapley (b. 1947), professor of Climate Science at University College London, proposed the construction of ocean pumps to pump water up from below the thermocline (a thin but distinct layer in a large body of fluid (as in an ocean or lake) or air (e.g., the atmosphere) in which temperature changes more drastically with depth than it does in the layers above or below) to "fertilize algae in the surface waters and encourage them to bloom" (Lovelock and Rapley 2007). The idea was to accelerate the transfer of carbon dioxide from the atmosphere to the ocean by increasing the synthesis of organic compounds from atmospheric or aqueous carbon dioxide and by doing so enhance the export of organic carbon to the deep ocean. The proposal was heavily criticised as essentially being a form of risky geoengineering/climate engineering, but at some point it may be one of the few, relatively benign ways to mitigate climate change. Lovelock en Rapley made their proposal as far back as 2007 and in the 16 years that have since passed hardly anything, apart from nibbling at the edge, has been done to address the problem.

There is 45 years of experience with the injection and storage of CO_2, and with abundant underground storage resources at our disposal, geological storage remains the easiest and most logical CO_2 mitigation solution. It involves injecting CO_2 captured from industrial processes into rock formations deep underground, permanently removing it from the atmosphere. There are many geological systems throughout the world that are capable of retaining captured CO_2. Almost all oil and gas production is associated with sedimentary basins, and the types of geologic formations that trap oil and gas (and also naturally occurring CO_2) are similar to those that make good CO_2 storage reservoirs. Any porous and permeable rock formation at a depth in excess of about 1 km is a potential candidate for CO_2 storage, if there is an overlying impermeable caprock to prevent the upward migration and leakage of the gas (English and English 2022).

Typically, effective storage sites have the following characteristics. The rock formations have enough millimetre-sized voids, or pores, to provide the capacity to store the CO_2. The pores in the rock are sufficiently connected, a feature called permeability, to accept the amount of CO_2 at the rate it is injected, allowing the CO_2 to move and spread out within the formation. And an extensive cap rock or barrier sits at the top of the formation to contain the CO_2 permanently.

Broadly speaking there are five main options for such CO_2 storage, as indicated in Fig. 10.1:

1. *Deep porous saline aquifer formations* (also called saline reservoir formations as they are different from drinking-water aquifers). These are deep water-bearing reservoirs and the largest available volumes for storage of CO_2.

2. *Depleted oil and gas reservoirs.* This option offers great potential, because once the oil and gas have been extracted, these well-characterized reservoirs offer an excellent storage solution with significant infrastructure already in place.

3. *Coalbed storage.* Coal formations can be used to store CO_2 in un-mineable coal seams or, in analogy to EOR for oil recovery, as part of enhanced coal-bed methane projects (see Chapter 11).

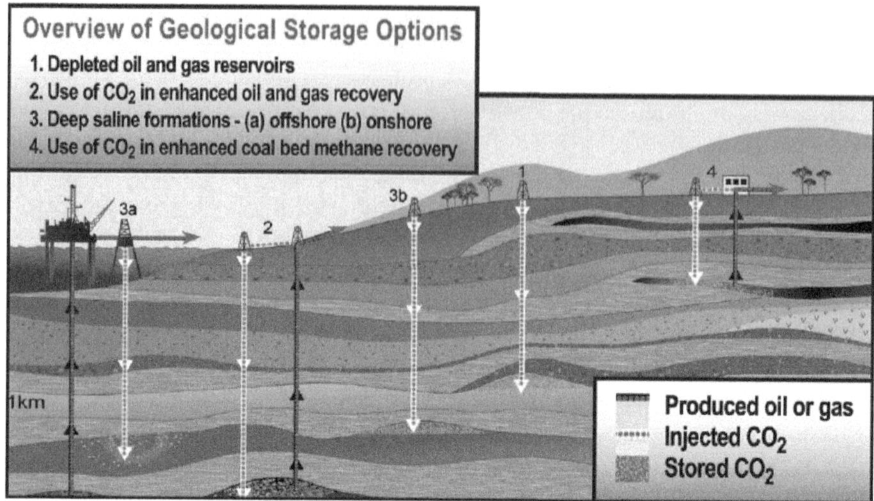

Fig. 10.1. Options for geological storage in sedimentary rocks (NAS 2019, p. 320).

4. *Other rock formations.* These include volcanic rocks, underground caverns, as well as rocks formed in the carbon mineralization discussed in Chapter 9.

5. *Storage as part of Enhanced Oil Recovery projects.* When used in such projects, the injected CO_2 remains in the subsurface as stored CO_2. This is very similar to option 2 but deployed before the oil reservoir has been completely depleted.

Geological storage is already a proven technology, effectively taking advantage of the same geological characteristics and processes that have sealed and trapped oil and gas in the subsurface over geological periods of time.

The above enumeration shows that the options include a mixture of permanent pure storage and utilization of CO_2 for oil or methane (natural gas) recovery in partially depleted fields. A final different method to be briefly discussed and unrelated to the ones mentioned above is deep ocean storage.

Saline aquifer formations and depleted oil and gas fields

The first two options, saline aquifer formations and depleted oil and gas fields, are porous reservoirs that would otherwise contain water, oil and gas. These two CO_2 storage options are at present considered the preeminent means of storing CO_2. This is due to their large storage capacities and the fact that some of the infrastructure needed for CO_2 injection is already in place as a result of the oil and gas industry. Moreover, due to historical investment by the oil and gas industry, there is usually a large amount of data available and a good understanding of the subsurface. To ensure that the CO_2 is in a dense form (a liquid or in a supercritical phase), it needs to be stored relatively deep. This is important for storage efficiency, as a higher density means that storage will be more effective. Depth is further important for storage security and leads to the need for a storage seal. The presence of such a storage seal, an overlying, thick, and continuous layer of shale, silt, clay, or evaporite (a water-soluble sedimentary mineral deposit formed from evaporation from an aqueous solution) is the single most important feature for a geologic formation to be suitable for geological storage of CO_2. It physically prevents the upward migration of CO_2 and takes care of the structural trapping of the sequestered CO_2.

These considerations have led to the widely accepted principle that CO_2 storage sites need to be deeper than 800 m, which is roughly the depth at which CO_2 becomes liquid. Approaching depths of around 1 km or more, we are in the domain of compacted and cemented rock which potentially contains low-permeable sealing units, like shales, faults and salt units. It is known from experience that at these depths natural gas has been trapped beneath geological seals for millions of years and hence they should equally be suitable for trapping CO_2 (Ringrose 2020, p. 10–20).

The essential issues of any CO_2 storage projects concern capacity (how much CO_2 can be stored, is there sufficient storage volume over the project's lifetime), injectivity (ease with which fluids flow through the connections between pores, which determines whether the CO_2 can be injected at a sufficient rate and cost effectively), and containment (will the CO_2 remain in the geological storage unit or are there dangers of leakage or migration to another geological formation). Containment is of course related to the mechanisms by which the CO_2 is trapped, which upon injection can occur either through physical or geochemical trapping mechanisms. Physical trapping mechanisms include capillary trapping caused by the interfacial tension between two fluids in a porous medium or residual trapping within the storage unit itself. This latter mechanism is well known as a limiting factor in oil and gas recovery (EOR, see below, tries to overcome this). These mechanisms take care of the initial trapping of CO_2 upon injection as geochemical trapping mechanisms take considerable time to come into play. However, geochemical trapping results in more permanent and secure storage of CO_2 and is therefore preferable to physical trapping. The two types of geochemical trapping are solubility and mineral trapping. Solubility trapping involves CO_2 dissolving in the local brine and becoming trapped as an aqueous component. The aqueous CO_2 then reacts with water to form carbonic compounds. Which compound depends on the pH of the brine. After solubility trapping, the CO_2 is no longer in a separate phase and so the CO_2 fluid is not subject to buoyant forces that drive it upwards, as the CO_2-saturated brine is denser than unsaturated brine, implying that the CO_2 is now securely stored within the geological formation.

Mineral trapping occurs when dissolved CO_2 reacts directly or indirectly with minerals in the geologic formation, causing the CO_2 to precipitate in the form of a carbonate. This type of trapping is attractive as it could immobilize CO_2 for very long periods. The process is however thought to be comparatively slow because it depends on dissolution of silicate minerals, so the overall impact may not be realized for tens to hundreds of years or longer (Benson and Cole 2008).

The most mature storage technique is sequestration in deep saline aquifers or other sedimentary formations. Supercritical CO_2 (density around 600 kg/m^3) may be injected and stored in such formations. Ideal reservoirs for this purpose are thick reservoirs located below a depth of 1 km with high porosity and high permeability ensuring a high storage capacity (i.e., 50–100 Mt CO_2/project). Suitable reservoir rocks with high porosity and permeability include sandstone, limestone, dolomite, basalt, or mixtures thereof, whereas a suitable caprock is made of shale, anhydrite (anhydrous (i.e., waterless) calcium sulphate ($CaSO_4$)), or carbonate rocks with low permeability. Long-term CO_2 sequestration and reservoir safety rely on the caprock being impermeable and on secondary trapping mechanisms to further immobilize the CO_2.

Sedimentary basins around the world have the potential to store 8,200 Gt of CO_2 in a conservative estimate and between 20,000 and 35,000 Gt of CO_2 in high estimates, while approximately 1,000 Gt of CO_2 capacity is provided by oil and gas reservoirs. It has been estimated (Schmelz et al. 2020) for example that the northeast/midwestern region of the United States and the mid-Atlantic offshore have a geological storage capacity to sequester over 500 Gt of supercritical CO_2 in onshore saline formations, between 150 and 1100 Gt in offshore saline formations, and 2.5 Gt in depleted oil and gas fields. The latter does not seem much in view of the enormous potential for storage in sedimentary basins, but the architecture and properties of such storage locations are well known as a result of exploration and production of hydrocarbons, and have often the added benefit that infrastructure for CO_2 transportation and storage already exists at such sites. In general, even conservative estimates for CO_2 sequestration capacity far exceed the cumulative anthropogenic CO_2 emissions since the Industrial Revolution, which amount to $2,035 \pm 205$ Gt of CO_2. Storage capacities therefore put no limitation on CO_2 mitigation, but whether a given capacity can be realized, depends on a number of factors. Uncertainties concerning the pore space usable for CO_2 storage include: the pressure build-up in the reservoir during injection; the quality of the seal, including possible old boreholes that penetrate the seal, other uses of the formation, and other restrictions on above-ground or surface resources. Assurance of long-term storage security can be enhanced by secondary trapping mechanisms that immobilize the CO_2, including dissolution, residual gas trapping, and mineralization. Before CO_2 injection starts, potential reservoirs are modelled using multiscale and multi-physics numerical simulations coupled to field observations in order to assess the ability of the reservoir to securely sequester CO_2. It is necessary to monitor the evolution of CO_2, the reservoir, and the caprock during and after injection, in order to detect leakage from the reservoir. Hence, suitable monitoring must be possible. In addition, competition with other vital resources, such as freshwater and land use needs to be avoided (Kelemen et al. 2019, p. 9–10).

The Norwegian Sleipner Saline Aquifer Storage Project (see Chapter 12 for a more extensive description) is the first commercial-scale CO_2 storage project. It commenced in 1996 with the injection of CO_2 into the offshore Utsira saline aquifer formation. This project shows that CO_2 storage in subsurface sedimentary reservoirs at a rate of 1 Mt/year is perfectly possible and safe. The well-tested and well-monitored Utsira formation has a capacity of 600 billion tons, which alone would suffice for a long time for all CO_2 to be captured globally. It concerns a 200–300 m thick massive Miocene sandstone formation, with several thin shales, overlain by a thick shale caprock, located at a depth of 800–1,000 m beneath the seabed. Seismic imaging is used to make the spread of the CO_2 injected into the formation visible and to model the dynamics of the CO_2 migration (Ringrose 2020,

p. 57). The experience with this project belies all those who keep shouting that CO_2 sequestration cannot be guaranteed to be safe and that carbon capture remains an unproven technology. The arguments used and regulations demanded for CO_2 are starting to look like the crippling demands made on nuclear power in the form of rules and conditions, clearly with the intention to block such path from the outset. In view of the fact that some CO_2 escaping from the reservoir is essentially the worst 'accident' that can happen, with just some money and effort wasted, which is an almost everyday occurrence in most fields of human activity, there is no reasonable justification for opposing carbon capture with storage in geological formations or depleted oil and gas reservoirs by demanding all too stringent guarantees. It goes without saying though that site monitoring is required to ensure safe CO_2 storage and confinement over long periods of time. Leakage can of course occur and, if it does, remedial action is required to stop or control such releases. A recent Spanish study showed that the sequestered CO_2 would remain deep in the subsurface for millions of years, even if the overlying low-permeability rocks were fractured, and that injecting billions of tons of CO_2 underground has a low risk of leakage back to the surface (Kivi et al. 2022).

Depleted oil and gas fields are the obvious candidates for storing the carbon dioxide that originated from burning that same oil and gas. The technology is virtually identical to sequestration in deep saline aquifers. An example of a project intending to use such reservoirs is the Porthos project in the Port of Rotterdam, which plans to bury over 2 Mt of CO_2 per year in such fields. The CO_2 is transported by pipeline to the coast, where it is compressed and then piped to depleted reservoirs in the North Sea. Another example is the CCS Ravenna Hub, a CO_2 storage site off the Adriatic coast of Ravenna in Italy. These and other projects will be discussed in more detail in Chapter 12.

Enhanced oil recovery

EOR also uses partially depleted oil fields but in this case the aim is to get some of the last oil out of the fields by injecting CO_2 into an oil well. It is a form of utilization of carbon dioxide, and its main aim is not (permanent) storage of CO_2. It will be discussed in more detail in the next chapter. This type of use and storage can at best be seen as a temporary solution so long as it is necessary to get more oil out of the ground, but it does not solve much of the CO_2 problem. It has been in use in the US since the 1970s. When CO_2 is injected underground for EOR, most of it, around 90 to 95 percent, stays there, trapped in the geologic formation where the oil once was. If the CO_2 comes from the right source and enough is buried, it could amount to substantial carbon sequestration. If the CO_2 that returns to the surface is separated and reinjected to form a closed loop, permanent CO_2 storage can be the result. Of course, the oil that is recovered in this way is used as a fossil fuel and in this process CO_2 will again be released, which in this book's philosophy, as will be clear I hope, must be captured. Today, in the US between 300 and 600 kg of CO_2 is injected in EOR processes per barrel of oil produced. Given that, when combusted, a barrel of oil releases around 400 kg of CO_2, and that around 100 kg of CO_2 is generated on average during production, processing and transport of the oil, EOR opens up the possibility for the full lifecycle emissions intensity of oil to be carbon neutral or even "carbon negative", the latter meaning that per barrel of oil produced more carbon dioxide is eventually sequestered (permanently stored away) than is released to the environment in the production and use (combustion) of the oil. This could even be improved when all or part of the carbon dioxide released in the latter processes is captured and stored. This is of course only true when the CO_2 used for injection originates from anthropogenic sources, such as power stations. In practice, however, most of the CO_2 injected in EOR projects originates from naturally occurring underground CO_2 deposits and not from carbon capture projects elsewhere. The reason for this is the absence of available CO_2 close to oil fields. Using natural sources clearly provides no benefit in terms of the emissions intensity of the produced oil. Such practice must be considered unacceptable (IEA 2018, p. 503).

Demand for CO_2 for use in local EOR operations is forecast to grow at least fivefold to 2030, and an increasing number of EOR projects intends to use CO_2 captured from anthropogenic sources, also outside the US. Examples are the Moonie Project in Australia developed by Bridgeport Energy, which targets injection of around 1 Mt of CO_2 per annum, sourced from power stations nearby, for CO_2-EOR and storage in southeast Queensland; the China Petroleum & Chemical Corporation, aka as Sinopec, which started constructing China's first 1 Mtpa CO_2-EOR project in Shangdong Province, capturing CO_2 from the Qilu Fertilizer Plant and injecting it into the Shengli Oilfield for EOR and CO_2 storage; and the Saudi Arabian oil company Saudi Aramco, which captures 0.8 Mt of CO_2 per annum at its Hawiyah Natural Gas Liquids plant and uses the CO_2 to demonstrate the viability of EOR at the Uthmaniyah oil field (Global CCS Institute 2021a).

While currently for EOR projects roughly only 80 Mt of CO_2 per year are injected into the ground, CO_2-EOR has the potential to sequester 30 Gt of CO_2, on the proviso of course that the CO_2 used in oil recovery is left in the ground after the oil is produced. Advanced CO_2-EOR methods, designed to optimize both oil recovery and CO_2 sequestration, could sequester over 90 Gt CO_2, and this could increase oil production compared to standard CO_2-EOR methods (Kelemen et al. 2019, p. 8).

In the next chapter we will come back to EOR when discussing utilization of CO_2.

Enhanced gas recovery

The next storage method I want to briefly discuss is coalbed storage, a technique similar to EOR but with the intention to recover methane gas from un-mineable coal seams, going under the name of enhanced coalbed methane recovery or enhanced gas recovery (EGR).

In nature, coal seams also contain gases such as methane, held in pores on the surface of the coal and in fractures in the seams. If CO_2 is injected into a coal seam, it displaces the methane, which can then be recovered. Coal has a higher affinity (about 2–13 times higher) for carbon dioxide than for the methane that is naturally found in coal seams. This property, called adsorption trapping, is the basis for the storage of CO_2, while the methane can be released (Mosleh et al. 2019, p. 493). Providing the coal will never be mined, the injected CO_2 will remain stored in the seam, hence the name coalbed storage. Coal seams that have never been disturbed can contain considerable amounts of methane (up to 25 m^3 per ton of coal). This coal typically lies between 300 and 1500 m below the surface, with total global reserves of more than 4,000 billion tons of coal at these depths. Typically 50% of the methane in the coal can be recovered using standard techniques. The methane is usually of high purity (> 90% by volume) and can be supplied more or less directly to a natural gas distribution system or used for power generation or heating.[4] If a gas-powered power station is built near such un-mineable coal seams, the methane recovered from the seams can be used to run the plant. The CO_2 resulting from the combustion can be captured and used in recovering further methane. This would result in a gas-fired power plant that would not emit any carbon dioxide into the atmosphere.

The IPCC in its 2005 report on CCS estimates that deep, un-mineable coalbeds have the capacity to store between 3 and 200 Gt of carbon, hence between 11 and 730 Gt of CO_2. An earlier estimate (Gale and Freund 2001) gives a figure of 150 Gt of CO_2.

Although distasteful to some and to some extent also to me, it must be realized that storage options such as EOR and enhanced coalbed methane have the added benefit of producing commercially useful by-products, in the form of oil and gas that can offset cost. In our over-commercialized world, where in the end cost is often the deciding factor, such benefits may be needed to get necessary developments on the rails.

Ocean storage

A final rather different method is ocean storage of CO_2, whereby the gas is intentionally injected into the deep ocean (at depths greater than 1,000 m), where it will remain isolated from the atmosphere for hundreds of years or longer. Various methods have been proposed, e.g., through a pipeline (running from the shore to the ocean depth) or from a ship, including injecting CO_2 as dry ice 'torpedoes' that would penetrate the sea floor. The CO_2 dissolves into the water and when injected at depth, it would take hundreds of years for the dissolved CO_2 to circulate back into the surface waters. Alternatively, at greater depth, CO_2 hydrates (in which CO_2 and water molecules form a crystal structure) are created which would sink to the bottom and collect on the sea floor. Also, at great depths, CO_2 is heavier than water and will sink into enclosed ocean basins to form a permanent 'CO_2 lake'. A further option is the injection of CO_2 into unconsolidated sediments in the top tens of metres of the seafloor, which can be seen as intermediate between geological and ocean storage.

As a general rule, the deeper the CO_2 is injected into the ocean, the longer it will take to get back to the atmosphere, but ocean storage has not yet been tested at a significant scale to date. Laboratory and field tests indicate that the impact of injected CO_2 on marine organisms in the immediate vicinity of the injection site is likely to be significant and the longer the injection goes on, the more widespread the impact.

---

[4]   Storing CO_2 in un-mineable coal seams, https://ieaghg.org/docs/general_publications/8.pdf.

The increase of atmospheric CO_2 over the past 200 years has already resulted in an increase in the concentration of oceanic CO_2 as the ocean absorbs a considerable amount (about 30%) of the CO_2 we emit into the atmosphere. It acts as a carbon sink. This explains the rising acidity of the ocean surface waters where pH has decreased by 0.1 pH units over the past 200 years, which seems little but is actually a lot and has a great impact on marine life. Direct injection of CO_2 into the ocean would greatly accelerate the acidification process, potentially resulting in significant unintended consequences to marine ecosystems. No wonder that the concept of ocean storage has met with strong opposition from NGOs and governments. Both national and international laws and treaties ban the dumping of industrial waste in the ocean, which is usually taken to include flue gases and anthropogenic CO_2. Although this has not been able to prevent the already colossal pollution of ocean waters, it makes it nevertheless unlikely that ocean storage will be an acceptable mitigation option in the foreseeable future, certainly since, as we have seen above, there is plenty of storage space available in dedicated geological sites that can relatively easily be monitored (Cook 2012, p. 58–59).

Storage costs

Generally, onshore storage is less expensive than offshore storage, and storage in depleted oil and gas fields is less expensive than storage in saline reservoirs.

A study, already quoted above (Schmelz et al. 2020), investigating total CCS costs for emissions from 138 electricity-generating power plants in the north-eastern and midwestern US with storage in subsurface geological formations, suggests that CO_2 emissions from coal-fired plants can be stored in this region at a cost of $52–$60 per ton and from gas-fired plants at approximately $80 to $90. Storing emissions offshore increases the lowest total costs of CCS to over $60 per ton of CO_2 for coal. This concerns total costs for capture *and* storage, where it should be noted that it only concerns capture from point sources. For storage alone representative costs range from $5 per ton of CO_2 in depleted oil and gas fields located onshore, to $18 per ton of CO_2 in an offshore saline reservoir. Finally, it is likely that more than 8 Gt of total CO_2 emissions from this US region can be stored at total cost of less than $60 per ton. This is much lower than the tax credit recently hiked up from $50 to $180 per ton of CO_2 sequestered, making CCS more than feasible in an economic sense.

Another study, also quoted before (Kelemen et al. 2019, p. 11), states that the cheapest storage methods with large capacity cost roughly $7–$30 per ton of CO_2 sequestered (storage costs only) and range from the injection of supercritical CO_2 into subsurface sedimentary formations to in-situ carbon mineralization via injection of CO_2-enriched fluids into on-land basalt or other on-land rock formations.

Storage of a total of 125 Gt of CO_2 would result in a capital input of approximately 1–4 trillion dollars in total and 10–50 billion dollars per year until 2100. For comparison, the total cost of the energy transition, i.e., decarbonizing energy by shifting to an energy infrastructure comprised primarily of renewables, required for < 1.5°C of warming was estimated to be 1.6–3.8 trillion dollars per year until 2050. From this it would follow that carbon capture and storage is a relatively cheap option.

This is also apparent from the case study for the Netherlands worked out in Chapter 3 where the avoidance of 16 Mt of annual CO_2 emissions from households is estimated to cost €245 billion (€15,000 per ton of CO_2) in capital costs for home insulation, solar panels and heat pumps (costs that will partly recur every 15–20 years) while Direct Air Capture of a similar amount of carbon dioxide would cost at most €3.2 billion per year (at €200 per ton). It will soon be realized that capital costs for the energy transition are prohibitively high, certainly when affordable storage options for electricity generated from wind turbines and solar panels fail to materialize.

Environmental aspects

Like any human activity CCS and Direct Carbon Removal will have impacts on the environment, both adverse and beneficial. The important question to answer is whether the beneficial impacts due to the reduction of carbon dioxide in the atmosphere exceed the unavoidable adverse impacts. CCS is obviously a greener way to operate power stations, while still ensuring an energy supply and allowing society time to make the transition to a low-carbon future. In addition, carbon capture, in the form of, e.g., Direct Air Capture, is absolutely necessary to lower the carbon-dioxide concentration in the air below the extremely dangerous levels we already have at present.

It looks as if people have been desperately searching for adverse impacts from CCS and DAC. Lacking any firm evidence for such impacts, it has led to falsely suggestive statements like "building and operating CCS technology *significantly* depletes natural resources. For example, *extra* energy is required for capturing the CO_2 and this is *mainly* responsible for the depletion of fossil fuels, whilst the infrastructure required to transport the CO_2 to storage is *largely* responsible for the increase in metal depletion."[5] This is a very disingenuous statement by the choice of suggestive adjectives (placed by me in italics). Of course CCS uses natural resources which is however completely different from '*significantly* depletes natural resources'. Neither does it require '*extra*' energy, where just 'energy' will do, while '*mainly* responsible for the depletion of fossil fuels' and '*largely* responsible for the increase in metal depletion' are just untrue statements. There are many areas of human activity that make a much larger claim on fossil fuels and/or metals than CCS does, and have much larger adverse impacts, such as the ridiculous numbers of private cars that are driving around. The statement quoted above brings CCS in the miraculous position of being both responsible for the depletion of fossil fuels and being a 'false solution' invented by the fossil-fuel industry to be able to continue the exploitation of fossil fuels. If there ever was a case of having your cake and eat it…! Moreover, who cares about the depletion of fossil fuels if we want to transit to a society in which fossil fuels, especially coal and gas, are no longer to be used? The passage in quotes can be rephrased as "building and operating *wind turbines* significantly depletes natural resources. For example, extra energy is required for *manufacturing such turbines* and this is mainly responsible for the depletion of fossil

---

5 https://ec.europa.eu/environment/integration/research/newsalert/pdf/297na5_en.pdf.

fuels, whilst the infrastructure required *to build wind turbines* is largely responsible for the increase in metal depletion," which is an equally true sentence, even more true, to which could be added that *wind turbines fail to lower the concentration of carbon dioxide in the air.* Such a passage you will however never see written, showing the negative bias with which carbon capture is approached.

That does not change the fact that transport and storage of CO_2 has some environmental impacts and we will briefly consider them here (in part adapted from Mosleh et al. 2019, p. 496–498).

1. Impact on water resources

During construction of pipelines spillage can accidentally happen and have a potential impact on surface water. In addition, discharges of waste water from a capture plant can be a source of contamination. Such impact is no different than from other industrial installations. Outflow water due to the injection of CO_2 into a brine-containing formation can contain toxic chemicals. For instance, water produced from coalbed methane recovery operations is typically contaminated with sodium, barium, bicarbonates and iron, and should be disposed of in a safe and proper manner.

2. Impact on air, ground and climate

As with any construction activity, there may be an impact due to production and suspension of dust, fuels used by machinery and shipping processes. Atmospheric leakage of CO_2 can occur after completion, have an impact on soil pH etc. All this applies equally to any other human activity; such impacts should be avoided by taking well-known appropriate measures.

Sulphur dioxide (SO_2) emissions from power plants are predicted to fall when CO_2 is captured, as SO_2 must also be removed after the fuel combustion stage. Particulate matter and nitrogen oxide emissions are expected to increase in line with the amount of the additional fuel consumed if no additional measures to reduce emissions are installed. Ammonia (NH_3) is the only pollutant for which a significant increase in emissions is expected to occur. Emissions of CO_2 in the EU would fall by around 60% by 2050 if CCS were implemented at all coal-based power generation plants. Implementing CCS at all coal, gas and biomass plants would result in net negative emissions—in effect removing CO_2 from the atmosphere. Overall, CCS is considered to be generally beneficial both in terms of climate change and air pollution (EEA 2011). In spite of the latter conclusion made in 2011, more than a decade ago, not a single power plant in the EU has been fitted with CCS, although as we will see in a future chapter some attempts have been made.

3. Induced seismicity

High-volume injection of fluids into the deep subsurface can have potential short-term seismic effects (small earthquakes) and long-term geochemical impacts. Such effects are also known from the extraction of natural gas, e.g., at the giant Groningen gas field in the Netherlands, and water injected for EOR has been reported to cause 38 cases of induced seismicity. This is the only serious and unique environmental impact from carbon storage and should be dealt with by properly choosing the sequestration site, e.g., at sea.

From the above it can be safely concluded that environmental impacts from carbon capture and storage are mild and can easily be dealt with.

The general conclusion from this chapter must be that compression, transport and storage of any CO_2 captured by CCS or DAC do not in reason pose any insurmountable environmental or other problems that might force us to refrain from such means of climate mitigation.

Chapter 11

Utilization of CO_2

Introduction

Instead of permanently storing the captured CO_2 in a suitable geological storage site like a used oil field or saline aquifer, it is also imaginable for it to be used to good effect in one way or another to produce economically valuable products or services, or transform it into useful chemicals and fuels. CO_2 has been used for a long time as a commercial input into a range of products and services and today around 230 Mt of CO_2 are used each year. The largest consumer is the fertilizer industry, which uses around 130 Mt per year in the manufacture of urea, an organic compound with chemical formula $CO(NH_2)_2$, mostly used as a nitrogen-release fertilizer. In second place stands the oil industry sector with a consumption of 70 to 80 Mt for enhanced oil recovery (EOR). In addition, it is widely used in the food and beverage industry, the fabrication of metal, for cooling, fire suppression and in greenhouses to stimulate plant growth. More than two-thirds of current global demand for CO_2 comes from North America (33%), China (21%) and Europe (16%), with the demand for existing uses expected to grow steadily year-on-year (IEA 2019, p. 5).

It is important to note though that CO_2 utilization may at best only complement mitigation, as CO_2 used is not the same as CO_2 avoided. It does not necessarily reduce emissions and quantifying resulting climate benefits is complex, requiring a comprehensive life-cycle assessment. In assessing the climate benefits, there are five key considerations: (1) the source of CO_2 (from natural deposits, fossil fuels, biomass or air); (2) the product or service the CO_2-based product or service is displacing; (3) how much and what form of energy is used to convert the CO_2; (4) how long the carbon is retained in the product; and (5) the scale of the CO_2 use (IEA 2019, p. 7).

The future market for CO_2-derived products and services is very difficult to assess and estimates in the literature vastly differ. Where an earlier IPCC report (IPCC 2005, Chapter 7) suggests that CO_2 utilization may not make a significant dent in reducing atmospheric emissions, a report by the Global CO_2 Initiative (GCI 2016, p. 5) suggested that by 2030 the mitigation potential of CO_2 utilization is about 7 Gt per year, approximately 20% of current-day emissions, a sizeable amount but still not sufficient, certainly if legacy emissions are taken into account. The IEA considers this latter estimate highly optimistic and it indeed looks that way as currently, just seven years away from 2030, we are still far from utilizing such volumes. The conclusion can in any case be that CO_2 use cannot and should not be considered as a replacement or alternative to permanent storage.

Although CO_2 utilization cannot be viewed as a game-changing strategy for emission mitigation, it nevertheless can help recycling carbon by producing useful products, especially if the energy input is supplied from renewable sources. CO_2 is a valuable source of carbon, which is the backbone of a wide range of industries, like the food packaging, refrigeration and beverage industries, and of products, including polymers and plastics, petrochemicals, pharmaceuticals, fire extinguishers, synthetic fibres, chemicals and fuels. So long as the carbon remains chemically bound in these products and materials, it is effectively blocked from entering the atmosphere and hence considered 'chemically stored'. The term carbontech, a subset of carbon use, not including EOR, refers to the wide variety of commercial products (e.g., building materials, chemicals, and fuels) made with captured carbon dioxide. Although the market for CO_2-derived products and services is difficult to assess, the climate NGO Carbon180,[1] which has fully committed itself to carbon removal from the air, estimates this market to be worth $1 trillion in the US alone, including everything from fuels and plastics to building materials and other industrial products. Fuels make up about 90% of that amount ($882 billion), building materials $101 billion, plastics $72 billion, wood-based panels $13 billion and chemicals $2 billion. At least 182 carbontech projects are ongoing in at least 14 different countries, including the US, Canada, Germany, China, and India.[2] Some applications of CO_2 use, such as fuels and chemicals, could grow to using multiple billions of tons of CO_2 per year, but in practice would compete with direct use of low-carbon hydrogen or electricity, which would be more cost effective in most applications.

Carbontech is a win-win for both business and climate and in the long run it could help offset the high cost of carbon capture. It can also play a key role in decarbonization by helping to reduce or eliminate emissions in hard-to-abate sectors, i.e., sectors in which it is hard to reduce CO_2 emissions like cement production and aviation. Electrifying aviation will be slow and difficult—but aviation fuels can be made of captured carbon today, significantly lowering the aviation sector's emissions. In 2018 Virgin Atlantic flew a Boeing 747 jet from Orlando, Florida to London, England, using jet fuel from LanzaTech[3] derived from CO_2 emissions from a steel mill.[4]

However, apart from the essentially old-hat usages in the fertilizer industry and for EOR, the history of carbon utilization (possibly followed by subsequent storage) has so far not been a great success story. And in fact, it should be recognized that, barring some great future discoveries, the scope for using captured or stored CO_2 is rather limited. Some use of carbon dioxide is of course rather obvious such as in fizzy drinks, but fairly useless as a climate mitigation method since as soon as the bottle is opened the CO_2 starts to be released into the atmosphere. To be useful for

---

[1]   https://carbon180.org/.

[2]   Carbon180, *Factsheet on Carbontech*; https://carbon180.org/fact-sheets (accessed 14 June 2022).

[3]   https://lanzatech.com/; Chicago-based LanzaTech traps carbon that would be emitted during industrial processes and uses bacteria to convert the waste gas into sustainable chemicals such as ethanol.

[4]   https://www.ecowatch.com/virgin-atlantic-recycled-waste-gas-flight-2609795725.html.

climate mitigation the used carbon dioxide must be stored in a durable product, such as the carbon dioxide captured from the air by trees remains buried for a long time in furniture that is eventually produced from the wood. Although not in volumes anywhere near emissions, a variety of existing applications, which exploit the chemical and physical properties of CO_2, is already commercialized. The origin of the CO_2 can be fossil fuels, biomass, industrial processes, air or natural underground deposits and its use can involve conversion or non-conversion as illustrated in Fig. 11.1 below.

In the following we will discuss some of these uses, starting with enhanced oil recovery.

Enhanced Oil recovery

EOR, also called tertiary recovery, is the extraction of crude oil from oil fields that cannot be extracted otherwise. Primary recovery is the initial stage in an extraction process for oil and gas, generally by placing increased pressure on the oil within wells to force the oil to the surface or using equipment to keep the pressure under control to prevent the spontaneous release of oil. Prior to the advent of such pressure control equipment in the 1920s, the uncontrolled release of oil and gas from a well while drilling was common and known as an oil gusher or wild well, or in the case of gas as a gas gusher. The typical recovery factor during this primary stage is just 5–15%. When pressure underground has become so low that primary recovery is no longer feasible, secondary recovery techniques must be used, such as additional water injection, which seek to force the oil to the surface by directly applying pressure. This can enhance recovery to 20–40%, after which a lot of oil, up to 60%, is still left underground.

There are three main methods of tertiary recovery, involving the use of heat, gas (CO_2, natural gas or nitrogen) or chemical injections. Here we are only interested in the method using CO_2 and in the US this use dates back to the 1970s. To date most activity is still in the US and Canada, with some projects being started up in Australia, China, Hungary, Saudi Arabia and the United Arab Emirates (Global

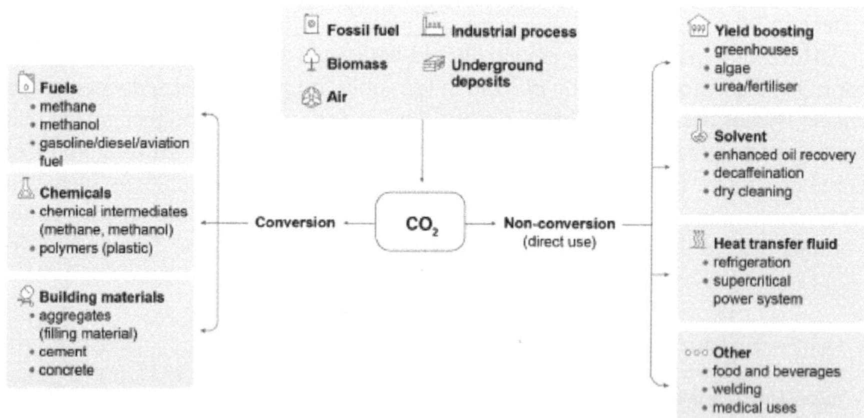

Fig. 11.1. Simple classification of pathways for CO_2 use (*from* IEA 2019).

CCS Institute 2021a, p. 62–66). When more oil fields get partly depleted in the near future, the demand for this technique will undoubtedly become more widespread.

The process involves injecting supercritical CO$_2$ into the oil reservoir, where it comes into contact with oil, increases the overall pressure and forces the oil upwards. The CO$_2$ may also dissolve in the oil, causing it to swell, forming a concentrated oil bank which is then pushed towards the production well, allowing the oil to be extracted. How this works has been illustrated in Fig. 11.2 below.

This technique can help extract 30 to 60% or more of a reservoir's oil but of course comes at a price. On the other hand, the increased extraction is an economic benefit with the revenue depending on prevailing oil prices. Onshore EOR has paid in the range of $10–16 net per ton of CO$_2$ injected. With oil prices at around $90/barrel, the economic benefit is about $70 per ton of CO$_2$. In 2015 the US Department of Energy estimated that 20 Gt of captured CO$_2$ could produce 67 billion barrels of economically recoverable oil, whereby 90% of this CO$_2$ could be sourced directly from industrial installations in the US (Hebert 2015; *Wikipedia* on Enhanced oil recovery).

EOR with CO$_2$ (CO$_2$-EOR) is one of the most successful EOR technologies and contributes to increasing the lifetime of fossil fuel reservoirs. It is the only form of large-scale, permanent carbon sequestration that currently makes a profit, a situation that can only change when a stiff price is attached to carbon emissions. When using

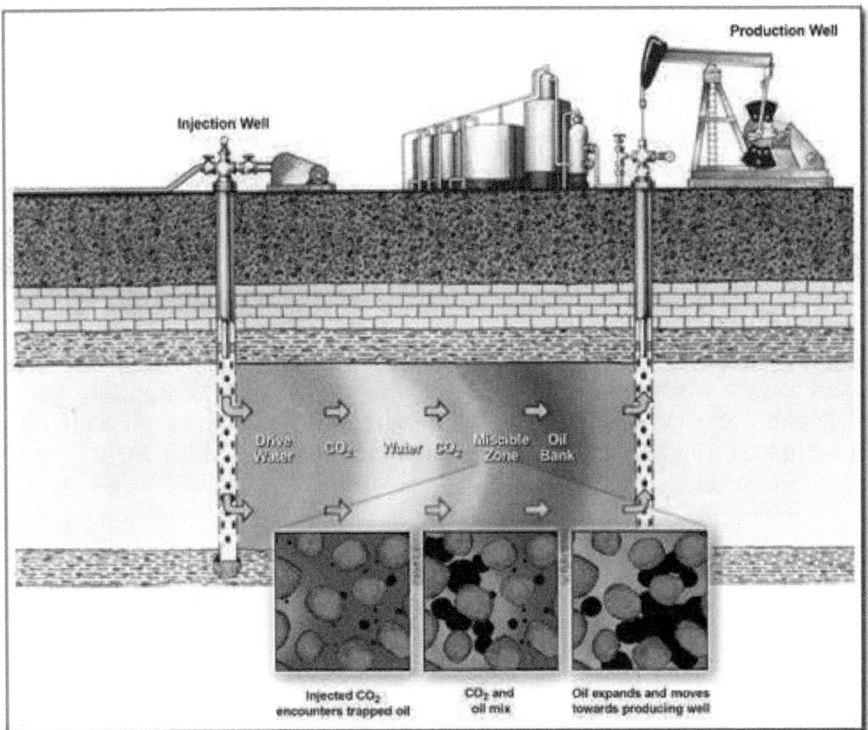

Fig. 11.2. Cross section illustrating how carbon dioxide can be used to flush out residual oil from a subsurface rock formation between wells (*from* NETL 2010).

CO_2 in enhanced oil recovery, some portion of the injected CO_2 remains below ground. If the CO_2 that returns to the surface is separated and re-injected, the result is permanent CO_2 storage.

CO_2-EOR has been implemented for over 50 years in oil fields in the Permian Basin (in Texas and New Mexico), using carbon dioxide from natural underground deposits in New Mexico and Colorado. In the US CO_2-EOR accounts for approximately 273,000 barrels of oil production per day (in 2020), slightly declining from a high of just over 300,000 barrels in 2017.[5] These are generated from 142 EOR projects, of which 80 are situated in the Permian Basin, responsible for 80% of these aforementioned 273,000 barrels. The US Department of Energy has estimated that full use of 'next generation' CO_2-EOR in the US could generate an additional 240 billion barrels (38 km^3) of recoverable oil, more than 30 years worth of total US annual oil consumption. For comparison, total US domestic oil resources still in the ground total more than 1 trillion barrels (160 km^3), most of it remaining unrecoverable (*Wikipedia*).

According to a 2010 report of NETL, the total volume of CO_2 consumed by EOR in the US up to 2010 has been about 560 Mt (mostly from natural, i.e., non-anthropogenic sources), which is not very much compared to current total US CO_2 emissions of about 5 Gt per year. But it shows that demand for CO_2 for EOR is not insignificant and could be an enabling catalyst for larger scale sequestration efforts.[6] It can of course only have an impact on climate mitigation if the CO_2 originates from carbon capture, either from point sources or from Direct Carbon Removal, and not from natural underground deposits. The origin is then indicated as "industrial" versus "natural". CO2-EOR can be a bridge to deploying CCUS on a larger scale as the revenues associated with selling CO_2 to an EOR operator can offset (part of) the cost of employing CCS in industries with heavy emissions.

More than 4,000 miles of CO_2 pipelines are in place and approximately 85 million tons of CO_2 are being injected underground annually for CO_2-EOR, 80% still originating from natural sources and only about 20% from industrial sources, but over time the supply of CO_2 from man-made sources is expected to grow significantly.[7] According to the Global CCS Institute, demand for CO_2 for use in local EOR operations is forecasted to grow at least fivefold to 2030 (Global CSI Institute 2021a, p. 45). In a 2019 report (Bobeck et al. 2019) the Center for Climate and Energy Solutions even claims that future CO_2 use in the US applying current state-of-the-art CO2-EOR techniques for economically recoverable oil is projected to be 10.7 Gt. This is however not borne out by the declining trend in EOR use in the US in recent years.

An example of a company providing industrial CO_2 for EOR is the Dakota Gasification Company, which produces a synthetic natural gas at its Great Plains

---

[5] Advanced Resources International, *The U.S. CO₂ Enhanced Oil Recovery Survey*, Oct. 2021; https://www.eoriwyoming.org/downloads/ARI-2021-EOY-2020-CO2-EOR-Survey-2021.pdf.

[6] NETL 2010, Carbon Dioxide Enhanced Oil Recovery: Untapped Domestic Energy Supply and Long Term Carbon Storage Solution (Strategic Center for Natural Gas and Oil (SCNGO), National Energy Technology Laboratory); available online at: https://www.netl.doe.gov/sites/default/files/netl-file/co2_eor_primer.pdf (accessed June 2022).

[7] https://www.c2es.org/document/understanding-the-national-enhanced-oil-recovery-initiative/.

Synfuels Plant near Beulah in North Dakota. Since 2000 the carbon dioxide it produces as a by-product is shipped via a high pressure pipeline to the Weyburn oil field in Saskatchewan, Canada where it is used for EOR and is permanently geologically stored. Dakota Gas has the ability to capture up to 3 million tons of CO$_2$ per year, reducing its emissions by about 45%.

The main barrier to taking further advantage of CO$_2$-EOR in the US has been an insufficient supply of affordable CO$_2$. There is a cost gap between what an oilfield operation can afford to pay for CO$_2$ under normal market conditions and the cost to capture and transport CO$_2$ from power plants and industrial sources. This is the reason why most CO$_2$ comes from natural sources as close as possible to the location where the CO$_2$ is used. Using CO$_2$ from power plants or industrial sources could reduce their carbon footprint (if the CO$_2$ is permanently stored). For some industrial sources, such as natural gas processing or fertilizer and ethanol production, the cost gap is small (potentially \$10–20/ton of CO$_2$). For other sources of anthropogenic CO$_2$, including power generation and a variety of industrial processes, capture costs are greater, and the cost gap becomes much larger (potentially \$30–50/ton of CO$_2$). This cost gap will of course always exist so long as "normal market conditions" mean that you can just let CO$_2$ escape into the atmosphere. That will always remain the cheapest option. It underlines the necessity of a sizable carbon tax or other form of carbon pricing. The low price of oil and gas, actually of energy in general, does not help either in this respect.

Apart from this, there are also serious environmental concerns connected with CO$_2$-EOR, as large quantities of water are pumped to the surface in this process. This water contains brine (a high-concentration solution of salt in water) and may also contain toxic heavy metals and radioactive substances, which can be very damaging to the environment, as well as to drinking water sources, if not properly controlled. Such reservoir fluids are separated at the surface, and commonly reinjected/recycled into the reservoir.

But, the point that mainly interests us here is that for this technique carbon dioxide can be used and that there apparently is quite a lot of experience with carbon capture, followed by utilization and storage (CCUS).

Enhanced gas recovery and enhanced coalbed methane recovery

Similar to EOR, enhanced gas recovery (EGR) is considered a promising technology to sequester CO$_2$ while improving recovery of difficult-to-recover natural gas or for recovering methane from un-mineable coal seams, as discussed briefly in the preceding chapter. Especially for coalbed methane and shale gas reservoirs, CO$_2$ enhanced coalbed methane (CO$_2$-ECBM) and CO$_2$ enhanced shale gas recovery (CO$_2$-ESGR) are attracting more attention in recent years because of these dual benefits.

CO$_2$ storage with EGR is based on differences in the physical properties of CO$_2$ and natural gas. Under atmospheric condition, CO$_2$ and CH$_4$, the main component of natural gas, are both in a gaseous state and mix easily. At typical reservoir conditions (depth > 800 m), however, their physical properties (in particular, density and viscosity (thickness)) are significantly different, which reduces the likelihood of the reservoir natural gas being contaminated by injected CO_2. Under such conditions CO_2 mainly stays in a supercritical state and has a density almost two orders of magnitude larger than natural gas. This causes gravity to separate CO_2 and methane, with the denser CO_2 sinking to the bottom of the reservoir. In addition, at reservoir conditions the viscosity of CO_2 is about an order of magnitude larger than of CH_4. These differences will cause natural gas to be displaced when sequestrating liquid-like CO_2 at the bottom of a reservoir.

From numerical simulations it can be concluded that this can be applied in depleted gas reservoirs in a technically and economically feasible manner. However, CO_2 easily forms preferential flow patterns (due to reservoir heterogeneity) which adversely compromise CO_2-storage and EGR performance. Preferential flow (also called non-uniform flow or non-equilibrium flow) refers to the phenomenon that a fluid or gas seeks its own preferred pathways in a porous medium, thereby bypassing large portions of the porous system. It can be slowed down by water trapped in the pores of sedimentary rocks (so-called connate water) or by injected water.

A few CO_2-storage and EGR research and demonstration projects have so far been conducted, of which the K12-B gas field project, located in the Dutch sector of the North Sea, is the first and most successful demonstration project for CO_2 storage in combination with EGR. The field has been producing gas since 1987. This gas has a relatively high CO_2 content (13%) and the CO_2 is separated from the production stream prior to the gas being transported to shore. The separated CO_2 used to be vented into the atmosphere but is now injected into the field at a depth of approximately 4,000 m. K12-B is the first site in the world where CO_2 is injected into the same reservoir from which it originated. CO_2 injection started in May 2004, simultaneously with extensive measurement programs. The main objectives of this ORC (Offshore Re-injection of CO_2) project were to determine the potential for EGR with CO_2 storage. The separated CO_2 has been continuously re-injected into the depleted gas field for more than 13 years, sequestering in total 100,000 tons of CO_2, while at the same time enhancing gas production. The success of this project provides useful field-scale knowledge and experience, with a view to commercial application.

Although the technical feasibility of CO_2 storage with EGR has been preliminarily proved by reservoir simulations and the mentioned demonstration projects, the technology is not yet fully developed or commercially ready. Other ongoing projects are the injection of CO_2 into the depleted Long Coulee Glauconite F Pool in south-eastern Alberta, Canada, and the Otway Project of CO_2 storage in a depleted gas field, located in the Otway Basin of Victoria, southeast Australia (Liu et al. 2022).

Industrial applications

CO_2 has long been deployed in many industrial scale applications, either directly or after conversion, including carbonated beverages, in fire extinguishers and refrigeration. More recent applications exploit its remarkable physical properties. Liquid CO_2 has very low viscosity and excellent wetting properties, and a density comparable to many other commonly used organic solvents. Supercritical CO_2 is also an excellent solvent and used in many extraction and separation processes in the food industry, and cleaning operations in the garment industry (e.g., dry-cleaning) as well as in microelectronics. Both liquid and supercritical CO$_2$ are considered 'green solvents' increasingly employed in a wide range of industrial applications including replacing organic solvents for precision coatings in car and furniture manufacturing, polymer production and processing, biomedical applications, catalysis, and decaffeinating coffee and tea (Gür 2022). Most of these uses do however not amount to CO$_2$ mitigation as in such use the carbon dioxide is released into the air and not permanently embedded in products.

New pathways involve transforming CO$_2$ into fuels, chemicals and building materials. Most of these chemical and biological conversion processes are still in their infancy and face commercial and regulatory challenges. The production of CO$_2$-based fuels and chemicals is energy-intensive and requires large amounts of hydrogen. The carbon in CO$_2$ enables conversion of hydrogen into a fuel that can be used, for example, as an aviation fuel. CO$_2$ can also replace fossil fuels as a raw material in chemicals and polymers. Less energy-intensive pathways include reacting CO$_2$ with minerals or waste streams, such as iron slag, to form carbonates for building materials (IEA (2019), p. 3) or as we have seen for carbon sequestration by enhanced weathering.

The GCI study, mentioned earlier (GCI 2016, p. 15), recommends further investment in four markets—building materials, chemical intermediates, fuels and polymers. Of these five product groups building materials have the largest potential for CO$_2$ mitigation, 3.6 Gt per year by 2030, hence more than half of the potential 7 Gt mentioned above. In contrast, the much more cautious IEA study considers the near-term potential of increasing the market to just 10 Mt of CO$_2$ use per year for each of these categories, plus CO$_2$ use to promote plant growth. This level of use would be almost equal to current CO$_2$ demand for food and beverages, and more than a factor of 100 less than the GCI estimate.

A company that pursues the use of CO$_2$ for plant growth is the American Infinitree LLC,[8] founded in 2014 to provide carbon-negative CO$_2$ enrichment solutions for enclosed agricultural applications, such as greenhouses where bottled gas and CO$_2$ generators are the most common methods to enrich the atmosphere with CO$_2$. These latter methods of course are carbon positive. Infinitree has developed a carbon-negative tool that utilizes an ion-exchange sorbent material to capture atmospheric CO$_2$ which can subsequently be discharged when desired to create a concentrated stream of CO$_2$. The material will extract CO$_2$ from relatively dry ambient air, so lowers the CO$_2$ concentration in the air. If exposed to high humidity, which is a typical greenhouse environment, or wetted directly, the sorbent material will release the CO$_2$ (humidity swing process) and be absorbed by plants. It does not require combustibles; is scalable to any level of enrichment; and sources a steady stream of CO$_2$ at ideal levels for greenhouse applications.

Building materials

The two main CO$_2$ utilization technologies for building materials are the use of CO$_2$ to cure concrete and CO$_2$ mineralization (see Chapter 9) for carbonate aggregates.

---

[8]   http://www.infinitreellc.com/.

The concrete industry, with a market size of $1 trillion, is viewed as one of the hard-to-abate sectors of industry as regards CO_2 emissions. Because of the enormous amounts of concrete used all over the world, concrete being the second most consumed material globally (after water) and responsible for 5–7% of global CO_2 emissions, this use of CO_2 could be relatively significant.

Concrete is a mix of aggregates, cement and water. Aggregates are coarse particles used in construction. They can be gravel or crushed stones or other similar materials. The aggregates in concrete consist for about 45% of coarse aggregates (stones/gravel) and for about 25% of fine aggregates (sand/crushed rock). The rest is water (18.5%), cement (10%) and air (1.5%). Cement accounts for most of the carbon embedded in concrete, as 860 kgs of CO_2 are embedded in one ton of cement.[9]

Concrete curing is the process of controlling the rate and extent of moisture loss from concrete during cement hydration (water absorption). Curing can be carried out by continuously wetting the exposed surface, thereby preventing it to lose moisture, e.g., by spraying the surface with water. Adding CO_2 to concrete during curing reduces the amount of cement and water required to produce equivalent-strength concrete and reduces emissions from cement production, in addition to the CO_2 being incorporated into the concrete.

Two North American companies, CarbonCure[10] and Solidia Technologies,[11] are leading the development and marketing of CO_2-curing technology. They inject recycled CO_2 into fresh concrete during mixing where it turns into the carbon mineral $CaCO_3$, making the concrete stronger. This reduces the carbon footprint without compromising performance, and technology costs are offset by cement savings as less cement is needed for the same strength concrete.

One hurdle to overcome is the existence of prescriptive standards, such as those of ASTM International (formerly known as the American Society for Testing and Materials), which sets standards all over the world to improve product quality, enhance health and safety, strengthen market access and trade, such that customers know they can rely on products. These prescriptions represent a significant challenge for advancing the use of CO_2-based construction materials. If such materials fail to match those specific requirements, they may not be accepted for use, even when they exceed performance levels of traditional materials.

The use of carbonate as aggregate does not face the same hurdles to market entry, but its cost is a significant barrier. Current gravel aggregate costs are typically near $50/ton depending on location, while technology developers say low-carbon aggregate might sell for $70 to $100/ton. Thus, it is unlikely that CO_2-based aggregate could soon be widely competitive purely on price, and instead would require some form of policy support. (Bobeck et al. 2019).

The Swiss company Neustark,[12] founded in 2019 as a spinoff of the ETH in Zürich, would not agree with this assessment. It presents itself as a company that

---

[9] https://www.concretecentre.com/Publications-Software/Concrete-Compass/Low-Carbon-Concrete. aspx.

[10] https://www.carboncure.com/.

[11] https://www.solidiatech.com/.

[12] https://www.neustark.com/; neustark is German for newly strong.

removes CO$_2$ from the atmosphere, stores it permanently in recycled concrete and cuts new emissions by reducing the use of cement. In each cubic metre of concrete about 10kg of captured CO$_2$ is permanently stored. Up to 20 kg of new CO$_2$ emissions per m³ of fresh concrete are avoided.

Recycled concrete aggregate can be carbonated, i.e., it can mineralize CO$_2$, when it is exposed to a concentrated CO$_2$ stream. This mineralized CO$_2$ is then permanently stored and, if the CO$_2$ is of anthropogenic origin, carbonating recycled concrete aggregate is a negative emissions technology. The anthropogenic source can for instance be biogas from a waste water treatment plant. Normally the CO$_2$ in such a gas is released into the atmosphere.

In case of mineralization, the CO$_2$ is collected, liquefied and transported to the mineralization plant located at the concrete recycling facility. The liquid CO$_2$ is evaporated before being fed at a high flow rate into containers with recycled concrete aggregate. The process duration is typically 2 hours. The CO$_2$ reacts with the cement phase of the recycled concrete aggregate to yield calcium carbonate. With this technology, the CO$_2$—originally taken up from the atmosphere by biomass—is stored in concrete aggregate instead of being released back into the atmosphere. Neustark's aim is to reduce the carbon footprint of the global construction industry by 2050 by 1 Gt of CO$_2$ per year (Tiefenthaler et al. 2021). The company does not seem to have its own capture technology, hence their approach essentially comes down to a method for utilizing or storing anthropogenic CO$_2$ and reducing the carbon footprint of the cement industry by recycling used concrete.

Other building materials that store carbon but at a much smaller scale include bioplastic cladding and insulation made from mycelium.[13] The Berlin start-up Made of Air[14] has developed a bioplastic made of forest and farm waste that sequesters carbon and can be used for everything from furniture to building facades. The material contains biochar, a carbon-rich substance made by burning biomass without oxygen, which prevents the carbon from escaping as CO$_2$. The recyclable material is 90% carbon and stores around two tons of CO$_2$ for every ton of plastic. It is however unlikely that such bioplastics will ever be more than a small niche market. It currently comprises only 0.2% of the global market and is unlikely to grow much as it costs a great deal more to use biomass than ordinary fossil fuels (Friedemann 2021, p. 82).

Another carbon-storing building material makes use of the excellent insulating properties of mycelium, which forms the root system of fungi. Among other things, it could soon be used to insulate and fire-proof buildings while sequestering carbon. The mycelium is grown on waste from the agriculture industry, sequestering carbon in the process. It grows fast and is cheap to produce in custom-made bioreactors.

A further example is olivine, one of the most common minerals on earth. We have met it in Chapter 9 as a carbon mineralization substance. It is capable of absorbing its own mass in CO$_2$ when crushed and scattered on the ground. It can be used as a fertilizer and replacement for sand or gravel in landscaping, while the

---

[13] https://www.dezeen.com/2021/06/27/carbon-negative-carbon-neutral-materials-roundup/.

[14] https://www.madeofair.com/.

carbonated version can be used as an additive in the production of cement, paper or 3D-printing filaments.

There are many more of such innovative ideas pursued by companies all over the world. Another example is the Australian company Mineral Carbonation International (MCI) which has developed methods to speed up the natural weathering process of mineral carbonation to create novel building materials. This process (CO_2-to-stone) sequesters CO_2 by reacting carbon dioxide with abundantly available metal oxides—either magnesium oxide (MgO) or calcium oxide (CaO)—to form stable carbonates.

For most of these applications it applies that they are not 'economical'. They cost more than the product they replace. So, as long as ordinary pecuniary aspects are still dominant, there will be little chance of success.

CO_2-derived fuels by electrochemical reduction

Producing fuels from CO_2 fits in with the macrotrend towards low-carbon fuels. CO_2 competes in this respect with petroleum-derived feedstock, as well as bio-based feedstocks such as sugar cane or other types of biomass. The latter, called biofuels, have the disadvantage that they compete with the food supply, which is not the case for the fuels considered here. Fuels represent one of the largest potential markets for CO_2 utilization technology given the global demand for greener fuel alternatives. In the US they make up 90% of the potential market for CO_2-derived products and services, as mentioned above.

The carbon in CO_2 can be used to produce methane, methanol, gasoline and aviation fuels. The production process involves the use of CO_2 in combination with hydrogen, resulting in a carbon-containing fuel that is easier to handle and use than pure hydrogen. Hydrogen is however highly energy intensive to produce. Low-carbon hydrogen can be produced from fossil fuels when combined with CCS, or through electrolysis of water using low-carbon electricity, e.g., from wind turbines during periods of low electricity demand. CO_2-derived fuels are particularly interesting for applications where the use of electricity or hydrogen is challenging, such as in aviation as electric intercontinental flying does not appear to be possible, at any rate not in the foreseeable future.

A large number of chemical and biological processes are suitable for manufacturing such fuels. The most technologically mature routes are the direct conversion of CO_2 (so-called hydrogenation where it is made to react with hydrogen) into methanol and methane, and indirect conversion whereby the CO_2 is first transformed into carbon monoxide (CO), followed by a synthesis step (Fischer-Tropsch process[15]) which then produces a range of other fuels (see Fig. 11.3). The rates of CO_2 use are high, with methanol requiring 1.37 tCO2 per ton of product, and methane requiring 2.74 tCO2 per ton of product, assuming 100% conversion efficiency. Unlike the chemical compounds making up fossil fuels, CO_2 is a very stable, non-reactive molecule which implies that large amounts of external energy must be supplied to convert it

---

[15] This name stands for a collection of chemical reactions that convert a mixture of carbon monoxide and hydrogen into liquid hydrocarbons; first developed in the 1920s in Germany.

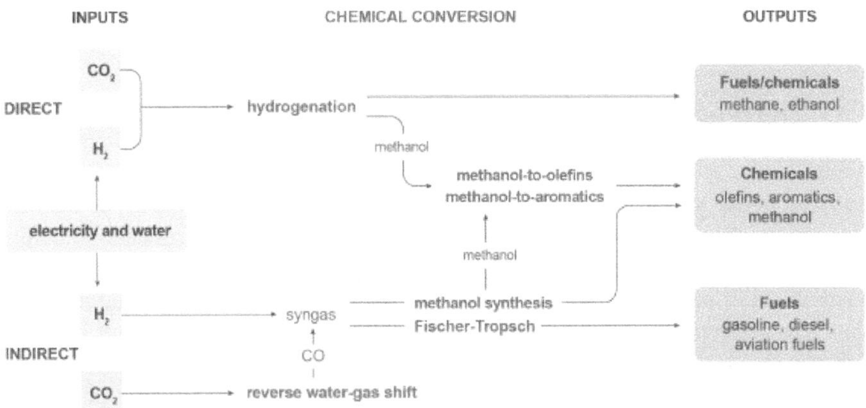

Fig. 11.3. Conversion route for CO$_2$ fuels and chemical intermediates (*from* IEA 2019).

into an energy-rich fuel. The most mature conversion pathways use energy in the form of hydrogen. In practice the overall conversion efficiency is around 50%, but differs per type of fuel. Methane is more energy-intensive to produce than methanol (IEA 2019, p. 41).

The conversion of CO$_2$ into useful fuels accomplishes two objectives in one stroke. It recycles CO$_2$ and transforms it into a useful product that closes the carbon cycle. If this conversion is carried out with energy from non-carbon sources such as nuclear, solar, wind or hydrothermal, the process can be carbon neutral. It is not carbon negative though, as burning the new fuels releases carbon dioxide, but if carbon sucked from the air by DAC is used, the net result is that the CO$_2$ concentration in the air will not be raised (hence carbon neutral). CO$_2$-to-fuels conversion is expected to play a major role in storing electrical energy in the form of fuels such as hydrogen, methane, ammonia, or methanol, and useful chemicals such as syngas (H$_2$+CO), CO or derivatives of formic acid (HCOOH). Estimated production costs of methanol and methane from CO$_2$ are currently 2 to 7 times higher than for their fossil counterparts. The chief cost factor is typically electricity, accounting for between 40–70% of the production costs. For electrosynthesis of fuels from CO$_2$ using renewable power to be able to compete favourably with fossil fuel-derived synthesis processes, the price of renewable electricity needs to fall below \$0.04/kWh and the electrochemical conversion efficiency needs to be > 60% (De Luna et al. 2019). Both conditions will almost certainly be satisfied in the near future.

According to Maarten Steinbuch, professor at Eindhoven Technical University in the Netherlands and a keen believer in e-kerosine (kerosine made from carbon dioxide and hydrogen) you need about 25 kWh in electricity to make one litre of e-kerosine, in addition to 3.3 kg of CO$_2$. In 2019 globally about 380 billion litres of kerosine were used, which means that you need more than 1 Gt of CO$_2$ and about 10,000 billion kWh (10,000 TWh) in electrical energy to meet the demand in e-kerosine.[16] This is close to half of total global electricity production (23,900 TWh in 2019). In spite of these colossal numbers Steinbuch believes that it can be done

---

[16] https://maartensteinbuch.com/2021/05/03/kunstmatige-kerosine-kan-dat-wel/.

and even within a decade. The real challenges in his opinion lie in the cost of the equipment for CO_2 electrolysis and DAC.

CO_2-derived chemicals, chemical intermediates and polymers

What applies to fuels also applies to other chemicals and chemical intermediates, such as ethylene, propylene and methanol. The carbon (and oxygen) in CO_2 can be used as an alternative to fossil fuels in the production of chemicals, including plastics, fibres and synthetic rubber. As with CO_2-derived fuels, converting CO_2 to methanol and methane is the most technologically mature pathway. The methanol can be subsequently converted into other carbon-containing high-value chemical intermediates, such as aromatics and olefins (organic compounds with chemical formula C_xH_{2x}, used to manufacture plastics), and used in a range of sectors, including health and hygiene, food production and processing. A special group of such chemicals, polymers (very large molecules composed of many repeating subunits), are used in the production of plastics, foams and resins. The carbon in CO_2 can replace part of the fossil fuel-based raw material in the polymer manufacturing process. Unlike the conversion of CO_2 to fuels and chemical intermediates, polymer processing with CO_2 requires little energy input, because CO_2 is converted into a molecule in an even lower energy state (carbonate). It is claimed that certain polymers can be made at 15 to 30% lower cost than their fossil counterparts, provided the CO_2 is cheaper than the fossil fuels-based raw material it replaces. Several production routes, e.g., for polyhydroxyalkanoates (PHAs; polyesters produced in nature by numerous microorganism) and polycarbonates, have already been commercialized for high-value products in niche markets (IEA 2019, p. 11).

By combining electrolysis with low-carbon energy sources, water and CO_2 can be converted into chemical feedstock with a minimal or even negative carbon footprint. Electrolysis can help alleviate the imbalance between the supply of electricity from intermittent sources (wind, solar, etc.) via various power-to-fuels scenarios, i.e., using electricity to produce fuels during periods of overproduction, an alternative to using the surplus electricity for the production of hydrogen via the electrolysis of water. One of the possibilities is to reduce CO_2 to CO, which has a wide range of industrial applications, used on a very large scale for the production of chemicals, either as a pure compound or as a constituent in synthesis gas (mixture of CO and H_2). Its use runs into millions of tons (Küngas 2020).

Such conversion processes, coupled to carbon capture, both from point sources and from the air, could potentially offer a sustainable route to produce many of the world's most needed commodity chemicals and address the gigaton challenges in reducing greenhouse-gas emissions. Capture and conversion are both energy intensive, and coupling the two processes can improve the energy efficiency of the system and reduce the cost of the reduced products. For instance, the cost of CO_2 transport and storage can be avoided as well as regeneration of the capture media. (Sullivan et al. 2021).

Production of sulphuric acid

The US technology start-up Travertine[17] (its name obviously derived from the form of terrestrial limestone (calcium carbonate) of the same name deposited around mineral springs) wants to use existing huge deposits of sulphate-containing waste to permanently sequester air-captured carbon dioxide in the form of stable carbonates, while producing at the same time sulphuric acid, H_2SO_4, reportedly the most used inorganic chemical in the world. Indeed, it is said (Chenier 1986, p. 45–46) that sulphuric acid production is a leading economic indicator of the strength of many industrial nations. There are various ways to produce sulphuric acid from sulphur, oxygen and water, the most common ones starting with burning sulphur to sulphur dioxide: $S + O_2 \rightarrow SO_2$, after which the sulphur dioxide is oxidized to sulphur trioxide in the presence of a catalyst: $2SO_2 + O_2 \rightarrow 2SO_3$. The SO_3 can either be directly dissolved in water to produce H_2SO_4 (strongly exothermic reaction meaning that a lot of heat is produced) or it can be absorbed into H_2SO_4 to form oleum ($H_2S_2O_7$), aka fuming sulphuric acid or pyrosulphuric acid, after which the oleum is diluted with water to form sulphuric acid: $H_2S_2O_7 + H_2O \rightarrow 2H_2SO_4$. As can be seen, in none of these processes any carbon dioxide is generated.

The process developed by Travertine first takes CO_2 out of the air via an alkaline solution in air contactors, similar to the procedure employed by Carbon Engineering, followed by an electrochemical process that uses the captured CO_2 to extract sulphuric acid from sulphate waste. Travertine's aim is to develop applications for fertilizer production and to recover critical elements from mines to help advance the transition to renewable clean energy.

The electrochemical process, aka electrochemical salt splitting and schematically represented in Fig. 11.4, involves recovering sulphate waste (containing, e.g., $MgSO_4$ or $CaSO_4$) generated from industries and upcycling it into sulphuric acid (e.g., $MgSO_4 + CO_2 + H_2O \rightarrow MgCO_3 + H_2SO_4$). The calcium and magnesium in the sulphate react with the CO_2 captured from the air, generating carbonate minerals that permanently sequester the carbon.

This procedure is as much a method for carbon capture as a means of utilizing captured carbon for the recovery of sulphuric acid from waste. It is a perfect and beautiful example of CCUS, carbon capture followed by utilization to produce sulphuric acid and store the carbon dioxide in stable carbonates. About half a ton of CO_2 is captured and sequestered for every ton of sulphuric acid produced. Today's global economy produces around 175 Mt of sulphuric acid every year by sulphur burning alone (the total including smelting, etc., is closer to 260Mt). If Travertine were able to replace conventional sulphur burning by its process via electrolytic sulphate upcycling, a maximum carbon mineralization of around 80 Mt CO_2/yr can be achieved.[18] (Or about 130Mt if all sulphuric acid were produced from sulphate waste in this way.)

---

[17] https://www.travertinetech.com/; Travertine is one of the 15 projects (as of 19 December 2022) accepted by *Frontier* as a (future) supplier of CO_2.

[18] Laura Lammers (Travertine), private communication 7.12.2022.

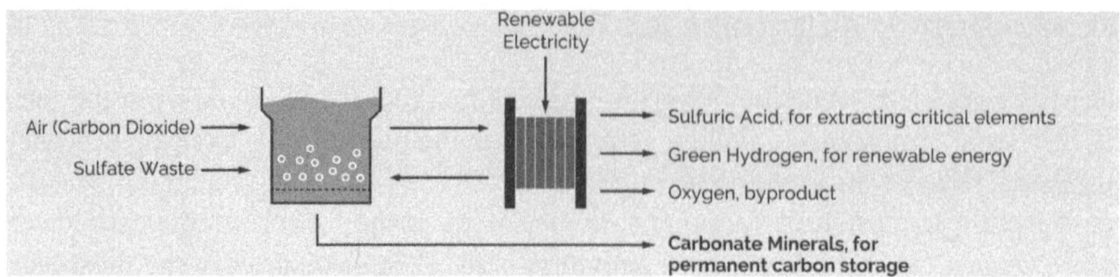

Fig. 11.4. Schematic representation of Travertine's electrochemical process (*source* Travertine).

Although useful, this is not (yet) the gigaton scale we are after, but, as the company says, its process recycles and reuses sulphate to eliminate a source of water and soil pollution while reducing the cost of industrial waste management, which in itself, apart from the carbon sequestration, is of course a very laudable aim.

Travertine draws CO_2 from ambient air, which as we have seen is a very dilute source of CO_2, containing just 0.82 grams of CO_2 per m³. They make it less dilute via an amine scrubbing process and pass the resulting pure CO_2 stream over or through the waste. To sequester one ton of CO_2 1.2 million m³ of air have to be passed through the air contactors in such a way that all the CO_2 in this volume of air is captured. The purpose of their electrochemical process is to speed up the reaction of CO_2 with the sulphate in the waste and separate the sulphuric acid from the resulting mixture. Gigatons of sulphate waste can only be cleaned up by gigatons of CO_2 and the challenge to get to the gigaton scale is the same as for Carbon Engineering or Climeworks. Still, the company claims that it is building a cost-effective, gigaton-scale carbon capture solution that co-produces sulphuric acid. Even if Travertine fails to achieve gigaton scale, it will be worthwhile as it can generate sulphuric acid, green hydrogen and oxygen. Given the magnitude of global greenhouse-gas emissions, it is unlikely that any single carbon sequestration strategy will achieve the desired reduction in CO_2 concentration. Each strategy will give a worthwhile contribution.

What are the costs? A ton of sulphuric acid currently (Dec 2022) costs about $300 per ton, more than double the price at the beginning of the year.[19] Capturing one ton of CO_2 via the Carbon Engineering procedure would cost probably something like $200, half of it being needed for producing one ton of sulphuric acid, which seems to leave plenty of room to make this economically viable.

Sulphuric acid is also an important agent of rock weathering. It has the potential to change rock weathering processes and accelerate the sequestration of CO_2. Dissolved CO_2 and H_2SO_4 reacting with carbonate and silicate minerals in sediments produce the bicarbonate ion HCO_3^- and release positively charged metal ions, such as Ca^{2+}, Mg^{2+}, Na^+, and K^+, and dissolved silica to solution.

Algae based products and novel materials

Algal growth in the ocean has been mentioned before as a storage possibility for CO_2, extracted by the algae from the atmosphere by photosynthesis. A novel application

---

[19] https://ycharts.com/indicators/us_producer_price_index_chemicals_and_allied_products_sulfuric_acid.

is algae cultivation for the production of commercial petroleum substitutes, whereby CO$_2$ is artificially introduced into closed systems to enhance algal growth.

According to the IEA, algae cultivation is still in an early stage of development. The main challenges include low yield rates, system sensitivity to impurities and the high energy requirements for processing algal products (IEA 2019, p. 63). The Global CO$_2$ Initiative is not optimistic either about the prospects for algal growth. Due to high downstream processing costs algae are not yet cost-effective. In 2009–2010 algae biofuel projects received over \$1 billion in funding, largely for development and pilot-scale testing, but investment began to dry up the year after. Intrinsic limitations in algae production and a weak business case for production at scale impeded further development. Novel materials, such as carbon fibres, have to this point received very limited development focus. Such products may potentially have a significant impact on CO$_2$ reduction, but there are great uncertainties as regards scale and the time it takes to take products to market (GCI 2016, p. 16).

The company Global Algae,[20] an Xprize Carbon Removal Milestone winner (see Chapter 9), is apparently undaunted by these assessments. Global Algae was founded in 2013 with the grand mission to change the world through technology, service, and compassion, by harnessing the unparalleled productivity of algae. Global Algae wants to exploit the properties of algae to provide food and fuel for the world, dramatically improving the environment, economy, and quality of life for people. To achieve this, large-scale algae farming needs to be made economically viable and energy efficient.

In their setup algae are cultivated solely on captured CO$_2$, directly absorbed from the atmosphere into open ponds so that no separate CO$_2$ processing system is needed. The algae harvesting system uses advanced membranes to achieve 100% harvest efficiency and to eliminate the need for secondary dewatering. Products are algae oil (seaweed oil) for fuel, polyurethanes, and omega-3 oil; silicon dioxide (SiO$_2$), aka as silica, for diatomaceous earth[21] applications; and protein meal for food, feed, and polymers. So, here we essentially have a combination of carbon capture and utilization by storing captured carbon in products.

The company's state-of-the-art algae laboratory and farming operations are located on the Hawaiian island of Kauai, nicknamed the Garden Isle, where it operates a 33-acre farm. Its corporate headquarters and algae equipment manufacturing facility are located in Southern California. It is now at the stage to scale up to a 160-acre farm (Farm 160) which will directly capture 7,500 tons of CO$_2$ per year. Part of the captured CO$_2$ will be sequestered for hundreds of years in the polymer products produced from the algae. The remainder is not sequestered as it will be released when using the fuel, food, and feed products, making this approach carbon neutral, not carbon negative.

In the way as presented by Global Algae, this farming is more a carbon-neutral procedure for growing protein than for capturing carbon, whereby the algae take

---

[20] https://www.globalgae.com/.

[21] Diatomaceous earth, aka as kieselgur/kieselguhr, is a naturally occurring, soft, siliceous sedimentary rock that can be crumbled into a fine white to off-white powder. It consists of the fossilized remains of a type of hard-shelled microalgae.

CO_2 out of the air at minimal cost. Large-scale algae farming for feed and fuel would greatly reduce CO_2 emissions. The company says on its website that 7,000 algae farms of about 10,000 acres[22] each would provide 25% of the global fuel supply and protein for feed and food to support 10 billion people. If one 160-acre farm captures 7,500 tons of CO_2 per year, the 70 million acres of these 7,000 farms would capture approximately 3.5 Gt per year, which is in the right ball-park of making a sizeable contribution to climate mitigation. Total land area required is just about 0.3 million km^2, 18% of US arable land area and less than 1% of global arable land area. Sounds fantastic, but I am not sure whether a diet consisting mainly of seaweed is something to look forward to.

The British company Seafields,[23] incorporated in April 2021, has probably also realized this scanty dietary appeal and therefore has a different plan. It wants to grow the seaweed Sargassum on a gigantic off-shore farm in the Atlantic, harvest it, pack it up in bales and sink these to the bottom of the ocean, locking away the carbon in the seaweed for ages. Sargassum is one of nature's best candidates for this. It is free-floating, has a high carbon-to-nutrient ration, doubles in size every two weeks and can grow to almost unlimited scale. To stimulate growth the Sargassum will be fertilized through a pumping system that causes nutrients to upwell from a few hundred feet below the ocean surface. The company believes that, since fertilized Sargassum grows very quickly, the farm will have a sequestration capacity of 1 Gt of CO_2 per year. Carbon accounts for around 30% of the dry mass of Sargassum, which means that with 7,500 tons of wet weight Sargassum sequesters 1,000 tons of CO_2. To get to the gigaton scale 7.5 billion tons of Sargassum need to be processed.

Just as the algae of Global Algae, Sargassum can also be grown and harvested for products to replace fossil fuels, including bioplastics, fertilizers and emulsifiers. The aim is to build a 1,000-ton pilot farm in the South Atlantic by the end of 2023 or early 2024. No details of costs are given, but revenue will be generated by selling carbon credits against the CO_2 stored in the sunken plants and from extracting valuable products from the Sargassum before it is being sunk. For this the company is working with carbon credit standards bodies to receive certification of its process.[24]

Final remarks

As regards new materials and new applications I am not sure whether we should rejoice in news items like the following: "Carbon recycling glamour has arrived at a boutique near you. Fashion retail giant Zara, based in Arteixo, Spain, has partnered with LanzaTech to use the biotech's carbon-capture technology to launch a collection of party dresses woven from pollutants. The 16-year-old biotech, first established in New Zealand, uses microbes to fix greenhouse gases produced by industry, agriculture or household waste for conversion to ethanol-based products. In the case

---

[22] I learned this figure of 10,000 acres from private communication with Global Algae.

[23] https://www.seafields.eco/.

[24] https://www.forbes.com/sites/erikkobayashisolomon/2022/06/07/seafields-an-innovative-ocean-based-nature-enhancing-solution-to-climate-change/.

of LanzaTech's collaboration with Zara, CO_2 is scrubbed from steel mill exhaust gases and converted to ethanol through fermentation. This ethanol is converted by Kolkata-headquartered India Glycols into ethylene glycol and then spun into a polyester yarn by Far Eastern New Century of Taipei, Taiwan. The resulting fabric, an alternative to petroleum-based polyester, is fashioned by Arteixo-based Inditex (which owns Zara) into clothing. LanzaTech has been engineering *Clostridium autoethanogenum*[25] using directed evolution to select for strains with high CO and CO_2 fixing efficiencies. These strains have been used by retailer Migros, based in Zurich, as feedstock to make 30% of the polyester polyethylene terephthalate (PET) contained in plastic bottles in its food and drink packaging, and by Zurich-based sports brand *On* to create ethylene vinyl acetate foam for running shoes."[26] All this sounds exciting and LanzaTech is for sure a company that does good work but, as the news item makes clear, with companies in Spain, India, New Zealand, Taiwan and Switzerland involved in the production of these dresses, before a dress can be bought anywhere a lot of material has to be flown all over the world, which undoubtedly undoes all the gains made by using carbon captured from the air.

A final use of CO_2 that has to be mentioned is for enhancing yields of biological processes, such as crop cultivation in greenhouses. The application of CO_2 with low-temperature heat in industrial greenhouses is the most mature yield-boosting application today, and can increase yields by 25% to 30%. The clear leader in the use of CO_2 in greenhouses is the Netherlands, with an estimated annual consumption between 5 and 6.3 Mt of CO_2, which sounds little but is still 3–4% of total annual Dutch CO_2 emissions. Of this amount, approximately 500 kt of CO_2 per year originates from external sources, mainly industrial plants, with the balance taken from on-site gas-fired boilers or co-generation systems. The replacement of these on-site systems with other industrial CO_2 sources or with CO_2 captured directly from the atmosphere could deliver climate benefits (IEA 2019, p. 13).

CO_2-based fuels and chemicals still cost several times more than conventional products, mainly due to the costs associated with hydrogen production. The availability of cheap renewable energy and CO_2 could change that. Polymers derived from CO_2 could be produced at lower cost than their fossil counterparts, but the market is relatively small. According to the IEA, the future prospects for CO_2 use will largely be determined by policy support. Many CO_2 technologies that use CO_2 will only be competitive when their mitigation potential is recognized in climate policy frameworks or where incentives for lower-carbon products are available. Building materials produced from CO_2 and minerals or waste avoid the costs associated with conventional waste disposal and can already be competitive today. Early markets for CO_2 use in concrete manufacturing are emerging, with CO_2-cured concrete being lower in cost and having superior performance compared to conventionally-produced concrete. The CO_2 used in building materials is permanently stored in the product, with additional climate benefits derived from lower cement input in the case of CO_2-cured concrete. They also offer larger emissions reductions than

---

[25] An anaerobic bacterium that produces ethanol (C_2H_6O) from carbon monoxide.

[26] Little black carbon-capture dress, *Nature Biotechnology* 40 (2022) 7; https://doi.org/10.1038/s41587-021-01199-6.

products that ultimately release CO_2 to the atmosphere, such as fuels and chemicals (IEA 2019, p. 3).

In conclusion, it is once more apt to note that all these means of CO_2 utilization will just be a drop in the ocean, although a useful and welcome drop. There is little chance that a sizeable part of the gigantic amounts of carbon dioxide that need to be captured, both from point sources and directly from the air, can be used in products to make a noticeable contribution to climate change. The problem that has been created in the last 100 years and has slowly crept upon us cannot be solved easily. All help is welcome and once CCS is applied widely so that CO_2 emissions are largely stopped and its concentration in the atmosphere is actually going down, we can take our time to clean up.

CHAPTER 12

CCS Projects and Hubs

Introduction

Each year the Global CCS Institute issues a report on the *Global Status of CCS*. Its latest report dates from early 2023 (Global CCS Institute 2022) and gives an overview of CCS facilities and projects in operation, in construction, and in advanced or early development in 2022. In their terminology that also includes DAC facilities and projects. As of September 2022 the capture capacity of all facilities and projects, both fixed-point capture and DAC, grew to 244 Mt of CO_2 per year from about 145 Mt per year at the end of 2021. This looks impressive growth, but it is just potential capacity while operational capacity in 2022 was just 42.5 Mt, only a few Mt more than in 2021. This is lightyears away from the over 5 Gt per annum needed by 2050 according to the IEA to limit global warming to 2°C. The report lists a total number of 197 facilities and projects (compared to 135 a year earlier), of which only 30, just 3 more than a year earlier, are operational (with a capacity of 42.5 Mt), 11 under construction, 79 in advanced development, 75 in early development and 2, accounting for 2.3 Mt, having suspended operations.

Until very recently the prospects for CCS did not look good at all. Many projects were started but then killed off in the development stage so that in recent years hardly any progress has been made. The vast increase in funding to billions of dollars recently announced by the Biden administration has brightened things up but there is still a very long way to go. Adequate funding is necessary of course but not a sufficient prerequisite.

In some sense it is confusing to speak of CCS as it does suggest that CCS projects are vertically integrated, i.e., consisting of a capture plant with its own dedicated downstream transport and storage system. Capture, transport and storage under the same roof. For some projects that is indeed the case but in general that is not a good situation, as the total chain of capture, transport and storage of CO_2 becomes dependent on the capturing facility (e.g., a power station, refinery or gas processing plant) whose core business has little to do with handling CO_2. The facility must be able to find a customer or storage location for its own CO_2 emissions and/ or venture far out from its own expertise into transport and storage. The suppliers of CO_2 for EOR generally just supply CO_2 to a single EOR project, which can be hundreds of kilometres away and for which they have their own (expensive) compressor and pipeline systems. CCS becomes much more viable if capturing facilities can outsource the transport and storage parts to a third party. As already noted in Chapter 10, there is a recent trend towards projects sharing transport and storage infrastructure: pipelines, shipping, port facilities, and storage wells (Global CCS Institute 2021a, p. 18; 2022, p. 45). Industries, e.g., power generation or oil refining facilities, are contracted to supply CO_2 to such network or hub, which is responsible for transport and storage, including finding a suitable storage location and constructing the pipelines and compressor stations. In this way the CO_2 supplier can concentrate on its core business and the shared transport and storage infrastructure makes smaller scale CO_2 capture projects viable.

No less than 20 of the 30 currently operational facilities are involved in EOR, some already for decades, the oldest dating back to 1972. Many of the projects listed as under development in the US plan to capture CO_2 from ethanol production facilities for dedicated geological storage and surprisingly few, just a handful, for EOR. This while the majority of the projects currently in operation is, as noted, for EOR. Each of these ethanol production facilities captures a relatively small volume of CO_2, supplied for transport and storage to a network as will be detailed below. The UK has 27 projects in development (none in operation), while the rest of the world contributes very little. China, for instance, in spite of being by far the largest emitter of greenhouse gases, has only 3 projects in construction and development with a maximum capture capacity of 3.3 Mt, contrasting starkly with its more than 10 Gt in annual CO_2 emissions. It has 3 operational projects with a capture capacity of 1.7 Mt per year. For comparison the maximum capture capacity of all operational US facilities and projects is about 20 Mt and more than 70 Mt for those in development, somewhat better but still very little compared to total US emissions of about 5 Gt.

Almost all projects concern CO_2 capture from a concentrated stream, either an exhaust stream from power stations, refineries and suchlike, from fertilizer production facilities, or from gas processing facilities that process raw natural gas to make it suitable for sale. There are very few and only small Direct Air Capture projects, which are still at an early stage of development. In respect of DAC, as of September 2023, the IEA stated that 27 DAC plants have been commissioned in Europe, North America, Japan and the Middle East. All of these plants are small-scale, capturing the pitiful amount of a little over 10,000 tons of CO_2/year with only a few commercial agreements in place to sell or store the captured CO_2, while the remaining plants are operated for testing and demonstration purposes. Six DAC projects are currently under construction, with the largest two expected to come online in 2024 in Iceland (36 kt CO_2/year) and in 2025 in the United States (500 kt CO_2/year, with plans to scale up to as much as 1 Mt CO_2/year). Plans for at least 130 DAC facilities are at various stages of development, some of the largest projects are in the US, UK, Norway and Iceland. If all of these planned projects go ahead and steadily capture CO_2 at full capacity, DAC deployment would reach around 4.7 Mt CO_2 by 2030; this is more than 500 times today's capture rate, but less than 7% of the 75 Mt CO_2 needed to get on track with the IEA's net-zero scenario.[1] We will discuss some of these DAC projects in the next chapter, together with the companies that carry them

---

[1] IEA, Tracking Report on Direct air capture (available at: www.iea.org/reports/direct-air-capture) (Accessed on 8 September 2021).

out. They deserve and need vastly more support if mankind wants to bring and keep down the CO_2 concentration in the atmosphere.

In the following we will discuss some CCS projects. We start our description with the EOR projects, active since the 1970s and still accounting for most of sequestered CO_2.

Enhanced Oil Recovery projects

The method and rationale of EOR has been set out in the preceding chapter as it essentially is a form of CO_2 utilization. Sequestration of CO_2 is at best a secondary benefit. The aim and purpose is to extract more oil. Most CO_2 used for EOR does not (yet) originate from the capture of anthropogenic carbon at point sources, such as power stations or refineries, but from natural CO_2 deposits. The double benefits of sequestering carbon dioxide from anthropogenic sources and getting more oil out of the ground are however increasingly being recognized. Those who believe that the only solution is to leave the oil in the ground probably don't see it that way but are unlikely to get their way. As noted in the previous chapter, the US Department of Energy (DOE) estimates that 20 Gt of captured CO_2 could produce 67 billion barrels of economically recoverable oil, enough for about nine years of current US consumption (7.2 billion barrels in 2021). It is unlikely that this opportunity will be allowed to slip by.

EOR is dominated by the US, both as regards projects that use CO_2 from anthropogenic sources and from natural deposits. Having been a pioneer in oil exploration the country was also faced with an early need to squeeze more oil out of (partly) depleted oil fields. Twenty of the 30 operational industrial facilities and projects worldwide listed by the Global CCS Institute that capture anthropogenic CO_2 supply to EOR projects that (intend to) sequester this CO_2. Eleven of these are in the US (Global CCS Institute 2021a, p. 62–66). Outside the US just 9 facilities are operational in supplying CO_2 to EOR projects, but that number is expected to grow in the coming years as more countries are faced with partly depleted oil fields.

Most of the EOR projects receive CO_2 from several sources, for instance the 80 active EOR projects in the Permian Basin (North and West Texas, and New Mexico) are supplied with CO_2 from four natural sources and two industrial sources that capture carbon: the Century Gas Processing Plant and the Val Verde Gas Plants, both in Texas.

Natural gas suitable for use by consumers (sales gas) is composed almost entirely of methane (CH_4). In addition to methane, raw natural gas usually contains a range of other substances, of which carbon dioxide is one. The purpose of natural gas processing plants is to remove the various impurities and produce pipeline quality dry natural gas. Sales gas is only allowed to contain a small percentage of CO_2. The surplus CO_2 is commonly vented into the air, but this practice is increasingly abandoned with the CO_2 being captured.

The first EOR project started in 1972, using a waste stream of CO_2 from several natural gas processing facilities in the Val Verde area of southern Texas, one of which is now called the Terrel Natural Gas Processing Plant. Instead of being vented, the CO_2 separated from the natural gas in the Val Verde gas plants was compressed and transported through the first large scale, long distance CO_2 pipeline to an oilfield several hundred kilometres away elsewhere in Texas.

The Century Plant processes high CO_2-content (more than 60%) gas from various gas fields in West Texas. The captured CO_2 is compressed and transported for use in Permian Basin EOR operations (Global CCS Institute 2016).

Apart from the Permian Basin, there are operational EOR projects in the Gulf Coast area (South Texas, Louisiana, Mississippi), in the Rockies (Montana, Wyoming, Utah, Colorado), in the Mid Continent (Oklahoma, Kansas), and in Michigan. They use CO_2 captured at a multitude of facilities, ranging from fertilizer production facilities and gas processing plants to ethanol and hydrogen production plants.[2] The Gulf Cost area includes the large-scale (1 Mt) Air Products Capture Project, which injects CO_2, captured at a hydrogen production facility in Port Arthur, Texas, into the West Hastings oil field (Núñez-López and Moskal 2019). Before it suspended operations the Petra Nova facility was also among them. This project has a special history which will be recounted in more detail below.

The activities in Michigan are run by Core Energy[3] using CO_2 captured from gas produced from the Antrim Shale gas fields. In 2008, Michigan produced its one millionth barrel of oil from CO_2-EOR; oil that would not otherwise have been produced.

Petra Nova: The first large U.S. power plant with CCS

Early in 2017 the Petra Nova project in Thompsons, Texas, became the first industrial-scale coal-fired power plant with CCS to operate in the US. It injects CO_2 captured at NRG's WA Parish Power Plant, southwest of Houston, into the nearby West Ranch oil field. One of the boilers at the WA Parish Plant was retrofitted with a post-combustion carbon capture system using amine-based absorption to treat exhaust emissions from this boiler.

The CO_2 is removed through a basic absorber-stripper system (see Chapter 7) and then compressed to a supercritical liquid. The CO_2 leaving the carbon capture plant is over 99% pure and the technology captures about 90% of the CO_2 in the exhaust gas after the coal is burned to generate electricity. The project is designed to annually capture approximately 33% of the CO_2 (or 1.6 Mt) emissions from the plant's boiler. The captured CO_2 is transported via an 82-mile pipeline to the West Ranch oil field, where it is injected for EOR. The joint owners of the Petra Nova project, together with the company that handles the injection and EOR, anticipated increasing West Ranch oil production from 300 barrels per day before EOR to 15,000 barrels per day after EOR. The CO_2 will eventually end up in sandstone in the Frio Formation of the West Ranch oil field, where it will remain stored about 1,500 m underground.

---

[2]   From the Global CCS Institute 2016 report it is difficult to connect a capture facility to a particular EOR project. That could have been done better. The survey provided by Advanced Resources International (https://www.eoriwyoming.org/downloads/ARI-2021-EOY-2020-CO2-EOR-Survey-2021.pdf) is better in this respect but contains errors.

[3]   http://www.coreenergyllc.com/.

Approximately $1 billion was spent to install the Petra Nova carbon emissions reduction system. Grant money totalling nearly $190 million was received from the DOE[4] under the Clean Coal Initiative, as well as a $250 million loan from the Japanese government. The increased oil recovery (from 300 barrels to 15,000 barrels per day) was expected to result in a net saving. However, when the project was first proposed, oil prices were very high (at $100 per barrel) and it was assumed that they would not drop. As of 2017, the oil price was about $50 per barrel and oil production at the field resulted in a net loss. In May 2020 Petra Nova's operators decided to turn off the CCS equipment, citing low oil prices, caused in part by the Covid-19 pandemic. The plant had also suffered frequent outages and missed its carbon capture goal by 17% over its first three years of operation. It has nevertheless captured close to 4 Mt of carbon dioxide in the three-year period it was in operation.

It illustrates the tight rope such EOR projects need to walk, certainly when making a profit is the ultimate motive. There is something basically wrong with the underlying philosophy of such projects, for if a profit must even be earned from trying to save the world by averting an irreversible climate catastrophe, then the world is beyond saving.

But, there is some light on the horizon thanks to the Inflation Reduction Act of the Biden administration, which provides major tax incentives to make a restart of the project viable and to give Petra Nova a second chance. Reuters reported on 14 September 2023[5] that the CCS installation has resumed operations.

The Kemper Project

A second failed project is the Kemper Project in Mississippi, aka as Plant Ratcliffe. It was intended to be an integrated gasification combined cycle (IGCC) facility with carbon capture, utilizing a technology to convert lignite coal—mined on the Kemper site—into natural gas (see Chapter 7). In the end the cost of the power plant totalled $7.5 billion, three times the initial cost estimate of $2.4 billion. These figures made it the most expensive power plant ever built, based on its generating capacity (528 MW). If it had become operational with lignite coal as intended, the project would have been a first-of-its-kind electricity plant to employ gasification and carbon-capture technologies at this scale. The plant missed all its targets and plans for 'clean coal' generation, resulting in carbon capture being abandoned in July 2017 and the company switching the Kemper project to burning only natural gas in an effort to manage costs. It is currently still operating as a gas-fired power plant.

The advanced coal-gasification technology to be used by the plant, called TRIG (transport integrated gasification), was supposed to work better with low-grade coal like lignite than the IGCC discussed in Chapter 7 and was developed in partnership with the DOE. It had to show that the economic benefits offered by the transport gasifier, compared to other systems, are preserved even when carbon-dioxide capture and sequestration methodologies are incorporated into the design. The Kemper Plant

---

4   https://en.wikipedia.org/wiki/Petra_Nova - cite_note-7.

5   https://www.reuters.com/business/energy/carbon-capture-project-back-texas-coal-plant-after-3-year-shutdown-2023-09-14/.

was planned to have 60 miles of pipeline to carry its captured CO_2, 3 Mt per year, to neighbouring oil reserves for EOR.

During the course of the construction legal issues popped up, then environmental issues, with environmental organizations adding their nefarious contribution with a number of red herrings, and finally political issues. But the main cause of its eventual failure, according to the Guardian newspaper (Kelly 2018), was that "top executives covered up construction problems and fundamental design flaws at the plant and knew, years before they admitted it publicly, that their plans had gone awry." The Kemper project ran aground due to poor market timing, poor management and mishaps that could have been avoided, but its failure still has long-running ramifications for the international response to climate change. It was a bad start for CCS, especially harmful as we cannot do without CCS to combat global warming.

EOR outside the US

A US-Canada transborder project and the first of its kind is the Weyburn-Midale Carbon Dioxide Project located in Midale, Saskatchewan, Canada, which started CO_2 injection in October 2000. It is the first project that used a man-made source of CO_2 for EOR. On injection of CO_2, oil production at the Weyburn field increased from 10,000 to almost 30,000 barrels per day and it continues to produce oil today, each ton of CO_2 increasing oil production by almost three barrels. In the process it sequesters about 3 Mt of CO_2 per year and over 30 Mt have been injected since the start of the project. The injected CO_2 is captured at the rate of 8,500 tons per day in the US at the Great Plains Synfuels Plant in Beulah, North Dakota, which is the only (lignite) coal-to-syngas facility in the US and in operation for 25 years. It is subsequently transported through a 320 km long transnational pipeline. Some of the injected CO_2 at both Weyburn and Midale is pumped back to the surface together with oil and water, then separated and re-injected. At the end of the EOR period, virtually all injected and recycled CO_2 will be permanently stored.[6]

Boundary Dam

The Boundary Dam project is the first commercial-scale power plant with CCS in the world to begin operations and is a flagship CCS initiative, although it can be questioned whether retrofitting such an old, albeit rejuvenated, lignite-fired power plant unit with CCS was a wise decision.[7] Boundary Dam is a Canadian venture operated by SaskPower. The CCS installation, which cost about US$1 billion, started operating in October 2014, after a four-year construction and retrofit of the relatively small, 160-megawatt generating unit 3 of the plant. Construction was no bed of roses as in 2015 'serious design issues' arose in the carbon capture system, resulting in regular breakdowns and maintenance problems that led to the unit being operational

---

[6]   https://ptrc.ca/projects/past-projects/weyburn-midale.

[7]   Boundary Dam was commissioned in 1959 and consists of six units with total original nameplate capacity of 813 MW and currently 531 MW. Some of the units have been shut down, but after decommissioning the 139 MW Unit 3 was replaced by a 160 MW unit in 2013.

for only 40% of the time. The captured CO_2 was sold for EOR, but due to the issues mentioned the contract obligations could not be fulfilled. Subsequent renegotiations significantly reduced annual revenues and seriously weakened the economics of the project.[8] Here too the mistake was made that the project was not set up with the primary goal of capturing carbon from power plant emissions but with the aim of making a profit by selling the captured CO_2 for EOR. The final project was smaller than the earlier plans which envisaged building a 300 MW CCS plant and were abandoned because of escalating costs.

Boundary Dam captures, transports, and sells most of its CO_2 for EOR, shipping 90% of the captured CO_2 via a 41-mile pipeline to the Weyburn Field in Saskatchewan. It captures 90% of CO_2 emissions, capturing 2–3 tons per day, and in addition reduces SO_2 emissions from the coal process by up to 100 per cent and NO_x emissions by 50%. CO_2 not sold for EOR is injected and stored about 2.1 miles underground in a deep saline aquifer at a nearby experimental injection site. It has so far captured over 5 Mt of CO_2 since full-time operations began in October 2014.[9,10] Capturing CO_2 costs money of course and it has been reported that it doubles the cost of the electricity generated.[11] As said before, the idea that capturing CO_2 would be profitable in the ordinary economic sense is ludicrous, just as sewage treatment plants cannot be made profitable, but it is certainly true that Boundary Dam is an expensive project. The costs of fitting or retrofitting power stations with CCS must come down considerably for this technology to become viable. At current prices in the EU ETS the roughly 0.5 Mt of CO_2 captured per year by the plant would be worth more than €80 million, far more than the $16–17 million price SaskPower allegedly[12] receives for a million tons of CO_2 pumped to the Weyburn Field. This again shows that CCS can only be viable if a sizable carbon tax is imposed.

Other projects outside the US

Apart from the projects discussed above, there are two further projects in Canada. Both are in Alberta and have to do with the Alberta Carbon Trunk Line (ACTL), which is a 240 km pipeline that collects CO_2 from industrial emitters, a fertilizer plant and a refinery, in Alberta and transports it to reservoirs across Central and Southern Alberta. The ACTL pipeline has been specifically designed to last for more than 100 years and at full capacity can transport up to 14.6 Mt of CO_2 per year. It has become operational in 2020. Liquefied CO_2 and syngas gathered from the sites is pumped into the pipeline and transported to the Clive Nisku and Leduc field reservoirs, owned and operated by Enhance Energy. The EOR fields are located at a depth of 1,800 m below ground. The project is estimated to capture and store more than 1 Mt of CO_2 per year once fully operational.[13]

---

[8] Nice try, shame about the price, https://www.economist.com/news/2014/10/03/nice-try-shame-about-the-price, and Wikipedia.

[9] https://ccsknowledge.com/blog/carbon-capture-and-storage-are-key-to-a-sustainable-transition.

[10] According to the SaskPower website, https://www.saskpower.com/about-us/our-company/blog/2021/bd3-status-update-october-2021.

[11] https://www.cbc.ca/news/canada/saskatchewan/carbon-capture-power-prices-1.3641066.

[12] Probably Canadian dollars, Wikipedia.

[13] https://www.hydrocarbons-technology.com/projects/alberta-carbon-trunk-line-alberta/.

One of the biggest EOR project is the Petrobas Santos Basin Pre-Salt Oil field project in Brazil. In developing its Tupi oil field in the offshore Santos Basin, Petrobras was looking for a CCUS solution to avoid the venting of associated natural gas with a high CO_2 content of 8–40%. Following a successful pilot project in 2011, Petrobras started commercial scale capture of CO_2 and EOR two years later. As of December 2019, Petrobras had reinjected 14.4 Mt of CO_2 into rock reservoirs. As new production units come on stream, Petrobras aims to reach a total of 40 Mt of reinjected CO_2 by 2025.[14]

A cute little project is the Hungarian MOL Szank Field CO_2 EOR project, the only EOR project in Europe. MOL Group is an international, Hungarian based, integrated oil and gas company (revenue about $20 billion in 2021). MOL Group operates the Szank oil field, which is a producing conventional oil field. Since 1992, CO_2 produced from natural gas processing is used for EOR at this field, sequestering between 59,000 and 160,000 ton of CO_2 per year.

China has also entered the EOR business with currently four projects in operation and a few in development. The Sinopec Zhongyuan CCUS captures CO_2 from a cracking facility at an oil refinery and uses it for EOR at the Zhongyuan oil field. It concerns sequestration of 0.12 Mt per year. The Karamay Dunhua project captures 0.1 Mt of CO_2 per year from methanol production and uses it for EOR at the Karamay oil field in Xinjiang, which began production in 1955. The field is developed by the China National Petroleum Corporation, which also runs the Jilin oil field EOR project, capturing CO_2 from a natural gas processing plant, which processes gas with a CO_2 content of 22.5% from the Changling gas field. It sequesters about 0.35 Mt of CO_2 per year. And finally in January 2022 Sinopec completed the construction of China's first megaton CCUS project, the Qilu-Shengli oil field CCUS, which will reduce carbon emissions by 1 Mt per year. CO_2 is captured from the Qilu Fertilizer Plant and injected into the Shengli oil field for EOR and storage.

Another noteworthy project in China concerns the carbon capture project at the Guodian Taizhou Power Station. It is one of the very few projects that capture carbon dioxide from a power plant (Petra Nova is another example). It is designed to capture 0.5 Mt of CO_2 per year to be used by Sinopec for EOR and will start operations in 2023.

The last EOR projects we mention are the Saudi Arabian demonstration project at Uthmaniyah in the Ghawar oil field and an Abu Dhabi CCS project. The Saudi Arabian project run by Saudi Aramco captures 0.8 Mt CO_2 per year at a gas plant and transports it via a 70 km pipeline to the Ghawar oil field. This huge oil field produced about 60–65% of all Saudi oil from 1948 to 2000. Cumulative production until April 2010 exceeded 65 billion barrels. In 2009 Ghawar produced an estimated 5 million barrels of oil per day (6.25% of global production). The field also produces about 60 million m³ of natural gas per day. After 60 years of production, the field is depleted and Saudi Aramco is going to start CO_2-EOR.

---

[14] https://www.ogci.com/case-study/petrobras-applying-carbon-capture-and-eor-at-scale-in-ultra-deep-waters-case-study/.

The Abu Dhabi project captures CO_2 from a steel factory in Mussafah and transports it to an oil field of the Abu Dhabi National Oil Company for EOR. It captures and stores about 1 Mt of CO_2 per year and has been in operation since 2016. It is being followed up by a phase 2 project which intends to capture 2 Mt per year by 2025 from natural gas processing also to be used for EOR at the same oil field, while a phase 3 could add a further 2 Mt of CO_2 per year from the Habshan and Bab gas processing facility by 2030.

Geological storage projects

There are only 10 operational CCS projects in the world that capture carbon from natural gas processing, chemical production or power generation solely or mostly for geological storage. The Boundary Dam Project discussed above is one of them, although most of its CO_2 is sold for EOR. Below we will discuss a few other projects that stand out, either through success or failure.

Quest CCS facility, Gorgon gas field and other projects

The Quest CCS facility, near Edmonton, Alberta, is a CCS project run by oil company Shell on behalf of the Athabasca Oil Sands project, which makes crude oil from the bitumen found in sand. This oil-making process requires hydrogen to make the oil lighter, which process releases carbon dioxide. Shell installed Quest to capture the carbon and pipe it away in liquid form to be injected and stored underground. Quest's technology is fairly simple and uses an amine solvent to capture the CO_2 from the process stream. The CO_2 is released from the amine by heating and then dehydrated and compressed. The compression reduces its volume by a factor of about 400, turning it into a very dense fluid. Since opening in late 2015, the facility has captured more than 6 Mt of CO_2 (about 1 Mt per year). The liquid CO_2 is transported by pipeline and stored 2 km underground into a layer of rock. Here, it will remain for thousands of years, hence it is truly permanently stored.

A second operational project, a joint venture of Chevron, Shell and ExxonMobil, is the Gorgon project in Australia which is developing the Gorgon and Jansz gas fields 130 to 200 km off the northwest coast of Western Australia. It is one of the world's largest liquid natural gas projects. The gas from the fields contains high concentrations of CO_2 and the CCS operations aim to capture 3 to 4 Mt of this CO_2 per year. The system started up in August 2019 and has a goal of storing 100 Mt of CO_2 over the life of the project, expected to be more than 40 years. The CO_2 is injected at the site of the LNG facility on Barrow Island 2,500 m underground in the Dupuy formation, a deep saline formation beneath the island.[15]

In the US, the Illinois Industrial CCS project is the follow-up of the Illinois Basin Decatur project which completed its goal of injecting and storing 1 Mt of CO_2 in 2014. It captures CO_2 from an ethanol production facility and aims to sequester 1 Mt

---

[15] https://australia.chevron.com/our-businesses/gorgon-project/carbon-capture-and-storage.

of CO_2 per year. One of its objectives is to demonstrate advanced CCS technologies at industrial scale facilities.

In the Middle East, CCS project activity is spread across Qatar, Saudi Arabia and Abu Dhabi. Around 3.7 Mt of CO_2 per year is captured at three CCS facilities, two of which were mentioned above under EOR projects. Even without further CCS activity, these projects could raise overall regional CO_2 capture to almost 10 Mt per year by 2030. Qatar Gas captures 2.1 Mt per year from the Ras Laffan gas liquefaction plant and expects to expand its capture rate to 5 Mt per year by 2025. It is the world's largest LNG liquefaction project, increasing Qatar's LNG production capacity by about 50%. The project will capture up to 3 Mt of CO_2 from the natural gas processing facility.

Schwarze Pumpe

A failed geological-storage project is the oxy-fuel combustion pilot plant with CCS which was part of the modern (2 × 800 MW) lignite-burning Schwarze Pumpe plant in Germany. The pilot plant was operated at 30 MW by Vattenfall. The idea was to compress, liquefy and store the resulting carbon dioxide in geologic formations. It was hailed as the world's first clean coal-fired power station, but when it came to commercializing the project and increasing its productivity, the company decided against it because they believed that the technology was not viable financially. It seems though that the true reason for abandoning the project was the short-sighted but nevertheless strong and in the end decisive opposition from the environmental movement, which considered and considers the phrase 'clean coal' as a 'false promise' and a contradiction in terms ('nothing can make coal clean'). It eventually resulted in Vattenfall selling the plant to a Czech firm. The plant is still in operation, burning lignite and annually belching about 10 Mt of CO_2 into the air. It occupies place 6 in the list of most polluting German power plants, the ten largest of which emit in total a little over 100 Mt per year.[16] The environmental movement had of course not intended this outcome but their actions can still be directly blamed for it and they should be held to account. It is certainly true that 'nothing can make coal clean' but, if there is no alternative to burning coal, we have a duty to find the cleanest way.

Germany still has more than 40 plants that run on hard coal that is imported, mainly from Russia, at any rate before the 2022 Ukraine war, and about 30 that run on lignite, which shows that its green image is largely fake (also illustrated by the fact that coal unseated wind power as the biggest energy contributor to the German electricity network in the first six months of 2021). If some of the subsidies for renewable energy (€25 billion per year) had been spent in equipping these plants with CCS (for around €40 billion in capital costs) from the moment the *Energiewende* was announced in 2011, it would have avoided more emissions than all the wind turbines and solar panels these subsidies have bought, cleaned the air and avoided thousands of deaths. Currently (2020) German coal-fired power plants still emit about 200 Mt

---

[16] https://www.dehst.de/SharedDocs/downloads/EN/publications/2020_VET-Report_summary.pdf?__blob=publicationFile&v=3.

per year, down from 350 Mt in 2013, and this will only stop in 2030; that is if the Ukraine war does not spoil the plans.

The Sleipner project

As mentioned in Chapter 4, a carbon tax has been levied in Norway since 1991. This taxation has resulted in the first successful offshore CCS plant at the Sleipner gas field with permanent geological storage. Sleipner is a natural gas field in the North Sea, about 250 km west of the town of Stavanger in the south of Norway, halfway between Stavanger and Aberdeen in Scotland. The water depth at this spot is 110 m. Two parts of the field are in production, Sleipner West (proven in 1974), and Sleipner East (1981). The field produces natural gas and light oil condensates from sandstone structures about 2,500 metres below sea level. Sleipner West is the world's first commercial CO_2 storage project.

At 9% the concentration of CO_2 in the gas from the Sleipner field is too high, as Norway only allows 2.5% before imposing production export quality penalties. Therefore, part of the CO_2 must be taken out before the gas can be sold. Using a chemical solvent, the CO_2 is removed from the gas at an offshore platform. The solvent is continually recycled in the process and the cleaned gas piped to land. The Norwegian state-owned oil and gas company Equinor (at the time called Statoil), which exploits the Sleipner field, chose not to release the surplus CO_2 into the air, but to capture and store it and thus avoid paying the Norwegian carbon tax.

After capture, the CO_2 is stored at the spot more than 800 metres below the seabed in a saline aquifer. The transport distance is minimal, no long pipelines are required. Although the word aquifer would suggest otherwise, saline aquifers are not underground bodies of water, but porous rock formations infiltrated with highly saline (salty) water. The pores of such aquifers are capable of absorbing large quantities of CO_2, which at the pressures at such depth has liquid-like density. Aquifers capped by an impermeable sedimentary layer, as is the case with the Utsira formation used in the Sleipner project, are ideal for storage, but this is not essential for long-term storage. If injected far enough from the aquifer boundaries, the CO_2 may eventually either dissolve into the aquifer water (hydrodynamic trapping) or precipitate as a solid carbonate mineral by reacting with the surrounding rock (mineral trapping) (Jaccard 2005, p. 199–200). Equinor captures about 2,600 tons of CO_2 every day from the gas produced at Sleipner. The way the CO_2 spreads underground at the storage site has been monitored by various research projects. The Utsira reservoir is continuously monitored using seismology, and comprehensive models have been developed for calculating how the CO_2 moves in the reservoir.

Amine scrubbing technology is used to remove CO_2 from the high pressure natural gas stream. The energy released by the amine treatment process runs two generators, yielding 6 MW of power, utilized on the platform itself. Following removal of the CO_2, the gas is piped to shore for export to continental Europe. The CO_2 is injected into the Utsira aquifer in a dense phase.

One million tons of CO_2 have annually been transported and stored 1,000 metres under the seabed since 1996. Operating costs are $17/ton of CO_2 stored and the project required an initial investment of $80 million. The Sleipner project is the CCS flagship and the one (and so far only) clear success story of CCS. It was the tax that provided the incentive for Statoil to act as it did. When the Sleipner CCS project was commissioned in 1996, five years after the carbon tax was introduced, the levy was $49 per ton of CO_2. Venting 1 Mt of CO_2 into the air in that first year 1996 would have costed $49 million, while as said $80 million was needed to set up the installation to capture, compress and inject the removed CO_2.

Without this tax the CO_2 burden of the atmosphere would have been more than 25 Mt higher. The costs for Statoil have been 25 × $17 million (operating costs) + $80 million = $505 million, while without CCS the tax would have amounted to $1,225 million (at the 1996 tax rate). Clearly a profitable business, and at the same time environmentally sound. What is also important to note is that the Sleipner CCS project has operated consistently near its capturing capacity (IEEFA 2022, p. 25–26).

Snøhvit CO_2 storage project

The Snøhvit (Snow White) project is almost a carbon copy of Sleipner. Like the latter it is an integrated CCS project, taking care of capture, transport and storage. Snøhvit concerns a gas field 140 km northwest of Hammerfest in the Barents Sea, hence much further north than Sleipner. It was the first major fossil-fuel extraction project on the Norwegian continental shelf without any surface installations. No platform or production vessel in the Barents Sea indicates where the field is located. The production facilities stand on the seabed, in between 250 and 345 metres of water.

The Snøhvit gas contains five to eight percent CO_2, which just as for the Sleipner gas is too high. Therefore the gas is piped from the field to the onshore separation plant on the island of Melkøya in the north of Norway. After separation the CO_2 is piped back to a formation at the edge of the Snøhvit reservoir, where it is stored 2,600 m beneath the seabed. A shale cap above the sandstone will seal the reservoir and ensure that the CO_2 stays underground without leaking to the surface. It is in operation since 2008 and captures and stores 0.7 Mt of CO_2 per year.[17]

CCS networks/hubs

As noted before, there has been a recent trend towards facilities sharing transport and storage infrastructure in a network: pipelines, shipping, port facilities, and storage wells. Emissions-intensive industries often develop in clusters due to the availability of feedstocks, infrastructure, such as ports and rail, a skilled workforce and specialist suppliers of engineering and other goods and services. That is why CCS networks or hubs centred on existing key industrial areas are emerging as a preferred model for CCS development (Global CCS Institute 2021a, p. 11). Networks can serve such areas by concluding contracts with industries to supply CO_2 to the network. The network then takes care of transport and storage, including finding a suitable storage location and constructing the pipelines and compressor stations. Such CCS hubs bring many benefits, including lower unit costs, reduced risk and the ability to standardize and scale up quickly. For emitters, the hub opens up CCS as a decarbonization option

---

[17] https://www.equinor.com/news/archive/2008/04/23/CarbonStorageStartedOnSnhvit.

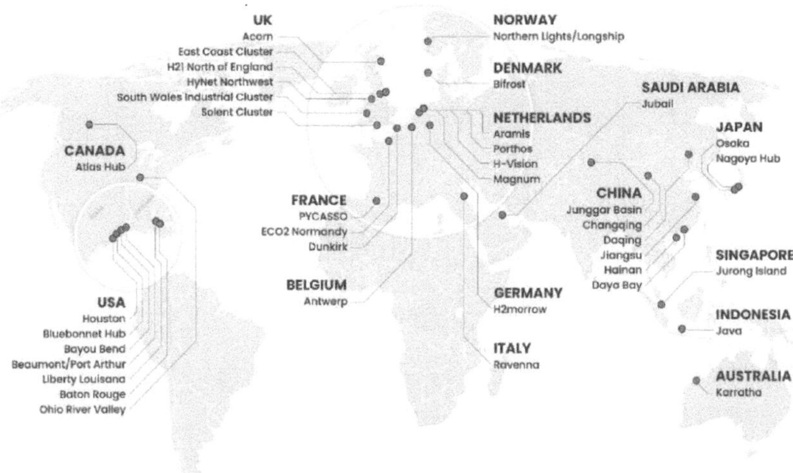

Fig. 12.1. Publicly announced emerging hubs as of early 2023 (*Source:* CCUS Play Book 2023 (OGCI)).

without having to take responsibility for building pipelines, drilling storage wells and without long-term liability for the stored CO_2. Networks are also making capture viable for smaller facilities. Figure 12.1 gives an overview of the various CCS hubs in development in the world, some of which we will discuss below, starting with the activity in Europe.

Northern Lights

The Northern Lights[18] project, a joint venture of Equinor, Shell and Total, is a CCS hub that will take care of transport and storage of CO_2 for a large number of potential customers up to 48 Mt of CO_2 per year, more than total current annual storage worldwide.

The EU has designated Northern Lights as a Project of Common Interest (PCI), a key cross border infrastructure project. It is a commercial ship-based CO_2 cross-border transport network, involving 18 companies and seven countries (Norway, Belgium, Finland, France, Germany, Netherlands, Sweden) and connecting European carbon capture initiatives with permanent carbon storage infrastructure under the Norwegian North Sea. As can be seen from Fig. 12.2 the activity on CCS in northern Europe is quite impressive and could make a real impact if allowed to go ahead.

Northern Lights is developing the infrastructure for transporting CO_2 by specially designed ships from capture sites across Europe to a terminal in western Norway for temporary storage, before transporting it by pipeline for permanent storage in a reservoir under the seabed. The construction of the infrastructure and facilities is well under way. An onshore receiving terminal is now being built in Øygarden, near Bergen in Norway, and will be connected to pipelines and wells that enable storage 2,600 m below the seabed.

---

[18] https://norlights.com/.

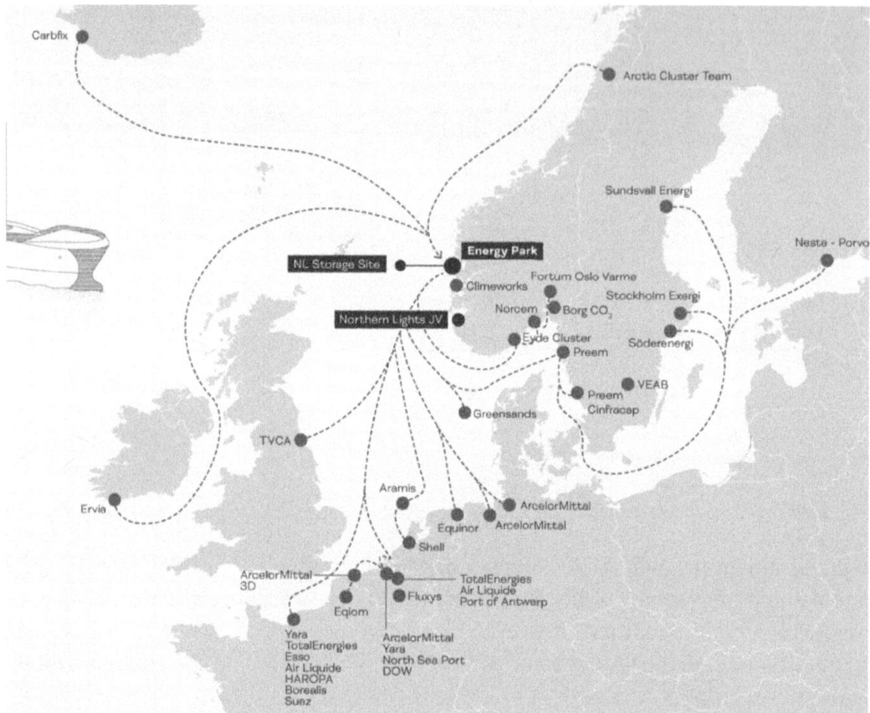

Fig. 12.2. The Northern Lights project bundling CO$_2$ storage efforts of various countries around the North Sea and the Baltic Sea. The routes on the map link up promoters and affiliates to the network (*source* Northern Lights (norlights.com)).

The first development phase of Northern Lights will be completed by mid-2024, establishing infrastructure to store 1.5 Mt of CO$_2$ per year in the Johansen Formation, located in the northern North Sea. As demand from industrial sectors in Europe grows, Northern Lights will increase storage capacity.

Athos project

It is unfortunately a common occurrence for a CCS project to be taken into development and then to be killed off at an early stage or later. An example is the promising Athos[19] project, a development in the Netherlands led by Gasunie, Energie Beheer Nederland (EBN), an entity owned by the Dutch state,[20] Port of Amsterdam and Tata Steel, which aimed to develop a transport and storage network in the *Noordzeekanaal* (North Sea Canal) industrial area. The canal connects Amsterdam with the North Sea at IJmuiden, enabling seafaring vessels to reach the port of Amsterdam. Tata Steel's

---

[19] Athos is an acronym of Amsterdam-IJmuiden CO$_2$ Transport Hub & Offshore Storage, but it is also the name of one of the musketeers in Alexandre Dumas book *The Three Musketeers*, as is its sibling project Porthos pursued in the Rotterdam harbour area (see below).

[20] https://www.ebn.nl/en/; EBN is involved as a non-operating partner in nearly all oil and gas projects in the Netherlands, generally having a stake of 40% in these activities.

IJmuiden plant would be one of the main CO_2 sources, separating CO_2 from blast furnace production gases. Athos only concerned the transport and storage of CO_2 captured in separate projects by the companies that committed themselves to using the transport and storage facilities provided by Athos. It was intended to transport CO_2 offshore for storage in depleted North Sea oil and gas fields or dedicated geological storage, with some CO_2 to be made available for greenhouse horticulture and other industries. The project was based on well-developed mature technology and designed to annually sequester 8 Mt of CO_2 from 2030 (almost 6% of total emissions in the Netherlands, hence in that respect a sizeable project).

It turned out to be short-lived when Tata Steel, enticed by possible billions in subsidies and encouraged by the familiar opposition from environmental organizations against CCS, decided to withdraw from Athos and said to accelerate the switch to steel making via green hydrogen (to produce direct reduced iron (DRI)). For DRI to be carbon neutral, a carbon neutral feedstock (such as hydrogen produced with renewable electricity) must be used in the steel making process. The technology is still in development and a number of demonstration plants are being set up in Europe. None are yet at a commercial stage.[21] For Tata Steel the money has to come from the Dutch taxpayer, much cheaper than itself paying for CCS in the Athos project (even with the subsidies for CO_2 storage), and the renewable energy has to come from wind turbines in the North Sea, which like the hydrogen factory still have to be built and will likewise not be paid for by Tata Steel. All this will at least take a decade or more, while the Athos network was expected to start operations in 2026. In the meantime Tata Steel in IJmuiden, with more than 12 Mt the largest CO_2 emitter in the Netherlands and also in other respects an extremely dirty and polluting industry, to such extent that the Dutch public prosecution service has started *criminal* proceedings against the company, can continue to harm the environment and the health of people living near the plant, without having to pay for this nefarious business.[22] In this connection it is especially galling that within the EU emissions trading scheme (EU ETS) Tata Steel was and still is allocated more carbon allowances than it needs, selling the surplus for an undeserved windfall profit.[23]

After Tata Steel's withdrawal, the other participants no longer deemed it worthwhile to continue with the project. This saga illustrates the steep mountain CCS has to climb before being accepted as *the* way forward for the world towards solving the CO_2 problem. Industry is often just not interested and will do anything possible, sometimes encouraged by environmental organizations, to save money by postponing to take measures. The main problem in my view is one of attitude. In the

---

[21] https://bellona.org/news/industrial-pollution/2021-05-hydrogen-in-steel-production-what-is-happening-in-europe-part-two#:~:text=Iron%20ore%20is%20reduced%20with,that%20steel%20can%20be%20produced.

[22] The Dutch criminal prosecution service has recently started a case against Tata Steel on the suspicion of having deliberately released toxic pollutants into the environment.

[23] According to the National Allocation Plan Tables of the EU, in 2020 Tata Steel was allocated 9.34 Mt (for two separate installations in IJmuiden, not the entire Tata Steel responsible for the 12 Mt mentioned in the text) while verified emissions in that year were 5.79 Mt, leaving 3.55 Mt for Tata Steel to sell on the market. Such practices with the resulting windfall profits were reported by the BCC already in 2016, but are apparently still possible (https://www.bbc.com/news/science-environment-35994279).

EU the situation has recently become more advantageous with the carbon price in the emissions trading system having risen to about €80. It may now actually become *profitable* to capture CO_2 and sell the resulting surplus emission allowances on the market.

Porthos case study

The Port of Rotterdam CO_2 transport hub and offshore storage project, Porthos,[24] is expected to be the first large scale CCS project in an EU member state (see Fig. 12.3) and is closely allied to Northern Lights. It is ideally located to transport CO_2 captured by industry in the port area via pipeline and store it deep underground in depleted offshore gas reservoirs. Porthos is not involved in the capture but only provides the transport and storage network.

The CO_2 emission sources are refinery facilities operated by Shell and ExxonMobil and blue hydrogen[25] plants operated by Air Liquide and Air Products in the Rotterdam port area. They will be responsible for capture. In December 2021 the definitive contracts with Porthos were signed for transport and storage of CO_2. From 2024, the companies will capture a combined 2.5 Mt of CO_2. In the above-mentioned list of the Global CCS Institute of CCS facilities they are included as four separate CCS facilities in advanced development. The Dutch government supports Porthos with a subsidy reservation of €2.1 billion for these four participants, which enables storage of 37 Mt over 15 years,[26] the equivalent of about €57 per ton. It constitutes the maximum amount that may be paid over this term of 15 years to bridge the difference between the €57 mentioned and the carbon price in the EU ETS. The total amount actually to be paid is expected to be significantly lower, since the ETS rate has steadily been rising recently. The EU ETS rate is rather volatile, but currently, February 2023, stands at more than €100 and has consistently been higher than €60 since October 2021.

There are a number of factors likely to further enhance the value of the Porthos project. As already noted in Chapter 10, the Porthos project has been oversized and is capable of handling 10 Mt per year, while the initial participants will only provide 2.34 Mt per year. CO2TransPorts, an EU common interest project, is working on how best to connect Porthos to the North Sea Port and Port of Antwerp where local industrial carbon capture clusters are being developed.

The business case for the Porthos project has been established and efforts are now focused on finalizing permits. Construction of the pipeline will begin shortly thereafter with operation anticipated in 2024. However, in the strange world we are currently living in, until August 2023 the project was in danger of being axed in view of the fact that an environmental organization had lodged an objection with the

---

[24] Porthos is an acronym, of course, of **Port** of Rotterdam transport hub and offshore storage, but is, no doubt not accidentally, also the name of one of the musketeers in Alexandre Dumas' book *The Three Musketeers*, as is the name of the cancelled Athos project.

[25] See Glossary for the various colours of hydrogen.

[26] https://www.porthosco2.nl/en/dutch-government-supports-porthos-customers-with-sde-subsidy-reservation/.

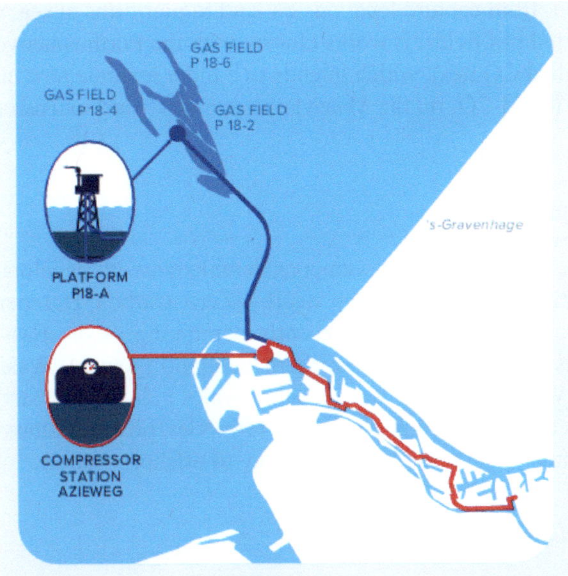

Fig. 12.3. The Porthos project in the Rotterdam harbour (*source* Global CCS Institute 2021a).

Council of State against the project. The project has been awarded an exemption for nitrogen emissions on the basis of the Nitrogen Reduction and Nature Improvement Act, which the said environmental organization has rightly argued is in conflict with European rules.[27] Hence the exemption should not have been granted. The Council of State upheld the complaint on 3 November 2022. The impact of the pipeline as regards nitrogen emissions should first be investigated. This resulted in some delay, but fortunately not that much as in August 2023 the Council of State decided that the project may go ahead as the impact of possible nitrogen emissions on nearby nature reserves 'was convincingly shown by the government to be insignificant'. Knowing the fickleness of courts of law, it was by a whisker that Porthos avoided the same ignominious fate as its fellow musketeer Athos. It would have been a tremendous blow to the efforts to get CCS off the ground in the Netherlands and surrounding countries if the permit for Porthos had been nullified. For once, common sense, a rare commodity in the climate change debate, has triumphed.

The failure of Porthos would also have left the third musketeer Aramis barely alive. Aramis CCS Network is a world-scale network in early development with a proposed capacity in excess of 20 Mt per year. As can be seen in Fig. 12.2, Aramis is part of the Northern Lights project. It intends to realize a new infrastructure for

---

[27] The Dutch government, under unremitting pressure from agricultural lobbying groups, has totally messed up the handling of nitrogen pollution in the Netherlands. Most building and infrastructure projects, including pipeline projects, result in the emission of nitrogen oxides. Dutch regulations for reducing nitrogen pollution are inadequate and contrary to European legislation, reason why the courts for some years now nullify permits for many building and construction projects granted on the basis of these inadequate rules. Before permits can be awarded, nitrogen pollution, which for the greater part is due to agriculture, must first be reduced.

transport of CO_2 from capture locations on land to platforms at sea, where it will be stored in depleted gas fields. It works closely with the Porthos network and another CO_2 project called CO_2next, which intends to build a separate terminal for receiving and supplying liquid CO_2 on the Maasvlakte, the westward extension of the Port of Rotterdam.

United Kingdom

In recent years, the UK has also seen considerable network development, including the Humber Zero network and the nearby Zero Carbon Humber and net zero Teesside networks—the latter two recently combining as the East Coast Cluster. More networks are underway in Northern Scotland (Acorn), Wales and England (HyNet North West) and South Wales (South Wales industrial cluster). All are based in areas with heavy industry—including oil refineries, power stations and natural gas processing plants—with reasonable proximity to offshore storage.

CCS Ravenna Hub

The CCS Ravenna Hub is a CO_2 storage site off the Adriatic coast near Ravenna in Italy. When realized, the project will have a final capacity of sequestering 10 Mt of CO_2 per year by 2030. The plan is to launch phase 1 in 2023, testing technologies in a full capture, transport and storage chain handling up to 100,000 tons per year. Phase 2, scheduled to start in 2027, aims to allow storage of 4 Mt of CO_2 per year, about half of it from three power stations and a hydrogen plant owned by ENI, and the rest from other emitters, including hard-to-abate industries in the region (cement, steel, fertilizer, glass etc). Interest from emitters in Italy and beyond has grown with the increase in EU-ETS carbon prices, and the European Commission's 'Fit-for-55' package of climate legislation. Total storage resource in the Adriatic is estimated to be 500 Mt, giving the possibility in subsequent development phases to increase storage capacity to more than 10 Mt per year, covering the decarbonization needs of additional clusters.[28]

China North-West

In China the China National Petroleum Corporation (CNPC) is setting up the country's first CCS hub, China North-West, in the Junggar Basin, an area with a high concentration of large-scale emitters with relatively pure carbon dioxide streams.

Pipelines and storage systems must be ready in 2025 for capturing 1.5 Mt of CO_2 per year from one of CNPC's own refinery facilities. By 2030 capture is expected to be 3 Mt per year; the hub will expand its transport infrastructure, taking CO_2 from hydrogen production, and from other potential customers, including cement, steel and power plants. The aim is to expand to 10 Mt per year by 2040.

---

[28] https://ccushub.ogci.com/focus_hubs/ravenna/.

CNPC's oilfield subsidiary will inject the CO_2. Initially it will be used for EOR to provide a commercial impetus and to develop the technology, but the plan is to move towards long-term geological storage. Transport may start with the use of tanker trucks, to be replaced by pipelines as the project scales up. China announced in 2020 that it was aiming for carbon neutrality by 2060, and a national emissions trading scheme started operating in July 2021. Although carbon prices in the scheme have started low, at a few dollars per ton, emitters are anticipating much higher prices by 2030 that will make CCUS a commercial proposition. CNPC is planning to build three additional hubs in China by 2030.[29]

Network projects in the US

In the last few decades a large number of biorefineries (converting biomass to energy and other products (such as chemicals)) has sprung up in the US. The biorefineries are run by various agri-businesses and produce ethanol from biomass, mainly corn. Ethanol can be used as a fuel instead of gasoline, or as a fuel additive. Its chemical formula is C_2H_6O. The refining process releases carbon dioxide. Most of these refineries are fairly small, in the sense that they cannot each by themselves capture CO_2, and transport it for storage to a geological or other storage site. Identifying and characterizing a storage location is an expensive business, requiring investment of tens to hundreds of millions of dollars. Various networks have been formed, one by Summit Carbon Solutions[30] and another one by Navigator CO_2 Ventures,[31] creating networks of pipelines spanning several US states, collecting the captured CO_2 at a number of participating biorefineries and transporting it to a geological storage site.

The Summit Carbon Solutions network, shown in Fig. 12.4, connects 32 biorefinery installations, where CO_2 is captured and transported for storage in a geological reservoir in North Dakota. The project will have the capacity to capture and permanently store up to 8–12 Mt of CO_2 every year. The starting date will be sometime in 2024.

The Navigator CO_2 Ventures network, structured in a similar way, is called Heartland Greenway and will provide a CO_2 transport and storage service to biofuel producers and other industrial customers in five Midwestern states (Illinois, Iowa, Minnesota, Nebraska and South Dakota) for a large part overlapping with the Summit Carbon Solutions work area. The Heartland Greenway project would comprise a CCS network spanning more than 1,930 km and transport CO_2 from biorefineries and other industrial facilities to a geological storage site in Illinois. Operations are expected to start in early 2025. Once fully expanded, Heartland Greenway will have the ability to capture and store 15 Mt of CO_2 every year, from more than 30 installations.

---

29 https://ccushub.ogci.com/focus_hubs/china-northwest/.
30 https://summitcarbonsolutions.com/.
31 https://www.navigatorco2.com/.

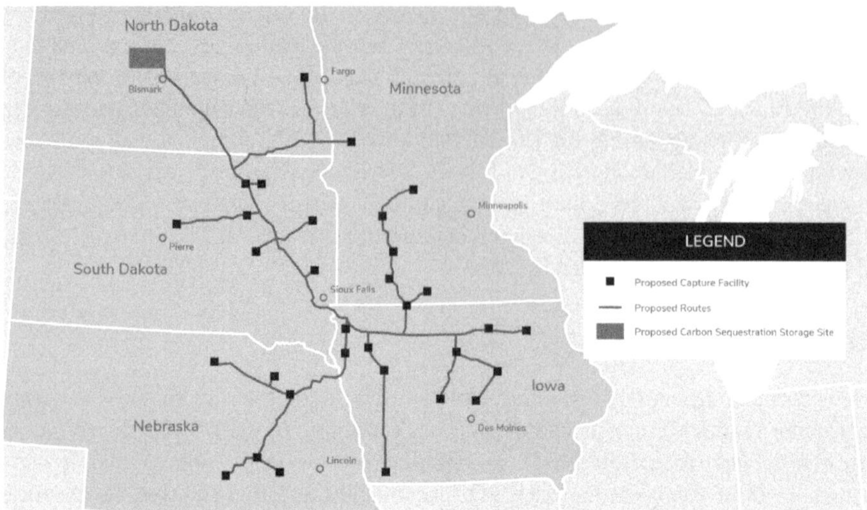

Fig. 12.4. CCS network of Summit Carbon Solutions (*from its website*).

Private companies

In addition to the projects sponsored by government or big industry there is a growing number of private companies that offer carbon capture services to industrial installations. An example is the UK company Carbon Clean,[32] founded as far back as 2016. The company is headquartered in London, with offices in India, Spain and the United States. As of October 2022, it has removed over 1.7 million tons of carbon from 49 facilities across the globe.

Their patented process, called CDRMax™, works similar to the amine scrubbing processes we encountered before: the solvent extracts CO_2 from the feed gas, after which the CO_2-rich solvent is heated in the heat exchanger using hot lean solvent from the desorber; followed by further heating of the CO_2-rich solvent within the desorber, where the carbon dioxide is released from the solvent. The CO_2-lean solvent passes back through the heat exchanger to the absorber for reuse and, once isolated, the CO_2 can be safely stored or converted into new products. The process can be used with source gases that contain CO_2 concentrations between 3% and 25% by volume. Carbon Clean claims to have made improvements, which are not revealed, apart from a low corrosion rate and smaller equipment, but which lower the costs of capturing CO_2 to $40 per ton, compared to $70 for conventional carbon-capture technology.

A second example is the Canadian firm Svante,[33] which has developed technology to capture carbon dioxide from flue gas, concentrate it, then release it for safe storage or industrial use, all in 60 seconds. It is tailored specifically to the separation of CO_2 from nitrogen contained in diluted flue gas generated by industrial

---

[32] https://www.carbonclean.com/.
[33] https://svanteinc.com/.

plants such as cement, steel, aluminium, fertilizer and hydrogen plants, which is typically emitted in large volumes, at low pressures, and dilute concentrations. Tailor-made nano-materials (solid adsorbents) are used with very high storage capacity for carbon dioxide. A sugar-cube sized quantity of their material has the surface area of a football field. These adsorbents have been engineered to catch and release CO_2 in less than 60 seconds, compared to hours for other technologies.

The latest 2022 report of the Global CCS Institute notes that firms such as Carbon Clean and Svante are good examples of capture technology development that is ideally placed for medium-scale applications, such as in the cement sector (Global CCS Institute 2022, p. 41).

Conclusion

As can be gathered from the above there is considerable activity on CCS, but actual results have so far been rather disappointing. Many projects are started up but go astray for various reasons, such as mismanagement, reluctance to cooperate, failure to get permits and dependence on oil price. A much greater effort is needed; gigatons of carbon dioxide need to be sequestered. That this can be done economically and safely has been amply demonstrated already for decades by various EOR projects but also by the Sleipner project and the Shell Quest project as far as geological storage is concerned. There is no reason to hold back. Fossil-fuel power plants all over the world should be (retro)fitted with carbon capture installations, pipelines should be laid and suitable storage sites identified. It looks likely that CCS's future involves international networks spanning multiple industrial clusters and storage sites. There is no time to lose, also because large infrastructure projects like CCS facilities or pipeline networks, usually take seven to 10 years from concept study through feasibility, to design, construction and finally operation. It is clear that fossil fuels will not go away any time soon. If anything, their grip on energy generation has hardly weakened in the last few decades in spite of the advance of renewable energy sources. It is greatly disturbing that promising projects such as the Athos project in the Amsterdam area are abandoned for the shaky prospect of a hydrogen-powered steel-making future for one of the participants and that environmental organizations are cooperating with polluters, allowing them to continue to belch carbon dioxide into the air for unrealistic and uncertain solutions. The best solution would have been if both the Athos and the hydrogen project had been pursued. The hydrogen would have been just as useful without a steel-making goal and once the Athos project would have been up and running other CO_2 emitting facilities could be connected to it, if Tata Steel or another participant thought it wise, convenient or cheaper to withdraw. It is clear that if a sizeable carbon tax, without any tax-free emission allowance, were levied on every single ton of CO_2 such industries emit, they will think twice before withdrawing from such a project; $100 per ton would set back Tata Steel a billion dollars per year, wipe out its profits and send it to an inevitable and long overdue closure.

Fortunately there are other network projects in Europe, the US and also in China that will probably fare better. It is inevitable that in future such CCS hubs will start a large role in point-source carbon capture. They can scale up faster, have lower costs and lower investment risks, and are more likely to gain government support than individual projects. The main disadvantage is their complexity with many stakeholders, which requires careful communication and alignment between partners, to avoid the frustrating failures experienced by countries, such as the UK, Norway and the Netherlands, in getting large-scale CCS off the ground. It is hoped that they have learned from these failures and will apply them to make CCS hubs a reality.

CHAPTER 13

Direct Carbon Removal Companies and Projects

Introduction

According to the IEA's report on Direct Air Capture (IEA 2022), there were as of November 2021 worldwide 18 operational Direct Air Capture (DAC) plants in Europe, the US and Canada, capturing the puny amount of about 8,000 tons of CO_2 per year. Fifteen of these plants are operated by the Swiss company Climeworks, two by the US company Global Thermostat and one by the Canadian company Carbon Engineering. In addition, there are 9 projects in development in various parts of the world, including several 1Mt CO_2/year capture plants, all of these with the involvement of Carbon Engineering. Most of the operational plants sell the captured CO_2 for use in industry. The latest plant to have come online, in September 2021, is a Climeworks facility (see below) capturing 4 kt CO_2/year for storage in basalt formations in Iceland. These operations will have to be vastly scaled up to make any impact at all on the roughly 800 gigatons of emissions that are still in the atmosphere (about half of the 1740 gigatons[1] that were emitted since 1751). More than half of these were emitted since 1990.

Many of the DAC projects are pilot plants and very small, some just a few tons of CO_2 per year, or still at a very early stage of development, which is the reason why the 2022 report of the Global CCS Institute only mentions one project. The institute lists only four DAC projects in development.

The two main challenges faced by DAC are cost and scale: how much does it cost to capture a ton of CO_2 from the air and how to do this at scale, i.e., capturing millions of tons of the stuff in a short time. Access to reliable and ample renewable electricity, on-site or proximate geological storage (combined with affordable CO_2 transport), onsite shared utilities, and sufficient land are all essential to achieving gigaton scale at low cost.

Apart from the three companies mentioned above and to be discussed below, which are the primary industrial developers of DAC, a large number of Direct

---

[1] https://ourworldindata.org/grapher/cumulative-co2-emissions-region?stackMode=absolute.

Carbon Removal start-ups have recently shot up, enticed no doubt by the enormous amounts of funding that have recently become available or were promised. This has also resulted in an astonishing diversification of the methods and combinations of methods used to capture CO_2.

Carbon engineering

Carbon Engineering[2] was established in 2009, in Calgary, Canada, and was the first commercial entity to pursue solvent DAC technology, i.e., using a solvent to capture CO_2 from the air. Its technology involves using and combining techniques that are widely deployed in industry. No new advanced technology is needed, but further research will almost certainly point the way to innovations that will improve current processes and technologies. Their capture procedure has been described in Chapter 7.

Carbon Engineering redesigned the absorption unit, tailoring it for air capture applications and making it suitable for processing large volumes of air. Tests of the air contactor and some of the units involved in the solvent regeneration at pilot scale gave promising results, prompting the company to plan the construction of a DAC plant that can capture 1 Mt of CO_2 per year. In setting up this DAC facility, the company works in partnership with 1PointFive,[3] a development company formed by Oxy-Low Carbon Ventures[4] and Rusheen Capital Management. Oxy-Low Carbon Ventures is a subsidiary of the oil company Occidental Petroleum. This latter fact will make it somewhat suspicious in the eyes of many environmentalists and deservedly so. More worrying in this respect is that according to news reports of mid August 2023[5] Occidental is going to buy Carbon Engineering for $1.1 billion, which sounds a lot but is actually small beer as far as Occidental is concerned. Whether this is a good thing remains to be seen. A few days earlier the U.S. Department of Energy announced that sites in Texas and Louisiana will get over $1 billion in federal grants for funding DAC plants, e.g., Occidental's proposed DAC plants in Kleberg County, Texas. The money is part of the $3.5 billion authorization for regional DAC hubs funded by Congress from the bipartisan infrastructure bill. On the other hand Occidental has a lot of experience in carbon capture, although not in Direct Air Capture, with its EOR operations. As part of these operations it has been permanently storing CO_2 for more than 40 years up to 20 Mt of CO_2 per year.

The Carbon Engineering facility will be located in the Permian Basin with a land footprint of approximately 100 acres and the captured CO_2 will be permanently stored deep underground in geological formations or used in EOR. The construction of the plant started in August 2022 and is expected to start up in 2024. Figure 13.1 shows a reference design.

Planning and engineering work for a second million-ton DAC facility began in Kleberg County, Texas in October 2022. The site is expected to provide access for the

---

2   https://carbonengineering.com/.

3   https://www.1pointfive.com/.

4   https://www.oxylowcarbon.com/.

5   Reuters, 16 August 2023.

Fig. 13.1. Rendering of CE's first commercial DAC facility in Texas (*Credit* Carbon Engineering Ltd).

potential construction of multiple DAC facilities that would be capable of collectively removing up to 30 Mt of CO_2 per year from the atmosphere for sequestration.

Carbon Engineering is so far the only company that has embarked on projects of that scale, with the exception of one by Global Thermostat (see below). It is also collaborating with Pale Blue Dot Energy, acquired in 2020 by Storegga,[6] to build a 1Mt DAC facility in North East Scotland (the Acorn Project)[7]. The facility is planned to become operational in 2026. Partners in this project are Storegga, Shell, Harbour Energy, a small independent oil and gas company, and North Sea Midstream Partners (NSMP), a gas processing and transport company. The latter company owns about 600 km of gas pipelines in the North Sea and supplies 20% of UK gas demand. The Acorn Project shows that fossil-fuel and related companies are becoming increasingly interested and involved in carbon-capture projects. This is an encouraging sign as their expertise in realizing and managing large-scale ventures is probably indispensable in scaling up carbon capture.

In the same area Storegga and Carbon Engineering are developing the Dreamcatcher project.[8] The proposed DAC facility will capture between 0.5 and 1 Mt of atmospheric CO_2 per year and store it safely and permanently deep below the seabed in an offshore geological formation. One of the locations being considered for this facility is in North East Scotland, with access to the Acorn Project.

Further projects in development are a commercial facility[9] in British Columbia (BC) that would produce 100 million litres of ultra-low carbon fuel each year out

---

[6] https://www.storegga.earth/; the company was established in 2019 as an independent developer of carbon reduction and carbon-removal technologies.

[7] https://theacornproject.uk/.

[8] https://carbonengineering.com/news-updates/uks-first-large-scale-dac-facility/.

[9] https://carbonengineering.com/news-updates/large-scale-commercial-facility-fuel-from-air/.

of atmospheric CO_2; the Kollsnes project[10] in Norway for capturing 0.5 Mt of CO_2 per year from the air, with a possible extension to 1.0 Mt, and permanently storing it deep below the seabed in a geological formation; and the Atmos Fuel project[11] in the UK, partnered with LanzaTech, British Airways and Virgin Atlantic, to recycle atmospheric CO_2 into ultra-low carbon jet fuel. In June 2022 Carbon Engineering and 1PointFive announced that they intend to embark on the construction of 70 DAC facilities worldwide by 2035, each with an expected capacity of up to 1 Mt per year.

Carbon Engineering also offers a carbon-emission offset service whereby an organization can reserve capacity from a facility that will capture and permanently store atmospheric CO_2 using their technology and geological storage. The aviation industry has shown an interest in this by Airbus pre-purchasing 100,000 tons of carbon removals per year over four years, as a means to directly offset its CO_2 emissions.[12]

Climeworks

Climeworks[13] is a direct air capture company based in Zurich, Switzerland, established in 2009 and a spinoff of the ETH in Zurich. Their solid sorbent technology has been in development since 2007. The first working prototype of its DAC technology was developed by 2013, and four years later it launched the first commercial DAC plant in Hinwil, Zurich, which had the capacity to capture 900 tons of CO_2 from the atmosphere per year. The capital cost of constructing the Hinwil DAC system was \$3–4 million, making up the largest part of the levelized capture price of \$500–600 per ton of CO_2 (McQueen et al. 2021).

As said above, 15 of the currently 18 operational DAC plants are operated by Climeworks in various European countries (Switzerland, Germany, Italy, Netherlands and Iceland). In 2019, it was the first to open a service offering carbon removal to individual customers.

Contrary to Carbon Engineering's liquid solvent system, Climeworks' system allows for modular CO_2 collectors. Modularity refers to the ability of a system to be partitioned into smaller, repeated systems. The modular design of these systems makes relatively rapid deployment possible, with plant construction taking as little as six months. The collectors are powered using renewable energy or energy-from-waste. They can be stacked to create machines of various sizes, as suited to the location and end-use of the carbon collected. Their grey emission, i.e., greenhouse-gas emission from sources and processes in the facility, is under 10%, which is a very low level of re-emission.

Climeworks' carbon collectors use a two-step process. First the air is drawn into the collector by a fan, where it is collected on the surface of a special filter material inside the collector. When the filter is full, the collector closes and the temperature

---

[10] https://carbonengineering.com/news-updates/partnership-dac-norway/?utm_campaign=CE%20 Announcements&utm_content=188638613&utm_medium=social&utm_source=twitter&hss_ channel=tw-1000093259207655425.

[11] https://carbonengineering.com/news-updates/ce-lanzatech-jet-fuel/.

[12] Carbon Capture Journal, Jan/Feb 2023, p. 7.

[13] https://climeworks.com/; see also Kolber 2021, p. 143 ff. about Climeworks and its Orca project.

Fig. 13.2. Part of Climeworks Orca plant in Iceland (*Credit* 2022 Climeworks).

is raised to 80–100°C, which releases the carbon. It is then collected in a pure and highly concentrated form, ready to be piped away for re-use or underground storage.

For underground storage Climeworks relies on the technology and expertise of its storage partner Carbfix (see below). Climeworks and Carbix collaborate, for instance, on the Orca DAC project in Iceland (Fig. 13.2) which came online in September 2021 and captures CO_2 from the atmosphere blending it with CO_2 captured from geothermal fluids for injection and underground storage in basalt rock formations. This is the first operational appliance of this type of storage, turning CO_2 into rock through mineralization. The plant aims to capture up to 4,000 tons of CO_2 per year, which although still very little makes it the world's biggest climate-positive facility to date. This captured carbon can be bought by people or companies to offset their own carbon footprint at the relatively high price of over €1,000 per ton, which has to come down by a factor of about 10 before most people will be happy to accept such offers. In spite of this high price, Microsoft, never short of cash, announced in January 2021 that it will pay Climeworks to help operate its machine in Iceland. Its contract with Climeworks is for the relatively small amount of 1,400 tons of CO_2 per year. Microsoft is not the only company taking such action. The US-Irish financial-infrastructure company Stripe has also supported the ventures of Climeworks. In 2020 and 2021 Stripe spent $9 million for such carbon removal (Joppa et al. 2021).

Climeworks is currently working on the construction of its next and largest direct air capture and storage plant, called Mammoth. Located in Iceland, right next to Orca, the facility will have the capacity to remove up to 36,000 tons of CO_2 per year. The plant is expected to start operations in early 2024 and is an important step in scaling up to megaton capacity by 2030 and gigaton capacity by 2050. In addition, the company is participating in three DAC Hub applications in the US selected for funding by the US Department of Energy and is exploring the development of large-scale direct air capture and storage projects in Kenya.

Climeworks' current business model focuses on selling carbon-dioxide removal services to individual and corporate customers. Additionally, the company explores supplying onsite CO_2 for renewable fuels and materials and carbon product processes, for example, as part of a newly formed consortium with Rotterdam-The Hague

Airport and the sustainable aviation fuel company SkyNRG, it intends to build a demonstration plant for fully renewable jet fuel (Bipartisan Policy Center 2021).

Climeworks is also developing a project with Norsk e-Fuel, headquartered in Oslo, to build the first European plant to generate renewable fuels from CO_2 and water.[14] In the e-fuel production process (renewable) electricity is used to produce hydrogen via electrolysis from water, after which the hydrogen is combined (synthesized) with CO_2 captured from air into e-fuel (syngas). This syngas can be processed further into other carbon neutral e-fuels, which can then be used in existing vehicles releasing the same CO_2 that was captured earlier.

In the Norsk e-Fuel project, the direct air capture technology of Climeworks and the electrolysis technology of the German company Sunfire GmbH,[15] a leading provider of Power-to-Liquid technology,[16] are combined to convert renewable electricity, water and CO_2 into syngas. Renewable fuels, such as jet fuel, are then produced through further processing and refining. The planning of the first plant, to be located in Herøya, Norway, is well underway and will start operation in 2023. Its initial production capacity of 10 million litres annually will be scaled up 10-fold to produce 100 million litres of renewable fuel by 2026. The plant will help reduce CO_2 emissions from industries such as aviation by 250,000 tons every year.

As far as energy consumption is concerned, Climeworks' procedure currently requires about 2,000 kWh per ton of captured CO_2. At 5 cent/kWh energy cost this will already amount to $100 per ton. This is obviously too high and must come down by something like 50% to become viable.

Carbfix

Carbfix[17] started back in 2006 and was incorporated in 2007, with founding partners originating from Iceland, France and the US. Its storage technology is based on the carbon mineralization processes discussed in Chapter 9, whereby the captured CO_2 is injected into the ground, where it is left to mineralize and turn into stone. Carbfix's mission is to become a "key instrument in tackling the climate crisis by reaching 1 Gt of permanently stored CO_2 as rapidly as possible." It has twice been awarded a $1 million funding prize from the Xprize competition for start-ups,[18] once with Heirloom[19] for "sequestering carbon in naturally occurring minerals and fluids through efficient exposure" and once with Verdox[20] for "electro-swing adsorption to fix CO_2 with 70% less energy". The Carbon Removal Xprize, funded by the Musk

---

[14] https://climeworks.com/news/making-unlimited-renewable-fuel-a-reality.

[15] https://www.sunfire.de/en/.

[16] See Glossary.

[17] https://www.carbfix.com/.

[18] In total 15 carbon removal projects scored $1 million in the Xprize competition (https://techcrunch.com).

[19] https://www.heirloomcarbon.com/.

[20] Verdox has designed an electric system that makes it easier to both soak up the CO_2 and squeeze it back out. The design of the capture device allows for gases to flow through with less resistance, making the soaking process more efficient. Instead of squeezing out the CO_2 with heat, only a specific voltage is applied to the capture material to release the CO_2.

Foundation, aims to reward novel methods to "pull carbon dioxide directly from the atmosphere or oceans and lock it away permanently in an environmentally benign way."

Carbfix takes the carbon captured at a source (power plant or other industry), dissolves it in water and injects it into the ground, into naturally occurring reactive rock formations of suitable composition (basalt). Here the carbon reacts with elements that are naturally present to form stable carbonate minerals, like calcium carbonate $CaCO_3$. This mineral makes up 4% of the Earth's crust, and its most common natural forms are chalk, limestone, and marble, produced by the sedimentation of the shells of small fossilized snails, shellfish, and coral over millions of years. These stable minerals provide a permanent carbon sink and over time turn to stone. This process was thought to take hundreds of years to occur, however, the Carbfix pilot project undertaken in 2012 showed that the desired result could be achieved in about two years. This makes their technology highly effective in permanently storing captured carbon.

Global thermostat

The third pioneering commercial DAC venture, Global Thermostat,[21] is based in the US. Founded in 2010, Global Thermostat currently has two pilot DAC plants with the potential of capturing 3,000–4,000 tons of CO_2 per year. It has partnered with ExxonMobil to expand its technology into a facility capable of capturing 1 Mt of CO_2 each year.

In the capture process, Global Thermostat uses amine-based (dry) chemical 'sorbents', which are bonded to honeycomb ceramic 'monoliths'. These ceramic monoliths, coated with the amine-based sorbents, soak up or absorb carbon from the surrounding atmosphere. This may be directly from air or concentrated emission sites like flues or smokestacks. A focus point of their process is the use of process heat to regenerate the sorbent after capture, with steam near 100°C up to roughly 130°C. Ideally, the steam used is generated by residual or waste heat from the operation of the site where carbon is being collected, which implies that no new energy is required for the process and operational costs are reduced. The end result is a 98% pure carbon stream, which can be transferred to locations where it is needed by other industries. The CO_2 can then be used for a number of possible applications: geological sequestration, biofuels, or non-fuel products like fertilizer or construction material. This approach makes carbon capture a profitable endeavour, rather than a financial liability for the emitting body. It also makes it a potential business operation for those who wish to capture atmospheric carbon and sell it to industries.

In a number of pilot plants in California and Alabama Global Thermostat has tested and refined its DAC and combined DAC-plus-flue-gas-capture systems over the last nine years. Its most recent pilot, in Alabama, consists of two containerized units. Each unit is capable of removing 2,000 tons of CO_2 per year. It has also begun constructing two additional 2,000-ton-per-year commercial demonstration plants

---

[21] https://globalthermostat.com/.

in Oklahoma and is launching a technology centre in Colorado, to support Global Thermostat's existing and emerging projects in collaboration with a number of interested parties.

In 2019, Global Thermostat signed a joint development agreement with ExxonMobil to evaluate the scalability of its DAC technology for large industrial use and climate-change mitigation, which will require CO_2 removal on the scale of billions of tons. Funding has been received from private family and individual investors, as well as from strategic investors, government entities and corporate partners.

Finally, in 2021 Global Thermostat signed an agreement with HIF,[22] a global e-fuels company based in Chile, the US, Australia and Germany, to supply its DAC equipment to HIF's synthetic fuels (synthetic gasoline) Haru Oni pilot plant in Magallanes, Chile. The unit can remove up to a maximum of 250 kg of carbon dioxide per hour from the atmosphere.[23]

Artificial trees

The artificial trees, developed by Klaus Lackner, are an application of the solid sorbent approach described in Chapter 7. It concerns carbon capture devices in the form of trees with plastic leaves that are 1,000 times more efficient in sucking up carbon dioxide from the air than true leaves of natural trees that employ photosynthesis. The carbon dioxide is captured in a filter, removed and stored. The 'leaves' of such a tree are actually 5-foot diameter discs (or tiles consisting of six 'leaves'). They look like sheets of papery plastic and are coated in a resin that contains sodium carbonate (Na_2CO_3), which pulls carbon dioxide out of the air and stores it on the leaf as a bicarbonate (baking soda with chemical formula $NaHCO_3$), according to the reaction:

$$Na_2CO_3 + CO_2 + H_2O \rightarrow 2NaHCO_3$$

To remove the carbon dioxide, the leaves are rinsed in water vapour and can dry naturally in the wind, soaking up more carbon dioxide. The leaves do not need to be exposed to sunlight for photosynthesis like a real tree and trees can be much more closely spaced.

In capture mode the tree extends to a height of ten metres and within twenty minutes its 'leaves' become saturated with CO_2 from ambient air carried over by the wind. It does not need any blowers or fans, making it a passive (the capture process not requiring any energy), lower cost and scalable capture solution. After saturation the tiles are lowered mechanically into a chamber at the base of the tree, where the CO_2 is pulled off the sorbent. The offtake flows from the chamber, after which the CO_2 is separated from the water, achieving purity of up to 95% CO_2. The remaining 5% is nitrogen. The tree then extends again to full height to repeat the process.[24]

---

[22] https://www.hifglobal.com/.

[23] https://globalthermostat.com/2021/04/global-thermostat-to-supply-equipment-needed-to-remove-atmospheric-co2-for-hifs-haru-oni-efuels-pilot-plant/.

[24] https://carboncollect.com/mechanical-tree/.

Although developed as long ago as 2008, the first prototype mechanical tree, as the device is called, has only recently, April 2022, been under construction at Arizona State University in collaboration with the company Carbon Collect Limited.[25] When operating continually, it is expected to collect up to 90 kg of CO_2 per day. Carbon Collect will begin deploying small-scale implementations of their trees (in clusters of 12) while preparing for mass production and larger scale deployment (up to around 120,000 units). One cluster of 12 trees would capture about 1 ton per day. The company plans to deploy farms with an annual capacity of up to 4 Mt of CO_2 per year per farm in the second half of the 2020s. One such farm of 120,000 units would require an area of 2–3 km^2; 250+ of such farms would be needed to capture one gigaton per year.

The company does not say anything about costs, but in 2012 Lackner said in an interview that his trees would do the job for around $200 per ton of removed carbon dioxide, dropping to $30 per ton as the project is scaled up (Vince 2012). At that price it starts to make good economic sense, certainly now that the carbon price in the EU ETS is already around $100 per ton of CO_2 (hopefully to be followed by other carbon pricing schemes around the world).

The companies discussed above were the main players in the DAC field until just a few years ago. Since then it is getting rather crowded in DAC with many start-ups entering the field, undoubtedly enticed by the upsurge in real and promised funding. Some of them will be discussed in the rest of this chapter.

Recent start-ups

In the last few years a flurry of start-ups have entered the Direct Carbon Removal scene, with often splendid ideas to capture carbon from the air, either by straight capture or in an indirect way. Important in several of these new developments is that they require much less energy input than the procedures discussed above. But cost is not all that matters. To be able to achieve the required gigaton scale is the real challenge.

Several of the start-ups are winners in the Xprize competition, already mentioned above and in Chapter 9. It is very difficult to assess how real and/or realistic the various approaches are. Often very little technical or other details are given, with some giving the impression that there is little more than a website (the only must for a company nowadays, it seems) and a couple of enthusiastic guys with a far-fetched dream. In the remainder of this chapter we will discuss some of them, without pretending to be exhaustive. Apart from a few that drew my curiosity for one reason or another, I will discuss those that have received an award from the Xprize foundation and/or have passed the selection process of *Frontier* (see Chapter 9) for inclusion in their portfolio of promising future suppliers of CO_2. Most companies are so young that they do not yet feature in any of the lists or reports mentioned above, and perhaps never will. Even if the majority of them fails and cannot make true their promises, it is gratifying to see that so many clever ideas are being tried out in practice, although I presume that the large amounts of money that are being made

---

[25]   https://carboncollect.com/.

available or promised for Direct Carbon Removal have something to do with this frenzied activity. Some have already disappeared, but if just one or two of the recent start-ups succeed in the coming decades in capturing CO_2 from the atmosphere at the gigaton scale required and at affordable cost, global catastrophe due to climate change can be averted.

Especially those approaches that make use of existing infrastructure that already moves huge volumes of air on a daily basis, combined with means of electrochemical capture, are promising from a cost perspective, as energy-wise they tend to be the most economical, although they may have difficulty in achieving the required scale and, if they do, in disposing of the captured CO_2. As far as scale is concerned, procedures based on any form of carbon mineralization seem to stand the best chance in this respect.

The first group of companies to be discussed use chemical methods similar to the ones employed by the pioneering DAC companies discussed above, but do this in a different setting or in slightly different ways. Then some start-ups using carbon mineralization or enhanced weathering will be discussed, and we finish with a miscellaneous group of companies, of which those focussing on CO_2 removal from the oceans are of particular interest.

Companies using existing infrastructure that displaces large volumes of air

Rail is one of the most efficient forms of large-scale land-based transportation. Trains when moving displace a lot of air and the braking system that slows and stops trains generates a substantial amount of energy, much of which is normally lost as heat. This energy can be utilized for capturing CO_2 in a DAC system integrated into the train and operating on the fly (regenerative braking).[26] This is what the very young start-up **CO2Rail**,[27] founded early in 2021, intends to do. They have developed a special rail car that can be coupled to freight or passenger trains running in regular service. During forward movement of the train this special rail car *draws in air* through one or more forward-facing intakes. At 111km/h, such intakes would supply over 10,000 m³/min of air to be processed for CO_2 removal in the collection chamber of the rail car by means of an electrochemical solid sorbent. The CO_2 is then stored in liquified form in the 15-ton reservoir of the car. Each rail car can capture and store approximately 25 tons of CO_2 per day. By being deployed only with trains already running in regular service, no external power is needed and no additional carbon debt is incurred from the operation. The system powers itself by using the significant amount of wasted energy produced in the braking system of (freight) trains. For carbon capture, the use of this energy amounts to a saving of 50–75% of the costs.[28]

The actual cost per captured ton of CO_2 is still to be revealed, and depends on the scale the system can be deployed and the special infrastructure needed for unloading

---

[26] See Glossary

[27] https://co2rail.com/; their procedure has been published in Bachman et al. 2022.

[28] https://www.advancedsciencenews.com/direct-air-capture-trains/.

the cars and transporting the captured CO_2 to the eventual storage location. The cars are unloaded twice daily at crew change or fuelling stops into a normal CO_2 tank car or into a pipeline.[29] Hence costs involve the transport of the CO_2 to the place of use or storage, and the infrastructure to be built at railway stations for shunting CO2Rail cars to sidings for unloading, maintenance and suchlike and in the form of CO_2 tank cars and/or pipelines. It is not easy to assess whether from a cost point of view this approach is more favourable than a DAC facility at sea close to a geological storage site and powered by a wind turbine, for instance.

CO2Rail claims that their approach can enable gigaton-scale CO_2 capture in an energy efficient manner, but it is hard to see how the process gets to that scale. The stated goal is to shove 30,000 m^3 of air through the collection chamber per minute. In Chapter 9 we have seen that 1 m^3 of air contains 0.82 grams of CO_2. So from 30,000 m^3 of air no more than 25 kg of CO_2 can be harvested; there is no more in them and to achieve this the process used in the rail car must be capable of processing 30,000 m^3 per minute. Is it possible for a sorbent to capture the CO_2 for such a mass of air, release it from the sorbent and feed the sorbent back into the system in a single minute?

In 2020 all US freight trains taken together travelled in total about 600 million km.[30] Assuming that they travel at 100 km per hour, they spent 6 million hours travelling these kms. This implies that 6 million $\times$ 60 $\times$ 25 = 9 Mt of CO_2 can be captured if one CO2Rail car travelled all these 6 million hours and captured CO_2 at 25 kg per minute. CO2Rail envisages each freight train to be equipped with 4 of its rail cars in 2030, 10 in 2050, and 27 in 2075, which would give the respectable yield of respectively about 36, 90 and 240 Mt of CO_2, solely from the US freight train operation.[31] This requires a lot of CO2Rail cars, at least 4, 10 and 27 times the number of locomotives, of which there are about 24,000 in the US, and a few spare ones. About a million CO2Rail cars would probably be sufficient to smoothly run the system.

CO2Rail promises impressive scalability, asset life cycles, and cost-effectiveness, but the global model it has presented[32] is not very transparent, has many assumptions that I am not sure apply to most railway systems in the world. Especially, the assertion that the approach is equally suitable *globally* for passenger trains and freight trains reflects poor knowledge of the structure and operation of passenger train transport in Europe and elsewhere. For a country like the Netherlands for instance, with a very dense (largely electrified) rail network with very frequent intercity and local trains, it is unthinkable that the required number of CO2Rail cars could be coupled to the trains. The entire rail infrastructure would have to be overhauled to be able to handle the 14 (!) CO2Rail cars that are supposed to be coupled to each train. In most cases they alone will be longer than the entire train and trains would become too long for the platforms. For Germany, a much bigger country, some CO2Rail cars can be

---

[29] Private communication with Eric Bachman of CO2Rail.

[30] https://www.bts.gov/content/class-i-rail-freight-fuel-consumption-and-travel.

[31] But by 2075 freight trains may virtually have stopped running, as they mostly transport coal from mines to power stations and coal is being phased out at a rapid rate.

[32] Eric Bachman, CO2Rail, private communication, April 2022.

imagined to be coupled to long-distance intercity trains, but for local networks the same would apply as for the Netherlands. It seems unescapable to conclude that scaling up carbon capture through this method to the gigaton scale required will be a tall order, but even 100 Mt would of course be a useful contribution.

The Finnish start-up **Soletair Power**[33] proposes to utilize existing ventilation in buildings for atmospheric carbon removal. The capture system is integrated with the ventilation system of the building and captures CO_2 from the air supplied to the ventilation system. The building becomes a CO_2 capturing machine and the proposed system essentially is an air purifying system that does not release the CO_2 to the outside.

For a standard ventilation unit handling 3.3 m³/sec, Soletair Power's building HVAC[34] integrated CO_2 capture system, called VIDAC 1.0, can capture 47 kg of carbon dioxide per day per module. With on average six modules per building, this comes down to about 300 kg/day for an average commercial building. This example again shows the mountain DAC has to climb, as for capturing 1 Gt in a year 60 million modules will be needed in 10 million buildings.

In addition to reducing the planet's CO_2, it is claimed that such a system also has the potential to create a localized supply of alternative energy sources. By integrating the captured CO_2 inside an electrolyser and synthesis unit, it can be converted to fuels or other hydrocarbons. A 1500 person building would be able to generate 1500 litres of fuel equivalent per day. Such use of the captured CO_2 on site or on a site nearby seems vital for the approach, as piping the CO_2 away or transporting the relatively small quantities (in gaseous form) over large distances from millions of buildings for permanent storage seems impractical.

Like CO2Rail the clever thing about Soletair Power's technology is that an existing air-displacement system (ventilation) is used which can greatly reduce the power required and provide other benefits.

The San Francisco based company Noya,[35] founded in 2020, has a similar idea and proposes to mobilise existing cooling towers to capture CO_2. A cooling tower is essentially just a box containing a set of showers moving water, and a huge fan moving air. As we have seen, one of the challenges in capturing CO_2 from air is to move large volumes of air through a capturing unit. Rather than wastefully venting water and air, tons of CO_2 can be captured simply by tweaking a tower's plumbing without impacting its original purpose. Noya handles the costs of the equipment, installation, and CO_2 distribution, but splits the profits generated from selling CO_2 with the cooling tower. It's like Airbnb for CO_2 removal. That they think big is clear from the fact that the company estimates that existing towers in the US could together capture 7–10 Gt of carbon dioxide per year, far more than annual total US emissions.

Cooling towers circulate water to remove heat from an industrial process or a commercial building and dump it into the ambient air. They come in various shapes and setups, e.g., in the form of rows of fans humming away on the rooftop of city

---

[33]   https://www.soletairpower.fi/.
[34]   Heating, ventilation, air conditioning.
[35]   https://www.noya.co/.

buildings as part of ventilation systems, or the hourglass-shaped concrete columns of power plants. The water they circulate can sponge up carbon dioxide from the atmosphere simply through contact of the water with air until saturated with the gas. Noya's trick is to add a chemical blend to the water that binds carbon dioxide and allows it to be removed from the system, in effect wringing out the sponge over and over. In Noya's system, the main water line at the bottom tub of a tower is diverted into an exterior processing unit that diverts the carbon dioxide, stores it in pressurized cylinders and returns the water otherwise unaltered. Noya expects the retrofit to catch 0.55–1.1 tons of carbon dioxide per day, year-round (Steiner 2021). This amounts to about 300 tons of CO_2 per year per tower. To capture 7 billion tons about 25 million towers will have to be fitted with Noya's capture equipment and the cylinders with CO_2 have to be transported to a site where the CO_2 can be sequestered, a nontrivial task.

Like Soletair Power and CO2Rail, having found a second use for existing infrastructure, Noya sidesteps some of the costs that typically come with newly-build DAC plants. It still seems a challenge though to achieve capture of gigatons of CO_2 as required to make a real impact, while the cost to transport the captured CO_2 to a storage site will be considerable.

Companies using (electro)chemical approaches

The Dutch start-up **Carbyon**,[36] founded in 2019, has developed a novel thin film DAC technology, which was awarded \$1 million in the Xprize competition for start-ups. It uses a solid-state sorbent, resembling the technology used by Climeworks and Global Thermostat, the main difference being the kinetics; the speed of the process.

Carbyon improves the kinetics by employing a fast swing process by means of a continuously rotating drum (1 cycle per minute) within a cylindrical carrier. The walls of the drum consist of commercially available woven fibres, with an open porous structure. The CO_2 adsorbing substance consists of an amine and a single atomic layer of this amine is deposited in the open pores of the fibre material (atomic layer deposition (ALD)). By coating the walls of the pores in this way, the CO_2 absorption surface area is greatly enhanced (1–3,000 m^2/gram of amine). This reflects what we noted in Chapter 7 that a material with a very high surface area can have lots of molecules stick to its surface.

The carrier consists of two chambers: an air chamber and a regeneration chamber. When the drum is in the air chamber, it is exposed to air and gets saturated with CO_2. After a minute or so the drum finds itself in the regeneration chamber of the machine where it is heated to 100–120°C to release the CO_2. The fast swing process is the key to lower the energy consumption as well as the cost of the machine. A fast swing process means that the cycle from the adsorption to the desorption stage and back is fast, in their case just a few minutes compared to hours for other procedures. Carbyon's target is a capture cost of just €50 per ton, which if realized would be one of the cheapest capture processes in existence. The fast swing process

---

[36] https://carbyon.com/.

requires 1,000 kWh/ton in energy while €25 is needed in capital expenditure due to a very lean modular design, hence electricity must be real cheap to stay at or around €50. Each stand-alone unit is 2 m in diameter and 4 to 5 m in height and can capture 100 tons of CO_2 per year, which is about three times as much as Lackner's mechanical trees discussed above. Farms of thousands of modules, similar to solar panel farms, are envisaged to achieve capture of 1 Mt per year, which subsequently needs infrastructure (compression stations and pipelines) to move the captured CO_2 to its eventual storage location.

The Californian company **Holy Grail**[37] founded in 2019 employs a system similar to the one used by Verdox, discussed in Chapter 7, a DAC system that resembles a discharging battery cell. Air flows through a positively charged cathode, ionizing CO_2 molecules that are transported from the cathode to the anode. No heat or water is required along the way, just (clean) electricity. This allows Holy Grail to build compact cells not much larger than a laptop that can be stacked together. They claim that this approach slashes the heat, water, and energy requirements typically associated with DAC, resulting in very low cost for stripping CO_2 from air, although the actual cost per ton of CO_2 is not (yet) revealed. The technology lends itself well to a perfectly modular approach that could just as easily scrub CO_2 at household level as at an industrial site with cells piled on top of each another, and many cases for use in between. Although the modular approach is useful, it seems farfetched and unnecessary to deploy capture systems at household level. It does not matter where the carbon is captured and it is much more economical to build a huge stack of modules close to a CO_2 storage location, since the obvious question that immediately springs to mind in case of household level capture is what a consumer deploying this system should do with the captured CO_2. Individual consumers do not have to capture their own emissions. They can just pay for having this done elsewhere.

Mission Zero Technologies,[38] based in London (UK) and founded in June 2020 as a spin-out from Deep Science Ventures, claims to have developed a process that captures CO_2 from the air with an ion-selective electrochemical separation process and concentrates it as a pure gas. The exact details of the approach have not yet been revealed. It concerns a modularized DAC technology that is projected to drop below the $100/ton price at commercial scales. The process is entirely electrically powered and does not require any heat. It involves three stages: solution-based air-contacting, an electrochemical separation method and a depressurisation step within a release chamber to produce CO_2 gas at ambient temperature. So the method is based on physical processes instead of chemical reactions, although undisclosed specific chemicals and substances are needed in the separation step. The chemicals can be recycled and, on a laboratory scale, no waste is generated during operation.

Mission Zero is involved in various projects including the Project Hajar in the Al Hajar mountains of Oman for a minimum 1,000 ton/year project combining its electrochemical technology and 44.01's peridotite mineralization approach. This project is a $1 million Milestone Xprize Award winner. For the company 44.01 see below.

---

[37] https://www.holygrail.ai/.

[38] https://www.missionzero.tech/.

In July 2022 Mission Zero Technologies won a £3 million contract from the UK government (part of a £54 million investment in DAC) to build a DAC pilot plant employing their technology. The plant will capture 120 ton of CO_2 per year using electricity from the grid. The project is carried out in partnership with Optimus,[39] a Scottish engineering and consulting company, and O.C.O. Technology.[40] The latter company was formed in 2010 and treats waste material by letting it react with CO_2 and water. Very little energy is required for the process, which relies upon the fact that waste material is often naturally reactive with carbon dioxide in the presence of water. Water acts as a solvent, allowing CO_2 and calcium atoms to react. If the conditions are carefully controlled, this process can be accelerated, taking place in minutes rather than years and resulting in the carbon dioxide being permanently converted into calcium carbonate. The idea is to supply the CO_2 captured by Mission Zero Technology to O.C.O. for this waste treatment process.[41] This makes it unnecessary to build pipelines to transport the CO_2 to a geological deposit, but it is doubtful that the right scale can be achieved.

Sustaera,[42] based in Cary, North Carolina, was also selected for a $1 million Xprize Milestone award. It is a spin-out of Susteon, which focuses on hydrogen production and CO_2 capture and utilization.[43] Like most of the other approaches, it claims to have developed a novel technology that allows the removal, replacement and reuse of carbon at a massive scale. It plans to remove a cumulative amount of 500 Mt of CO_2 by 2040. To remove 1 Mt per year less than 100 acres of land are needed and, when applied at scale, costs will be less than $100 per ton. Another factor differentiating the Sustaera's DAC system from other approaches is the use of natural minerals repurposed as CO_2 capture sorbents and the carbon-free energy powering the DAC systems. The start-up uses a modular component design which allows it to rapidly scale up the new technology. The sorbent can be regenerated with renewable electricity at relatively low temperatures (85°C). All this does not sound very spectacular and resembles very much other sorbent or solvent based CO_2-capture approaches. I have not been able to find any details about their procedure that give any confidence that this can easily be scaled up, but I suppose that the Musk Foundation had more information before they awarded a $1 million prize to this approach.

Skytree,[44] with offices in Amsterdam, Netherlands, and Somerville, MA, USA, is a spinout of the European Space Agency (ESA) and builds on research carried out by ESA on the Advanced Closed Loop System to recycle carbon dioxide on

---

[39] https://optimusaberdeen.com/.

[40] https://oco.co.uk/.

[41] More details can be found in Mission Zero Technologies: Phase 1 report: https://assets.publishing. service.gov.uk/government/uploads/system/uploads/attachment_data/file/1075300/mission-zero-technologies-d4.8-project-phase-1-report.pdf.

[42] https://www.sustaera.com/.

[43] https://susteon.com/, Susteon, an acronym made up from Sustainability for Aeons and launched in 2018, is developing a novel DAC technology that (i) uses alkali-based sorbent materials, (ii) directly integrates renewable sources of electricity, and (iii) does not require new manufacturing or special material supply chains for scale-up.

[44] https://skytree.eu/.

the International Space Station into oxygen. It has developed a tool for onsite CO_2 capture from the air.

In the ESA system carbon dioxide is trapped from the air as it passes through small porous plastic beads made from a unique amine developed by ESA. The beads soak up the CO_2. Steam is then used to extract the carbon dioxide and process it to create methane and water. Electrolysis splits the water back into oxygen while the methane is vented into space.

Skytree favours decentralized DAC, i.e., capture the carbon dioxide where you need it. However, the trouble with carbon dioxide is that we need relatively little of it compared to the amounts that need to be scrubbed out of the atmosphere. Each Skytree unit is effectively working in the same way as a tree would suck CO_2 out of the air, only faster and more effectively based on the ESA system described above, but contrary to a tree it does not store it in wood. So, why should I want such a CO_2-capturing unit on the building I live or work in? To whom should I deliver the CO_2 I have captured and how? Not all of us are running a greenhouse where the CO_2 might come in useful. Most of the captured CO_2 must be tucked away safely deep underground, never to be seen again.

I have not been able to find out either how Skytree manages to meet the challenges of cost and in particular scale through their approach. ESA has apparently evaluated a number of sorbents that are useful in capturing CO_2 from air. Where and how Skytree's approach is better than any of its competitors that use amine or alkali scrubbing procedures is not clear.

Carbon mineralization approaches

Mineralizing CO_2 is a well known natural process. As discussed in some detail in Chapter 9, it involves the formation of solid carbonate minerals,[45] like calcium carbonate or magnesium carbonate, through the reaction of CO_2 with rocks rich in calcium and magnesium. In general, since the carbonates formed are stable, it represents the safest storage mechanism as regards leakage. A number of start-ups have sprung up that try to utilize and speed up such processes (enhanced weathering). They differ in the 'vehicle' they use to catch the carbon dioxide in the air and turn it into the stable carbonate. Some of these companies actually do not intend to themselves catch the CO_2, but offer a service to store carbon dioxide in stable carbonates. Let's discuss a few of them.

Calcite Carbon Removal,[46] a technology developed by 8 Rivers Capital[47] which calls itself a full-service Net Zero solutions provider, is named for the calcium carbonate crystal called calcite. It was also awarded $1 million in the Xprize competition and is part of the *Frontier* portfolio. Calcite exists since 2019 and has developed its technology in collaboration with MIT, receiving also some US government funding. Its process captures CO_2 directly from air and sequesters

---

[45]   A carbonate mineral is a mineral containing the carbonate ion CO_3^{2-}. There are a huge number of them, for a list see the *Wikipedia* Carbonate mineral page.

[46]   https://8rivers.com/calcite/.

[47]   https://8rivers.com/; 8 Rivers has also developed the Allam-Fetvedt Cycle described in Chapter 8.

it underground. It has the aim, common as we have seen to many carbon removal approaches, to remove over a billion of tons of CO_2 for less than $100 per ton.

The Calcite process passes ordinary air containing about 420 ppm of CO_2 across calcium hydroxide, $Ca(OH)_2$, traditionally called slaked lime, in a large warehouse, whereby CO_2 from the air is absorbed into calcium carbonate crystals. The process is similar to how concrete sidewalks dry and absorb carbon in the process. CO_2-depleted air with < 315 ppm CO_2 is then returned to the atmosphere. The created calcium carbonate is fed into a kiln to regenerate the calcium hydroxide to be reused for capturing CO_2, and the CO_2 is injected underground for permanent storage. The procedure is very similar to the one employed by the company **Heirloom Carbon**.[48] The innovative Calcite process, it is claimed, enables rapid carbon uptake at large scale and low cost, through the use of simple equipment, abundant feedstocks, and optimized chemistry. From the short description provided, it does however not seem that innovative, nor very spectacular, and it is far from clear, as is also the case for Heirloom, that the procedure is fast enough to be able to capture and eventually sequester the billions of tons of CO_2 required. The crux in both methods is that enough calcium hydroxide must be made available at low cost. In Heirloom's procedure the CO_2 is extracted from the carbonate in electric kilns, powered by renewable electricity, where the calcium carbonate decomposes into calcium oxide (CaO), commonly known as quicklime or burnt lime, and CO_2 at temperatures near 900°C:

$$CaCO_3 \rightarrow CaO + CO_2. \tag{1}$$

The pure CO_2 stream is diverted to be stored, while the quicklime is slaked with water (hydration). Quicklime is unstable and, when cooled, spontaneously reacts with CO_2 from the air until, after enough time, it will be completely converted back to calcium carbonate unless hydrated. The hydration process is exothermic and presents a suitable opportunity to recover some of the heat spent in reaction (1):

$$CaO + H_2O \rightarrow Ca(OH)_2. \tag{2}$$

The resulting calcium hydroxide ($Ca(OH)_2$) is moved through a passive air contactor, where it soaks up CO_2 from the ambient air through the process:

$$Ca(OH)_2 + CO_2 \rightarrow CaCO_3 + H_2O, \tag{3}$$

again recovering part of the heat spent in reaction (1), making the combination of the three reactions above energy neutral. These three chemical reactions are typical for any carbon mineralization process.

Once saturated, the carbonate is cycled back through the kilns to extract the CO_2 and the process is repeated. In Heirloom's procedure, to carbonate the material (reaction (3)), thin layers of $Ca(OH)_2$ powder are spread out on large area trays. These trays are stacked into vertical, tiered structures to minimize land area requirements,

---

[48] https://www.heirloomcarbon.com/.

all this while enabling maximum air-to-sorbent contact. The use of tiered structures also renders the process highly modular.

The result, Heirloom claims, is a procedure which requires relatively low capital and operating expenditure, which at scale will enable a cost well below \$100 per ton of CO_2. On its website it simply states that they will be "removing one billion tons of carbon dioxide by 2035". But, achieving such scale might well be the main issue, as the process consists of three rather slow chemical reactions and quite a lot of handling, resulting in the entire procedure probably taking up considerable time, even if the carbonation rate (reaction (3)) is speeded up. The fastest rate consistently achieved by the process they say is equivalent to 630g of CO_2 removed per 1 m^2 of exposed contactor area every 2.5 days (McQueen et al. 2022). This implies that for capturing one ton of CO_2/year more than 10 m^2 of contactor area are needed. If this is indeed the case, a staggeringly huge number of modules will have to be deployed for capturing sizeable volumes of CO_2.

Parallel Carbon[49] also presents direct carbon capture based on mineral carbonation paired with an electrochemical process, similar to some of the approaches discussed above, e.g., Mission Zero Technologies and in particular Calcite and Heirloom Carbon.

Calcium hydroxide ($Ca(OH)_2$) captures CO_2 through ambient carbonation to create solid calcium carbonate ($CaCO_3$), exactly as reaction (3) above. The CO_2 is then liberated from the carbonates to regenerate the hydroxide using aqueous, acid-base reactions facilitated by electrolysis. Fresh $Ca(OH)_2$ is loaded onto trays and racked. Thin granule layers passively carbonate over hours/days. Fully-reacted $CaCO_3$ is recovered and the cycle closes.

The CO_2 will be utilized to create infrastructure materials from highly alkaline materials, by-products and wastes, while some of it will be stored via underground injection. Parallel Carbon rather optimistically envisages to store more than 2 Gt of CO_2 from the air by 2040. It is hard to see whether their procedure is better or worse than the ones of Calcite and Heirloom, as the details of the various procedures are still shrouded in mystery. It is waiting for the first pilot projects to bring some more clarity.

For the companies discussed above it is essential that they can cost effectively (re)produce their hydroxide capture medium. We will now pay some attention to companies that try to find this capture medium in nature. In the end their aim is the same, namely locking CO_2 from the air in stable carbon carbonates.

The very young company **Lithos Carbon**,[50] founded in March 2022, intends to apply enhanced weathering by spreading basalt powder on croplands where it will react with rainwater, capturing atmospheric CO_2 and converting it into dissolved bicarbonate, HCO_3^-, and in addition releasing nutrients to the soil. The dissolved bicarbonate is washed out to rivers and streams and further to the ocean. Eventually, dissolved bicarbonate reacts to form calcium carbonate minerals that are permanently deposited on the ocean floor. Three tons of basalt can capture 1 ton of CO_2, but the

---

[49] https://www.parallelcarbon.com/.
[50] https://www.lithoscarbon.com/.

basalt, being rich in nutrients such as phosphorus, magnesium and calcium, also acts as fertilizer and will increase crop yields and reduce the need for high-cost artificial fertilizer. However, a lot of crushed basalt, 3 gigatons, and great areas of land are needed to capture 1 gigaton of CO_2. The study referred to in Chapter 9 (Strefler et al. 2018) assumed that at most 150 tons of basalt powder can be spread on 1 ha of land (15 kg per square metre), capturing 50 tons of CO_2, implying that 0.2 million km^2 of land is needed for capturing 1 Gt of CO_2, about 2% of the total area of the US. Lithos intends to work with farmers to optimize both crop yield and carbon capture, to regenerate top soil and add nutrients, and to monitor the leakage of the bicarbonate into the river network and from there to the ocean. Mining, crushing and transporting the basalt will result in CO_2 emissions undoing some of the gains made by this method of enhanced weathering. There may also be potential environmental and human health risks depending on the crushed materials used and where they are applied (Choi et al. 2021).

The effect of enhanced weathering on agricultural output will be one of the main factors determining the success of the method. Thus, the potential effects of enhanced rock weathering on agricultural productivity need more research, not only regarding enhanced weathering as a CO_2 removal method but also regarding alternative fertilization methods in the face of dwindling phosphate rock resources.

In spite of being less than one year old, Lithos is already part of the *Frontier* portfolio, a sign that some people apparently have a great trust in its method.

The Canadian company **Carbin Minerals**,[51] affiliated with the University of British Columbia and although just founded in September 2021, already earned $1 million as an Xprize milestone award winner and inclusion in *Frontier's* portfolio of promising projects. Like several other recent start-ups (e.g., Carbfix and Travertine), it focuses on unlocking the potential of mine waste for removing carbon dioxide from the air through carbon mineralization, enabling mines to become massive carbon sinks, while producing the metals needed to drive the clean energy transition.

They claim to have techniques that increase the rather slow natural reaction rates by a factor of three or five so that a single site can sequester hundreds of thousands of tons of CO_2 per year. The company has committed itself to removing 1,000 tons of atmospheric CO_2 by 2024 and demonstrating a pathway to removing mega- and gigatons in the future.

Rich stores of nickel and cobalt are found in ultramafic[52] rocks. The mining and grinding of these rocks to powder to recover these metals greatly enhances their CO_2 sequestering potential. According to an initial test at a mine in Northern British Columbia, they are able to absorb roughly 0.72 ton of CO_2 per ton of nickel mined, offsetting nearly all CO_2 expected to be emitted per ton of nickel. The test also showed that only about 10% of the magnesium hydroxide mineral in the ultramafic rocks responsible for sponging atmospheric CO_2 was consumed during the test,

---

[51] https://carbinminerals.ca/.
[52] See Glossary.

indicating that the tailings would absorb more of the greenhouse gas with longer exposure to the air.[53]

Peridotite is a dense, coarse-grained igneous rock, meaning that it is formed through the cooling and solidification of magma or lava and consisting mostly of the silicate minerals olivine and pyroxene.[54] It reacts with CO_2 and water to produce calcite, the common carbonate mineral $CaCO_3$, we have already encountered above. In fact this process has occurred at enormous scales throughout history, as witnessed by large streaks of calcite piercing peridotite deposits. Peridotite is normally found miles below sea level, but on the easternmost tip of the Arabian peninsula, specifically on the northern coast of Oman, tectonic action has raised hundreds of square miles of the stuff to the surface, which theoretically could hold billions of tons of CO_2. That is where the company **44.01**,[55] an Omani start-up, saw an opportunity.

The technology of 44.01 intends to speed up the natural mineralization process by taking captured CO_2 and accelerating its reaction with peridotite underground (as in enhanced weathering). The company does essentially the same as Carbfix and also works with Climeworks as its Direct Air Capture technology partner, so the company 44.01 is actually a little out of place in this chapter as it does not itself capture any CO_2 from the air, at least not yet, but only provides a safe method for storing CO_2 in a geological deposit on the Earth's surface.

The captured CO_2 is dissolved in water to create CO_2 charged water. This is then injected into peridotite rock formations underground where it mineralizes. The mineralization process normally takes years as CO_2 and water interact with the rock. It is a natural process that does not need any energy to be applied. By injecting water with a higher CO_2 content than you would get in the atmosphere the process is speeded up.

In the second quarter of 2023 44.01 will start injecting CO_2 on a commercial scale and the aim is to have mineralized one gigaton of CO_2 by 2040. A source of CO_2 is however needed for this. As things stand at present, Climeworks will not easily be able to capture 1 Gt of CO_2 from the air any time soon, nor will any of the other DAC companies. That is why in 2021 44.01 announced plans to build its own DAC plant in Khazaen Economic City in Barka, Oman, solely powered by renewable energy.

Another company intending to capture and store CO_2 by carbon mineralization is **Paebbl**,[56] founded in late 2021 and based in Stockholm, Helsinki and Rotterdam. Actually, like 44.01, they only store CO_2 captured in some way by others, e.g., by Climeworks. Olivine rock is grounded and fed into Paebbl's mineralization unit, together with captured CO_2, and out comes CO_2-storing mineral filler, a carbon-negative raw material. They claim to have found a way to accelerate the natural process, which takes roughly 100 years, by a factor of more than a million, i.e., to one hour, with minimal energy use. At the end of 2023, their pilot plant in Rotterdam

---

[53] https://www.metaltechnews.com/story/2022/04/27/mining-tech/carbon-capture-mining-tech-wins-xprize/921.html.

[54] Silicate minerals are minerals containing silicon (Si) and oxygen atoms. Dunite, which we met in Chapter 9, is a type of peridotite rock.

[55] https://4401.earth/; the number 44.01 happens to be the molecular mass of CO_2 (44.01 g/mol).

[56] https://paebbl.com/.

is scheduled to make per day about 300 kg of this filler material, which is intended to be scaled up by a factor of 10 every 12 to 24 months.

Ocean-based approaches

A few companies use an indirect route, via the sea, to draw carbon dioxide from the sea and through this enhance the ocean's power to draw CO_2 from the atmosphere. Drawing CO_2 from the ocean is easier than from air as the concentration in the top layer of the ocean is 150 times larger.

The Project **Vesta**,[57] born out of the climate think-tank, turned climate do-tank, Climitigation[58] and founded in 2019 in San Francisco, intends to draw carbon from the oceans by adding carbon-removing sand (ground olivine) to the water. Instead of Direct Air Capture Vesta uses the name Coastal Carbon Capture. Its goal is a cost figure as low as $10 per captured ton of CO_2 (Temple 2020). According to Vesta's website, the process captures 20 times more carbon dioxide than released in the extraction and transportation processes of the olivine. If olivine is deployed on just 0.1–0.25% of global shelf seas, 1 Gt of CO_2 could be removed from the atmosphere. One ton of olivine sequesters 1.25 tons of carbon. A volume of 7 km^3 spread over the world's 2% most tidally active areas would offset all of humanity's annual CO_2 output. Whatever the merits of this approach, the thinking is on the right scale of billions of tons of CO_2. Any other scale is insufficient.

A pilot project is currently underway in Puerto Plata, Dominican Republic, to determine the approach's ecological safety, such as the risks from the release of trace metals like nickel contained in the ground olivine.

Captura,[59] founded in 2021, is also a winner of a $1 million Xprize award and part of the *Frontier* portfolio of promising future CO_2 suppliers. The company is led by a former employee of Carbon Engineering where he was CEO of DAC, and employs established electrochemical concepts to indirectly remove CO_2 from the atmosphere via direct ocean capture, using only seawater and renewable energy, producing no by-products.

The company claims to be developing cost competitive CO_2 capture and sequestration technology for extracting CO_2 from oceanwater that is scalable to Mt/year-Gt/year. The technology has been developed at Caltech. A plant is built in the ocean, preferable at a location where captured CO_2 can be stored in a geological deposit, and a flow of ocean water is passed through it. This flow of water is treated to remove its carbon dioxide content via electrodialysis (an electrochemical process to separate dissolved products from a solution by means of an electric potential difference and semipermeable membranes) and subsequently returned to the ocean.

The process starts by purifying part, less than 1%, of the ocean water drawn into the plant, turning it into pure salt water. Using renewable energy, this pure salt water is split by electrodialysis into an acid and an alkali base, after which the acidic part is added to the original flow of ocean water through the plant, triggering a chemical

---

[57] https://www.vesta.earth/.

[58] https://climitigation.org/.

[59] https://capturacorp.com/

process that draws the CO_2 out. This CO_2 is captured as a purified stream, ready for sequestration or utilization. The flow of acidic, decarbonized ocean water left in the system is recombined with the alkaline base to neutralize the ocean water flow and is returned to the ocean. This decarbonized water sits in the top layer of the ocean and will draw down an equivalent quantity of CO_2 from the atmosphere.

The advantages that immediately spring to mind and make this a promising approach are that no sorbents are needed, there are no by-products and no capital costs for air contactors, that sequestration is possible on site deep under the ocean, and that energy costs are lower as seawater contains per volume about 150 times more CO_2 than air. Captura claims that costs will remain below \$100 per ton of CO_2 and that the process is easily scalable as the oceans are vast. The technology has been fully demonstrated at Caltech with imported ocean water. In summer 2022 Captura launched its first ocean-based, stand-alone pilot facility in Newport Beach, CA.

Ebb Carbon[60] wants to accelerate the ocean's ability to remove CO_2 from the air, while at the same time making the water less acidic. When CO_2 enters seawater from the air, it goes through a series of chemical reactions, producing a number of hydrogen ions—which lower the alkalinity of the water—and bicarbonate, HCO_3^-, which stays in the water. About one third of the CO_2 generated from anthropogenic sources is absorbed by the oceans, resulting in the ocean water slowly having become less alkaline or more acidic.

With a team of researchers at Stony Brook University, Ebb Carbon runs a project called SEA MATE (Safe Elevation of Alkalinity for the Mitigation of Acidification Through Electrochemistry). In the SEA MATE process the salt (NaCl) is taken out of the sea water and separated into hydrogen chloride, HCl, and sodium hydroxide, NaOH. The HCl is sold for other uses while the NaOH is controllably returned to the ocean, resulting in an increase in ocean alkalinity and in the storage capacity of dissolved carbon. Additional atmospheric CO_2 can be absorbed and safely stored as oceanic bicarbonate.

The net effect of SEA MATE is the reversal of ocean acidification along with the net removal of carbon dioxide from the atmosphere. By harnessing the ocean's natural process for storing CO_2, the system is less energy-intensive and less costly at scale than comparable forms of carbon removal. Energy requirements are estimated at $1,500 - 2,000$ kWh of electricity per ton of CO_2, to be supplied by an offshore wind turbine.

One of the problems to worry about is that the HCl removed from the ocean must be disposed off in some way. The market for this stuff is not large enough to scale SEA MATE to the gigaton level. New markets or applications of HCl must first be found. Moreover the HCl produced by SEA MATE is very dilute (3–6% by weight) while the stuff sold on the market has 32% concentration by weight.

Biomass-based approaches

Instead of installing mechanical capturing and separation equipment to draw carbon dioxide directly from the atmosphere, many approaches exploit the fact that plants

---

[60]   https://www.ebbcarbon.com/

do this already via photosynthesis. A large portion of the CO_2 captured by plants is re-released when plants decompose (or are burned through deforestation and wildfires). Interrupting this cycle to store the biomass in a durable format can act as a permanent form of carbon removal. Once the CO_2 has been turned into plant matter, biomass conversion technology can be used to sequester the carbon. Several companies have devised methods to lock up large amounts of carbon in biomass or biochar. Biochar has been discussed in Chapter 9. There are many companies in the biochar business. The ones on the *Puro.Earth*[61] register sell carbon credits for prices varying between €100 and €535 per ton of CO_2. In Chapter 11 we also already met two companies processing biomass (Global Algae and Seafields) whose procedures are a mix of carbon utilization and storage. We will discuss a few others here.

InterEarth,[62] a Western Australian start-up, proposes gigaton-scale carbon removal at less than $100 per ton. Their approach focuses on the use of degraded land with groundwater five to ten times saltier than seawater. They plan to coppice trees (cut in a way enabling regrowth) they plant on plots, then bury (or otherwise permanently store) the wood, and repeat the process. Transport costs are minimized by growing trees right next to burial sites, while the land used is almost incapable of supporting life. By reinstating fast-growing native species that are able to survive in harsh conditions, InterEarth hopes to restore biodiversity and potentially revitalize the soil over the course of multiple planting cycles. The main barriers are transport costs, confirming that the biomass does not decompose, and that burying makes sense over other use cases. Such an approach can provide permanent removal: if the environment allows for only anaerobic decomposition (by bacteria, not fungi), the bacteria are incapable of processing and degrading the wood. Therefore, the key is avoiding oxygen exposure, with salt (found in InterEarth's target environment) enhancing the process, although achieving scale will be a challenge.

The American company **Atmocean,**[63] incorporated in 2006, researches ideas related to those of Lovelock and Rapley discussed in Chapter 10, to transport carbon dioxide from the atmosphere into the deep ocean. Atmocean proposes to use wave action for both the generation of clean water and carbon sequestration, while their OSCAR project, or Ocean Surface CArbon Relocation, is an ocean pump designed to send carbon rich surface water into deeper water where it is sequestered for decades to centuries. Initially, the idea was to pump deep colder water towards the surface in an attempt to reduce the temperature of warm surface waters with the ultimate goal of reducing the strength of hurricanes. Later the system was redesigned as an up-welling system to bring nutrient rich water up into the photosynthetic zone, thereby promoting algae growth, which is similar to the proposal by Lovelock and Rapley.

The US company **PHYCO$_2$ LLC,** founded in 2008, is using flue gas from the T.B. Simon gas-fired power plant of Michigan State University to grow algae via bio-sequestration in a bioreactor filled with algae, water and nutrients, and light provided by an LED system. The grown algae will be converted via a chemical

---

[61] https://puro.earth/CORC-co2-removal-certificate/.

[62] https://www.inter.earth/.

[63] https://atmocean.com/.

cascade process into several chemicals, including an amine that will be used for the bulk of the CO_2 capture process at the power plant.[64]

Brilliant Planet[65] is a British company that wants to store carbon by growing vast quantities of microalgae (diatoms) in open-air pond-based systems on coastal desert land using seawater as feedstock and source of algal microorganisms. Their pilot project is in Morocco. The process itself is essentially solar-powered—because the algae are effectively powered by the sun—but also needs pumps to move seawater around. It starts with sea water intake from fairly deep below the surface at a depth of 50 to about 200 m where light intensity is just 0.1% to 1% of that at the surface. No freshwater is used. In the ponds the microalgae in the seawater start to bloom, drawing CO_2 from the seawater, which is partly replenished by the water drawing CO_2 from the atmosphere. The algae grow very quickly and very large volumes of seawater have to be moved around. The estimate is that 22,000 tons of seawater are needed to capture one ton of CO_2. So, trillions of tons of seawater have to be pumped around to achieve the necessary gigaton scale of CO_2 capture. That makes it very important for the pumping system to be extremely energy efficient. Gravity feeds down through most of the system from one pond into the next, but for getting the water out of the sea pumps are needed. Once the pond is full the algae are separated from the seawater, which is returned to the sea where it will quickly draw further CO_2 from the atmosphere. The biomass is dried very quickly in the desert and the dry salty biomass is buried at a place that will stay dry. One hectare of land is needed for the ponds to capture 100 tons of CO_2 per year. These 100 tons can be compressed into about 62 tons of biomass. A thousand hectare farm will capture 100,000 tons of biomass which after compression can be buried in two hectares of land. The dryness and saltiness guarantee that the biomass, if left undisturbed, remains stable for centuries. This is obviously a weak point of the approach as the company can in no way guarantee that the land will not be disturbed in future, let's say in two to three hundred years, even if it currently concerns virtually worthless desert land.

Brilliant Planet expects that the cost per ton of CO_2 captured, which at present is about $200, can be brought down to below $100 when operating at scale. To achieve that scale, e.g., 1 Gt, huge areas of land will be needed.

Climate Robotics,[66] founded in 2020, a venture-backed start-up based in Houston, Texas, develops automated agricultural machines designed to fight global climate change with advanced robotics and artificial intelligence. Its robots are designed to efficiently generate biochar to sequester carbon and improve soils, starting with urban land. The company's machines identify degraded or blighted tracts of land in urban and peri-urban areas, and then convert the biomass waste on-site efficiently into biochar, enabling users to dispose of waste organically to boost nutrient and water retention and help remediate contaminants while permanently sequestering atmospheric carbon. The company's philosophy is that "Earth's soils

---

[64] https://phyco2.us/; http://www.biodieselmagazine.com/articles/2516491/phyco2-algae-pilot-program-with-msu-reaches-new-milestones.

[65] https://www.brilliantplanet.com/.

[66] https://www.climaterobotics.com/

contain more than 14 times more carbon dioxide than the atmosphere. We can safely store much more CO_2 in the ground, buying time to develop carbon free energy. Using existing technologies, it is possible to sequester all of man's carbon emissions deep in the ground for thousands of years."

Storing carbon dioxide in carbonated materials

Finally we have to mention methods to store carbon in carbonated building materials. They are primarily ways to store carbon, not to capture it, although there is of course overlap. No new ways to capture CO_2 are involved in this. That is why we have discussed this in Chapter 11. The use of wood in furniture and other applications is an example, but in them the carbon is only stored for a relatively short time. Such carbonated building materials also concern concrete-like building elements manufactured for instance from steel slag (waste material from the steel industry) instead of traditional cement, and the approach to mineralize CO_2 in concrete pioneered by the Swiss company Neustark (see Chapter 11). Basically, it is CO_2 negative concrete that removes more CO_2 than its production emits. During the hardening phase CO_2 is chemically bound and mineralized permanently into the building element. Benefits of the method include easy measurability and storage of CO_2 for good. It can give a useful contribution as do other bio-based construction materials but to achieve the scale needed will be very hard.

Concluding remarks

A variety of companies involved in Direct Carbon Removal were passed in review in the above. There are many more, but I hope that the selection I made is to some extent representative and illustrative. It is gratifying to see that so many approaches are going to be tested in the near future. Particularly interesting are those that make use of existing infrastructure that processes large volumes of air but as indicated they may have difficulty achieving the right scale and, if successful, there is the added difficulty of removing the carbon dioxide to its eventual storage location. Of the companies reviewed above this applies to CO2Rail, Soletair Power and Noya.

All approaches claim cost effectiveness and scalability, where so far only a single company (Carbon Engineering) has indeed built a facility that can capture per year 1 Mt of CO_2 from the air. Most approaches will have a hard time achieving the required scale. This especially applies to those working with solvents or sorbents (Climeworks, Carbon Engineering, Global Thermostat, Carbon Collect, Carbyon, Sustaera, Skytree) that must be heated to release the captured carbon, or related electrochemical methods (Holy Grail, Mission Zero). However, they all might make a useful contribution in a game where we must seize every opportunity that presents itself in order to lower the CO_2 concentration in the air or, as a minimum, to prevent it from rising further and reach catastrophic levels in a few decades from today. If they all fulfil their promises we will have nothing to worry about as by the middle of the century more CO_2 will be taken out of the air than goes in. The most promising as regards scalability seem to be the various mineralization (enhanced weathering) approaches (Heirloom Carbon, Calcite, Parallel Carbon, 44.01, Vesta) and the algae farming approaches, including those (Global Algae, Seafields) discussed in Chapter 11. Especially Global Algae is very interesting as it could also solve other problems at the same time.

Outsiders are Captura and Ebb Carbon, by taking the detour to tackle the CO_2 concentration in the air by emptying the top layer of the oceans of CO_2, and Neustark (also discussed in Chapter 11) that concentrates on the hard-to-abate concrete sector. If easily scalable, the Captura and Ebb Carbon methods are very promising, also from a cost perspective. The same applies to the biomass-based approaches.

Research is continuing and will undoubtedly result in improvements of existing techniques and in (entirely) new approaches. This is for instance borne out by a recent paper by scientists at MIT who claim to have developed a new and cheap method for removing carbon dioxide from the ocean (Kim et al. 2023).

CHAPTER 14

Costs and Economics

Introduction

In this chapter we will discuss another important question concerning carbon capture, perhaps the second most important one after whether carbon capture is necessary and technically possible on a large scale. The latter we have discussed in the preceding chapters. Now it is time to look in some detail at costs, although I want to stress that in the face of an existential threat, as climate change undoubtedly is, we should not be too preoccupied with economics.

Estimates for current and future costs of CO_2 capture and storage have considerable uncertainties. Studies of capture and compression costs for power stations completed ten years ago averaged around \$95 per ton of CO_2. Comparable studies completed in 2018/2019 estimated these costs could fall to approximately \$50 per ton of CO_2 from 2025.

As shown in Chapters 12 and 13 the experience with operational carbon-capture projects is rather limited. Very few projects are up and running and those that are running are generally small scale. Several attempts have been made to set up large-scale capture ventures, but only a few have met with success for various reasons. No fossil-fuel power plant equipped with CCS has operated for any length of time, possibly except for the Boundary Dam power plant in Canada, which is the first commercial-scale power plant with CCS in the world and as such one of the flagship CCS initiatives. Other large-scale successful projects are the CCS plants in the natural gas processing sector at the Norwegian Sleipner and Snøhvit gas fields. For the Sleipner field we quoted rather low operating costs of \$17 per ton in Chapters 6 and 12. But apart from that number, which is not representative for the costs of carbon capture from power plants or for scrubbing CO_2 out of the air, there are very few hard and reliable data from practical experience that can give an insight in the cost of carbon capture and we must mainly be satisfied with model calculations.

Let us start with the 2005 IPCC Report on Carbon Dioxide Capture and Storage already mentioned in Chapter 4. In chapter 8 *Cost and Economic Potential* the report discusses carbon capture from point sources, like power stations, and states in this respect: "The cost of employing a full CCS system for electricity generation from a fossil-fired power plant is dominated by the cost of capture. The application of capture technology would add about 1.8 to 3.4 dollar cent per kWh to the cost of electricity from a pulverized coal power plant, 0.9 to 2.2 cent per kWh to the cost of electricity from an IGCC coal power plant, and 1.2 to 2.4 cent per kWh from a natural gas combined-cycle power plant. Transport and storage costs would add between –1 and 1 dollar cent per kWh to this range for coal plants, and about half as much for gas plants. The negative costs are associated with assumed offsetting revenues from CO_2 storage in enhanced oil recovery (EOR) or enhanced coal bed methane (ECBM) projects. Typical costs for transportation and geological storage from coal plants would range from 0.05–0.6 dollar cent per kWh." Reading this, one remains baffled by the fact that, in view of the alleged 'existential threat' from climate change and global warming, CCS has not been forcefully pushed by the IPCC (and others) in the last fifteen years. Although the increase in price is nontrivial, as a kWh of electricity from, e.g., coal only costs a few dollar cents, and will of course hurt competition and have an effect on growth, one would have thought that is not the point at issue here. Energy from fossil fuels is now more expensive than from renewable sources,[1] but the use of fossil fuels is nevertheless not going away soon, even if renewable energy becomes dirt cheap. Renewables alone cannot yet ensure reliable uninterrupted power supply anywhere, let alone globally. If they could, the transition would have been completed by now. Transitioning to a carbon-free economy will take time. In the meantime we must mitigate the negative consequences from using fossil fuels. Sacrificing a little growth by paying a few dollar cents more for averting an 'existential threat' is for sure a small price to pay. An average household in the Netherlands, for instance, consumes 2,740 kWh of electricity[2] per year for domestic purposes, so the IPCC analysis implies that an *existential* threat, not just any ordinary threat but a threat to existence itself, comparable to what Ukraine is experiencing currently from the Russian onslaught, could be averted by paying an annual amount of some $80, about 10% of total electricity costs charged by a typical utility company, including all taxes and surcharges. Such amount is trivial in view of the threat. Why has this not happened? The IPCC adds that, without an explicit policy that substantially limits greenhouse-gas emissions to the atmosphere, CCS systems are unlikely to be deployed on a large scale. Such policy could be some sort of carbon tax coupled to a cap as in the EU ETS, which can act as a powerful incentive for emitters to install a CCS installation to avoid having to pay the tax or exceeding their allocated emissions allowance. The IPCC then continues: "With greenhouse gas emission limits imposed, many integrated assessment analyses indicate that CCS systems will be *competitive* with other large-scale mitigation options, such as nuclear power and renewable energy technologies. Most energy and economic modelling done to date suggests that the deployment of CCS systems starts to be significant when carbon prices begin to reach approximately $25–30/tCO2. They foresee the large-scale

---

[1]   The Dumat Al-Jandal wind power park in Saudi Arabia was connected to the Saudi grid in August 2021 and supplies electricity at $0.0199/kWh, a stupendously low price that no fossil-fuel plant can beat. The park consists of 99 4.2 MW wind turbines, so has a total nameplate capacity of slightly over 400 MW. This price does not include any costs for backup to make sure that the grid will always be able to provide energy, e.g., when there is no or insufficient wind.

[2]   https://www.nibud.nl/consumenten/energie-en-water/#:~:text=Elektriciteitsverbruik,is%202.730%20 kWh%20per%20jaar; the average rate for electricity paid in the Netherland is about 25 eurocent per kWh (price July 2021), with taxes and surcharges accounting for the bulk of this price.

deployment of CCS systems within a few decades from the start of any significant regime for mitigating global warming." The choice of the word *competitive* in the above citation is telling and rather odd in the face of an existential threat. Ordinary competition mechanisms between the various options are hardly suitable, one would think, to decide the issue. It shows that, at any rate at that time, the IPCC was still mainly thinking in economic terms, i.e., what does it cost instead of what do we have to do to stave off disaster. But, the carbon price range mentioned above has already been achieved for some time now in the EU ETS and with the cap being continuously lowered we should soon see an upsurge in CCS systems if the above prediction has any truth in it.

This IPCC analysis dates from 2005 and concludes that costs are considerable but far from high. Nevertheless, when discussing the costs of CCS it is generally assumed from the outset that the costs will be high. There is no shortage of commentators who through the years, without much investigation into the matter, have decried CCS as too expensive and hence not a viable tool for mitigating climate change. The question 'too expensive for what' or 'compared to what' is hardly ever addressed. Now that an increasing number of respectable organizations like the IPCC and the IEA argue that carbon capture is indispensable for curbing CO_2 emissions and achieving net zero, this attitude is slowly starting to change. The IEA, for instance, has stated that "the idea that CCUS is "high cost" ignores the bigger picture" and that "achieving net-zero goals will be virtually impossible without CCUS."[3] The preceding chapters have made clear, I hope, what this bigger picture is and that carbon capture is a necessary tool. If not already necessary now, Direct Air Capture, which makes no sense without CCS, will undoubtedly become an absolute necessity as soon as CO_2 concentration levels in the atmosphere hit 500 ppm or more, a value which at present emission rates will be reached around mid century. In this connection, the Global CCS Institute showed already in 2017 (Irlam 2017) that the cost of CCS on several industrial applications is far below what many would expect given the repeated claims that CCS is 'too expensive'.

In many cases in which carbon capture or, in particular, CCS is rejected as too expensive, this is done by coupling the cost question to the so-called social cost of carbon. This, as we have seen in Chapter 6, is at best a shady concept and can be manipulated at will by fiddling with the discount rate.[4] Such purely theoretical calculations have become largely irrelevant now that carbon prices in market-oriented systems like the EU ETS have risen from around €35 in March 2021 to over €85 per ton of CO_2 at the end of the year and now in February 2023 to more than €100,[5] as can be seen in Fig. 6.1, lifting the 'market' cost of carbon far above most values quoted for the 'social' cost of carbon.

The EU ETS price has the practical consequence that an installation that emits more than its allocated (permitted) carbon dioxide emissions will have to buy extra

---

[3] https://www.iea.org/commentaries/is-carbon-capture-too-expensive.

[4] The Obama administration introduced the first estimate for the social cost of carbon at $43 per ton. The Trump administration estimate was $3–$5 per ton, and the Biden administration estimate is around $51 per ton.

[5] https://tradingeconomics.com/commodity/carbon, accessed 22 February 2023.

permits and pay the market price for this. It also implies that those who emit less, e.g., by capturing the carbon emitted and store it somewhere safe, can make a profit by selling their permits, if the costs for CCS stay below €100.

I have already noted that to many CC(U)S is unattractive as it only involves costs and no revenue, this in stark contrast with renewable sources of energy, like wind turbines and solar panels. But such comparisons are beside the point. CC(U)S is not meant to generate energy but is a necessary part of a survival strategy for mankind on this planet, in the same way as putting (sulphur dioxide) scrubbers on smokestacks was necessary to solve the problem of acid rain at the end of the previous century.[6] Many environmentalists, who are opposed to carbon capture as it would possibly extend the use of fossil fuels, compare it with renewables, partly to be able to conclude more forcefully that we should switch to 100% renewable energy, in spite of it being abundantly clear by now that this goal will not be achieved in the short term, if at all, while the dangers from continuing emissions and ever increasing atmospheric levels of carbon dioxide are there for everyone to observe and experience. This comparison with energy from renewable resources is apparently an irresistible mantra. The September 2022 very negative report of the Institute for Energy Economics and Financial Analysis (IEEFA) again states that [CCS] "is not cost competitive with renewables and storage as a climate-change mitigation option for the power sector" (IEEFA 2022, p. 11). Apart from the fact that this statement is incomprehensible (CCS not competitive with storage?) it is also irrelevant. It is akin to saying that pigs are not competitive with cows as far as milk production is concerned.

Nevertheless, the cost of carbon capture remains of course a key issue and has many dimensions. Some CO_2 capture technologies are already commercially available now, while others are still in development, which contributes to a large range in costs. If the 2005 IPCC analysis discussed above had arrived at a ten times higher cost or if the power load needed for CCS is too high, i.e., if a too large part of the energy generated by a fossil-fuel fired power plant is needed to capture, transport and store the CO_2 it emits, it is obviously not a good procedure. At least, not yet! It may still become necessary if we are getting really desperate a few decades from today, and gladly pay three times as much for a kWh of electricity to make sure it is clean.

The cost of carbon capture can be split into the costs of:

- CO_2 capture at a source of emissions or from the air—generating a CO_2 stream of over 95% purity by volume from a (flue) gas stream or a very dilute source like air;

- CO_2 dehydration and compression/liquefaction, depending on the transport method;

---

[6] Acid rain was (and is) due to the burning of sulphur-containing coal and is still a considerable problem in China and Russia. Many coal-fired power plants use flue-gas desulphurization to remove sulphur-containing gases from their stack gases. This typically removes 95% or more of the SO_2 in the flue gases.

- CO_2 transport by pipeline, ship or truck;
- CO_2 injection, monitoring and verification of its storage.

As we have seen, there are essentially two separate approaches to carbon capture: CCS applied to carbon emitting installations like power plants and certain carbon-intensive industries, e.g., iron and cement producing factories, and Direct Carbon Removal, such as Direct Air Capture. Power plants and other installations produce flue gases with a high concentration of carbon dioxide (either post- or pre-combustion) and can be (retro)fitted with CCS installations, while Direct Carbon Removal involves the capture of CO_2 from ambient air in which the concentration of CO_2 is much lower. This makes capturing CO_2 directly from the air currently the most expensive approach, but it will nonetheless play an essential role in carbon removal as it is the only way to lower the CO_2 concentration in the air. Both approaches are necessary, with Direct Air Capture making no sense if no CCS is deployed. DAC would perhaps not have been necessary if fossil-fuel power plants and other installations had routinely been provided with CCS installations from 2005 onwards, the year of the publication of the above-mentioned IPCC report when the CO_2 concentration in the air was still 'only' 380 ppm. In any case, the task would then have been much easier. In addition, there are the costs of compression and transport of the carbon dioxide to the eventual storage location, and finally injection at that location, which the two approaches have in common. These costs can also vary greatly on a case-by-case basis, depending mainly on CO_2 volumes, transport distances and storage conditions. Hence, it is hard to quote a single cost figure for carbon capture; and the various components all have their separate cost structure. In the following we will separately discuss costs for pre- and post-combustion capture of CO_2, the case of bioenergy and carbon capture (BECCS), where carbon capture seems to be an accepted practice and which is especially loved by the IPCC, and finally Direct Air Capture, which is still very much in the process of proving itself.

Pre- and Post-combustion capture

Applications of CCS on power plants are typically characterised as "clean coal" technology. CCS however has uses beyond fossil-fuel combustion for energy, as illustrated by a multitude of other industries. Even when viewed as a coal technology, coal-fired generators fitted with CCS emit around 90% less CO_2 than unabated coal plants, including even the most efficient new coal plants using ultra-supercritical technology. So it does what it promises to do, almost completely stop CO_2 emissions from such plants into the atmosphere.

The cost of CCS on several industrial applications is far below what many would expect, while the costs of CCS for coal-fired power generation are poorly understood and so tend to be misused or misinterpreted, and there still is a poor but growing understanding of life-cycle emissions from gas-fired power generation, including from upstream fuel processing, and the role CCS can play there (Irlam 2017).

As we have seen, the benchmark technology for post-combustion carbon capture is amine scrubbing. The *theoretical* minimum energy requirement for such capture of CO_2 is 100 kWh per ton of CO_2 but in practice it is considerably more. It is expected that improved solvents and process configurations can bring this so-called parasitic power demand *in practice* down to about 200 kWh per ton of CO_2, equivalent to about 20% of typical power plant output (Rochelle 2009).

At \$0.05 per kWh, energy costs per ton of CO_2 captured would then be \$10 and total costs for a plant annually emitting 3 Mt of CO_2 would be \$30 million per year. As mentioned in Chapter 6, in 2020 worldwide coal combustion was responsible for 14 Gt of CO_2 emissions. To capture this 14 Gt would cost \$140 billion per year, a lot of money but not an amount to get worked up about as it is just 0.165% of total global GDP of 85,000 billion in 2020. Moreover, such CCS systems will only be needed so long as coal-fired power plants are still felt necessary, hence in the transition period to a sustainable energy system based on renewable sources. Assuming a capacity factor of 50%, a 500 MW coal-fired power plant generates about 2.2 TWh per year in electricity. Hence the CO_2 capture costs would add 1.4 dollar cent to the price of 1 kWh of electricity, not negligible but not an amount that destroys the world, and certainly worth it if it saves us from a climate apocalypse. It should be realized though that this is just the energy cost for capture and does not include capital, compression and transport costs. Some would argue that this would make energy generated from coal *uneconomical* since renewable sources can generate electricity at a far lower price. That is certainly true, but the argument is only relevant if renewables were capable of generating sufficient electricity to cover all demand at any time and place, which they aren't. If they were, coal plants would be shut down at a much faster rate than they are now. The 'market' would rapidly strangle them all. The number of installations generating renewable energy is still far from sufficient and will remain so for several more decades, while in addition electricity grids are far from ready to handle the corresponding huge volumes of intermittent electricity. The storage problem still remains to be solved and society has to be thoroughly transformed to make the transition to 100%-renewable energy possible. As things stand at present, it is therefore more appropriate to compare CCS cost figures with the amounts needed for renewable energy installations or for making society sustainable and this only as regards the price for avoiding the emission of CO_2, such as the Urgenda scheme in the Netherlands to make all homes sustainable, mentioned in Chapter 3! Or with the upfront costs mentioned by Griffith and Jacobson to carry out such sustainability exercises in the US.

The IEA's Sustainable Development Scenario, presented in its *World Energy Outlook 2018*, defines a pathway where 15% of the world's emissions reductions between now and 2050 are delivered by CCS, with roughly 20% of coal capacity being equipped with carbon-capture technology by 2040.[7] This equates to 2,000 large-scale carbon-capture facilities being deployed by 2050—around 100 facilities commissioned each year. This amounts to a colossal effort in view of the fact that there currently are only around 20 commercial, pitifully small, CC(U)S operations worldwide. The capital requirement would be around \$650–1,300 billion, roughly

---

[7]   The report says that retrofitting thermal power plants with one of the three main carbon capture routes—post-, pre- and oxyfuel combustion—appears to have only a small impact on their operational flexibility, provided that the capture systems are designed properly (p. 232).

between \$325 and \$650 million per installation. For ease of calculation I will assume in the following that the capital costs for a typical CCS installation are \$500 million.[8] The financing has to come mainly from the private sector where most of the world's liquidity is locked up. (Global CCS Institute 2021a, p. 50.)

Looking specifically at carbon capture, the cost can vary greatly by CO_2 source, from a range of \$15–25/t CO_2 for industrial processes producing "pure" or highly concentrated CO_2 streams (such as ethanol production or natural gas processing, e.g., the Sleipner project) to \$40–120/t CO_2 for processes with more "dilute" gas streams, such as cement production and power generation.[9]

A number of rather old studies (Singh et al. 2003; Roussanaly et al. 2013; Li et al. 2016; NETL 2007 and 2010; Rubin and Zhai 2012; Holt and Booras 2009) have investigated the cost of CO_2 capture and retrofitting coal and natural-gas fired power plants especially as regards amine scrubbing and arrived at amounts varying between \$32 and \$114 per ton of CO_2, with gas-fired plants being more expensive (\$74–\$114) than coal-fired plants (\$32–\$76). Despite the higher efficiency of gas plants, the emissions penalty is higher than for coal plants due to the lower concentration and amount available for capture. An extensive update (NETL 2019, p. 15) by NETL for the US Department of Energy in 2019 arrived at the conclusion that, at a capital cost of \$400–\$500 million per unit, commercial technology can capture carbon at roughly \$58 per ton of CO_2. CO_2 capture, including transport and storage, adds between \$31 and \$69 per ton of CO_2 to the levelized cost of electricity depending on the technology and fuel used. Not-surprisingly capital costs form the largest part of this. The cost for transport and storage remains below \$10 per ton.

These costs are all below the current price of carbon in the EU ETS quoted above. The IEA has stressed that CCUS can be a cost-efficient strategy to tackle emissions from existing coal- and gas-fired power plants. Around one-third of today's coal and gas plants were built only in the last decade; retrofitting with CCUS can allow them to continue operation and avoid the much higher cost of early retirement when these plants become stranded assets.

Any carbon-capture technology necessary involves energy penalties as illustrated in Fig. 14.1, although it is also claimed that it can be powered with "free or low-cost, widely available residual low-temperature (85°C) heat" (Chichilnisky and Bal 2019, p. 281).[10] As an illustration of the ongoing developments in this field, a March 2021 paper (Jiang et al. 2021) by researchers at Pacific Northwest National Laboratory studied the use of a single-component water-lean solvent, which can capture CO_2 without high water content (water-lean) and is much less viscous than other water-lean solvents. It showed that particularly the heat of vaporization can be significantly reduced by replacing aqueous amine with a water-lean solvent.

---

8   If German subsidies for renewable sources of energy in the last ten years (€25 billion per year) had been used to equip Chinese or other (their own, for instance) coal-fired power plants with CCS, 1.1 Gt of CO_2 emissions per year could have been avoided (more than twice total German emissions).

9   IEA, https://www.iea.org/commentaries/is-carbon-capture-too-expensive.

10   On p. 249 this publication claims that there is "enough residual heat in a coal power plant that it can be used to capture twice as much CO_2 as the plant emits, thus transforming the power plant into a "carbon sink"."

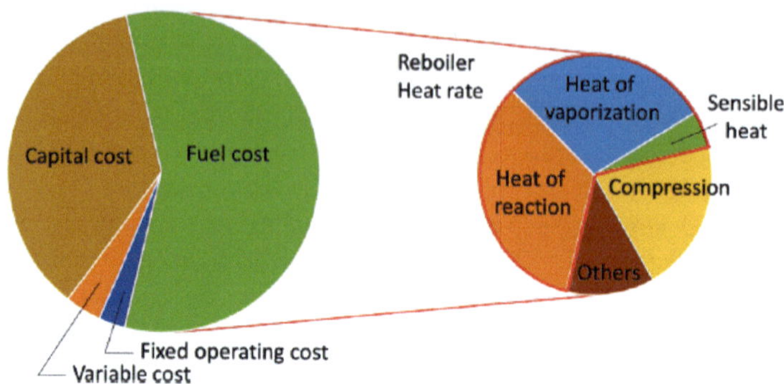

Fig. 14.1. Cost breakdown of post-combustion carbon capture using aqueous amine solvent (from Jiang et al. 2021).

The solvent used, called EEMPA, can absorb CO_2 from power plant flue gases and later release it as pure CO_2 for as little as \$47 per ton. Tests will be continued at increasing scales and the solvent's chemistry will be further refined, with the aim to reach the goal set by the US Department of Energy of deploying commercially available technology that can capture CO_2 at a cost of \$30 per ton by 2035.[11]

In 2020 a large scale study was carried out (Schmelz et al. 2020) in which the costs of carbon capture with storage/injection into subsurface geological formations were modelled for emissions from 138 electricity-generating power plants in the north-eastern and midwestern United States. This region has 1308 emission sources, both industrial and electricity generating ones, that produce 604 Mt of CO_2 emissions per year, more than 10% of total US emissions. The 138 largest point sources of CO_2 emissions from electricity-generating power plants (82 coal fired and 56 gas fired) accounting for nearly 390 Mt of CO_2 emissions are used in the analysis. The analysis suggests that CO_2 emissions from coal-fired power plants can be captured and stored at a cost of \$52–\$60/ton, whereas the cost for emissions from natural-gas-fired plants ranges from approximately \$80 to \$90/ton. Storing emissions offshore increases the lowest total costs of CCS to over \$60 per ton of CO_2 for coal. Finally, it is said to be likely that more than 8 Gt of total CO_2 emissions from this region can be stored for less than \$60/ton, including all costs (capital, capture, transport and storage). I would say let's get started!

All this concerns post-combustion carbon capture, extracting the carbon dioxide from the smokestacks of power plants or industrial installations. In Chapter 8 we have also considered pre-combustion and IGCC. In these two, much less mature, approaches the carbon dioxide is removed by first turning the coal into a synthesis gas before the latter is burned for generating electricity. The Office of Fossil Energy and Carbon Management of the US Department of Energy, whose mission is to minimize the environmental impacts of fossil fuels while working towards net-zero emissions, says about pre-combustion: "Compared to post-combustion technology, which removes dilute CO_2 ($\sim$ 5–15% CO_2 concentration) from flue gas streams and

---

[11] https://www.pnnl.gov/news-media/cheaper-carbon-capture-way#:~:text=RICHLAND%2C%20 Wash.,compared%20to%20current%20commercial%20technology.

is at low pressure, the [water-]shifted synthesis gas stream is rich in CO_2 and at higher pressure, which allows for easier removal before the H_2 is combusted. Due to the more concentrated CO_2, pre-combustion capture typically is more efficient but the capital costs of the base gasification process are often more expensive than traditional pulverized coal power plants. Today's commercially available pre-combustion carbon-capture technologies (…) cost around \$60/ton to capture CO_2 generated by an integrated gasification combined cycle (IGCC) power plant."[12]

There are some unsuccessful or underestimated cost examples of implementing CCS pre-combustion projects. The Kemper power plant in Mississippi, discussed in Chapter 12, is a clear example of how expensive it can be to implement pre-combustion technology, as the cost of the power plant totalled \$7.5 billion, three times the initial cost estimate of \$2.4 billion. It should be noted that this concerns the pre-combustion technology of the project, not the carbon capture part. But if you are looking for a stick to hit a dog, it is in general easy to find one. The Kemper project has been used as such, without explicitly saying so, in the announcement on the *endcoal.org* website[13] in 2018 that "New renewables cost data closes off coal power's CCS escape route." It quotes a 2017 Lazard[14] estimate that the levelized cost of electricity (LCOE) from an IGCC plant capturing 90% of CO_2 will be \$231 per MWh (not including transport and storage), of which the website says that it is far more expensive than energy from solar and wind. This is true and at such a high price burning coal or gas with CCS will certainly not be *competitive* (a word I would like to avoid in this discussion as we are beyond the stage of solving the climate problem by hard-nosed capitalist competition). However, the website is misleading as it only quotes the high-end Lazard estimate, based on the experience with the above-mentioned Kemper project; \$203 of the \$231 quoted above are capital costs, which completely got out of hand in the Kemper project. The low-end estimate in the Lazard report gives an energy cost of \$96 per MWh which is in the same ball-park as the values quoted for renewables (considerably cheaper actually that the \$187 quoted for rooftop residential solar PV). However, apart from the attempt to mislead the reader by only showing one side of the coin, the main objection against the quote on the *endcoal.org* website is that CCS is not meant to generate energy or to be competitive, but to bring down CO_2 emissions and to lay out a viable pathway to net zero, climate mitigation *and* a world powered only by renewables. So long as coal- and gas-fired power plants are being built and operated, this apparently in spite of being more expensive than renewable options, and not enough *reliable* energy from wind and solar can be generated, CCS will have to be deployed. That will of course cost money and will have an effect on the price of energy, you can't have your cake and eat it. When the website suggests that the results of the Lazard analysis imply that coal plants with CCS have lost the race, one wonders which race is meant here. The race is not for producing the cheapest electricity but for keeping the planet liveable.

---

[12] https://www.energy.gov/fecm/science-innovation/carbon-capture-and-storage-research/carbon-capture-rd/pre-combustion-carbon#:~:text=Today's%20commercially%20available%20pre%2Dcombustion,cycle%20(IGCC)%20power%20plant.

[13] Website no longer exists; is now https://globalenergymonitor.org/.

[14] https://www.lazard.com/media/450337/lazard-levelized-cost-of-energy-version-110.pdf.

There is still over 2 million MW of coal energy generating capacity operating[15] in the world, supplying more than 30% of global electricity. It just can't be ignored and dismissed in such a flippant manner! The comparable figures for wind energy are 830,000 MW of generating capacity, supplying 5.3% of global electricity,[16] also reflecting the (still) much lower capacity factor of wind turbines.

Moreover, another race, the race for improving the CCS process has hardly begun. In its latest analysis[17] Lazard quotes $159 per MWh for the price of electricity (the LCOE) from a coal-fired power plant with 90% CO_2 capture (excluding transport and storage) and $88 for a certain type of IGCC with carbon capture. These figures compare well with solar but are higher than for wind. A proper comparison is however only possible when the intermittency problem of these renewable sources has been solved. It might well be that the solution for this lies in back-up fossil-fuel power plants with carbon capture to keep the electricity grid stable.

A post-combustion CCS project that was at any rate in principle successful, was the Petra Nova project, also discussed in Chapter 12, designed to reduce carbon emissions from one of the boilers (654 MW nameplate capacity) of a coal burning power plant in Thompsons, Texas. Some may say that it was a failure since it was shut down prematurely in 2020, citing low oil prices during the Covid-19 pandemic. In the Petra Nova project the carbon dioxide was captured at 99% purity, and after compression piped about 82 miles to the West Ranch Oil Field for enhanced oil recovery. If oil prices are too low, enhanced oil recovery can obviously not be made profitable. As already pointed out in Chapter 12, the main aim of CCS should be avoiding carbon-dioxide emissions, not making oil extraction from almost depleted oil fields profitable. The CCS installation for the project cost approximately $1 billion to install (twice the average quoted above) and was put into operation early in 2017. It was designed to capture 1.6 Mt of CO_2 per year. To make it dependent on the profitability of enhanced oil recovery was short-sighted, as such profitability is highly uncertain with the comings and goings of periods of oil glut.

The Department of Energy has backed a number of projects spending a total of $1.1 billion from 2010 to 2017 on nine projects, of which only three, including Petra Nova, can be called successful. The other two are a hydrogen production plant at La Porte in Texas, the CO_2 of which is also used for enhanced oil recovery, and an ethanol plant in Illinois, which injects CO_2 into dedicated geological storage, which of course is the preferred option for getting rid of the stuff.

A comparison of post-combustion, pre-combustion and oxy-fuel combustion, including IGCC, has recently been made in the journal *Sustainability* (Kheirinik et al. 2021). It concluded that "pre-combustion is the most expensive process to implement and operate over its life span, as it requires considerably more overall investment than a pulverized-coal power plant with oxy-fuel CCS technology (≈ 1.6 times more). This is mainly due to the complexity of the process and the operation units required for the successful implementation of capture technology. The amount

---

[15] Carbon Brief, https://www.carbonbrief.org/mapped-worlds-coal-power-plants/.
[16] 2020 figure *Our World in Data;* https://ourworldindata.org/electricity-mix.
[17] https://www.lazard.com/perspective/lcoe2020.

of fuel required for the IGCC power plant (pre-combustion) was slightly more than the pulverized-coal power plant with post-combustion and oxy-fuel, which, again, validates the initial point made that, from a financial perspective, the pre-combustion power plant requires a significantly higher cost to construct and operate." The Kemper power plant (see above) is a clear example of how pre-combustion technology could be expensive to implement, but in that particular case other aspects, e.g., management problems also played a significant role. "Although the oxy-fuel process proved to be more economic in terms of LCOE (levelized cost of electricity), capital, and investment costs, it has failed to operate on a commercial scale. If a company is looking to implement one of the three CCS technologies to existing operational plants, post-combustion technology could be the most attractive option. This will cause less disruption to the existing operation and a lower loss of revenue due to shutting down and revamping the plant. It requires less total investment and indirect costs, including utilities and salaries. Moreover, due to the maturity of such technology compared with other technologies, such as pre-combustion, it would be safer. Therefore from a financial standpoint, post-combustion stands as the better option."

The calculated LCOE for oxy-fuel combustion with and without CCS amounts in British pounds to £0.095/kWh and £0.053/kWh, respectively, implying that CCS would almost double the price of electricity. It is consequently no surprise that for the costs of CO_2 avoided the paper gives the rather high values of £60.4 per ton of CO_2 for IGCC, £124.7 for post-combustion and £206.6 for oxy-fuel combustion. Especially the last number is very high indeed. These numbers are in stark contrast with the numbers quoted above and with the results of earlier calculations in the literature.

In 2015 Rubin and collaborators (e.g., Rubin et al. 2015) made an extensive analysis to assess the costs of CO_2 capture and storage (CCS) for new fossil fuel power plants and to compare the results with the costs reported in the 2005 IPCC Special Report on Carbon Capture and Storage. Costs had gone up, which is not surprising, as hardly any large-scale CCS projects had been or were being carried out at that time to establish a learning curve. For the CO_2 capture costs they arrive at (the rather low values of) \$35–57/t CO_2 and for the costs of CO_2 avoided, excluding transport and storage, at \$45–73/t CO_2. These values are more in line with the \$60 quoted above.

In summary, a cautious conclusion may perhaps be that capital costs for a CCS installation for a coal-fired power plant will be about €500 million per installation, resulting in total costs for carbon capture and storage of around €60 or \$60 per ton of CO_2 captured/avoided, significantly less than the current price of carbon in the EU ETS. There are differences between post-combustion, pre-combustion, and IGCC but these are not likely to be very significant. Lacking any large-scale demonstration projects and/or power plants with CCS it is not (yet) possible to provide any really reliable numbers.

Recently, in the Biden Administration's \$1.2 trillion *Infrastructure Investment and Jobs Act*, \$10 billion has become available for carbon capture projects, to be spent over five years for demonstration projects and R&D in this field. Of this amount \$3.5 billion is allotted to support technologies for DAC and another \$3.4 billion for commercial-scale demonstration plants and large-scale pilots of technologies for capturing CO_2 from power plants and industrial facilities (CCS). The DOE is required to spend $2.5 billion on six commercial-scale carbon capture demonstration projects: two for coal power plants, two for natural gas plants, and two for large industrial emitters such as cement, steel, and chemical manufacturers. (Kramer 2022a.) The Department of Energy had only asked for about half a billion for CC(U)S for the current financial year, so it will be hard pressed to wisely administer this huge amount, but it is gratifying to see that CC(U)S is at last taken seriously. We can only hope that the debate will not be dominated by profitability arguments and other purely economic considerations, but by climate change and lowering CO_2 concentrations in the atmosphere.

Bioenergy and carbon capture

Carbon capture seems to be seen as almost self-evident in the case of the burning of biomass, at any rate for large plants. BECCS is the generally accepted form of bioenergy with carbon capture; its main appeal being that it can result in negative emissions. The lowest cost trajectories for achieving the 2°C target typically include massive deployment of BECCS, to avoid the even steeper costs associated with relying on emission cuts alone. However, as for instance pointed out by a 2019 report of the US National Academy of Sciences (NAS 2019, p. 3), BECCS at this scale requires more feedstock than is available from biomass waste. For example, 30 million to 43 million hectares are required to raise BECCS feedstocks for each Gt/y of negative CO_2 emissions. Thus, 10 Gt/y from BECCS requires almost 40 percent of total global cropland to be used for raising BECCS feedstock, clearly an unattainable and undesirable prospect.

In its 2005 report on CCS the IPCC is not yet certain of the potential for BECCS, as it states: "there is considerable interest in some regions of the world in the use of biomass to produce energy, either in dedicated plants or in combination with fossil fuels. One set of options with potentially significant but currently uncertain implications for future CO_2 sources is bioenergy with CO_2 capture and storage. Such systems could potentially achieve negative CO_2 emissions" and "based on the available literature, it is not possible at this stage to make reliable quantitative statements on [the] number of biomass energy production plants that will be built in the future or the likely size of their CO_2 emissions."

As far as costs are concerned the report says: "At present, biomass plants are small in scale (< 100 MWe). Hence, the resulting costs of capturing CO_2 are relatively high compared to fossil alternatives. For example, the capturing of 0.19 MtCO2 yr^{-1} in a 24 MWe biomass IGCC plant is estimated to be about 82 $/tCO2, corresponding to an increase of the electricity costs due to capture of about 80 $/MWh. (…) significantly larger biomass plants could benefit from economies of scale, bringing down costs of the CCS systems to broadly similar levels as those in coal plants. However, there is too little experience with large-scale biomass plants as yet, so that their feasibility has still not been proven and their costs are difficult to estimate."

In later reports the IPCC seems to have changed its mind. In its 2018 special report on *Global Warming of 1.5°C*, BECCS is presented as essential for realizing mitigation targets in three of four pathways the IPCC presents. BECCS is modelled to deliver up to about 20 Gt CO_2 sequestration per year, requiring several million square km of land.

The Sixth Assessment Report (AR6) of the IPCC states that bioenergy with carbon capture and storage (BECCS), and afforestation/reforestation are the dominant carbon removal options used in climate stabilisation scenarios implying large requirements for land and water. All analyzed pathways limiting warming to 1.5°C by 2100 with no or limited overshoot include some use of such removal to offset anthropogenic CO_2 emissions and the median of CO_2 removal across all scenarios was 730 Gt of CO_2 in the 21st century. The required scale can vary from 1–2 Gt of CO_2 per year from 2050 onwards to as much as 20 Gt of CO_2 per year and would play a pivotal role in limiting climate warming to 1.5°C or 2°C.

BECCS has however a number of drawbacks: "sequestration potentials from BECCS depend strongly on the feedstock, climate, and management practices. If woody bioenergy plants replace marginal land, net carbon uptake increases, enriching soil carbon. On the other hand, replacing carbon-rich ecosystems with herbaceous bioenergy plants could deplete soil-carbon stocks and reduce the additional sink capacity of standing forests. Furthermore, wood-based BECCS may not be carbon negative in the first decades, initially emitting more CO_2 than sequestering. BECCS has several trade-offs to deal with, including possible threats to water supply and soil nutrient deficiencies. Deployment of BECCS at the scales envisioned by many 1.5–2.0°C mitigation scenarios could threaten biodiversity and require large land areas, competing with afforestation, reforestation and food security."

A 2020 analysis from the Oak Ridge National Laboratory in the US, exploring the potential supply and cost of BECCS under a range of feedstock, logistics, and power generation scenarios, concluded that BECCS in the US has a total technical potential to sequester about 181 to 737 million tons of CO_2 annually. In terms of cumulative potential, the US has a technical potential to sequester up to 46 Gt of CO_2 by BECCS by 2100. Average prices range from \$42 to \$137 per ton of CO_2 depending on cost accounting, power generation system, and biomass logistics system (Langholtz et al. 2020).

I just repeat the opinion already stated before that burning biomass is only a good idea when that biomass would anyway go to waste. To cultivate biomass just for the sake of burning it for generating electricity is not a good idea, even if the resulting carbon dioxide is captured. The main objections are that it will aggravate air pollution and will lay a too heavy claim on scarce land resources. Even burning coal with CCS would be a better option.

Direct air capture

In the preceding chapter we have met a multitude of startups that almost all promise to capture CO_2 from ambient air at \$100 per ton or even less. So far these promises are only numbers on paper, none of them have been tested in practice, and must therefore be taken with a pinch of salt. The only three companies (Climeworks, Carbon Engineering and Global Thermostat) that actually have had some success in capturing CO_2 from the air tell a slightly different story. Climeworks offers a carbon-offset service for the price of €1 per kg, which I suppose reflects the costs they actually make to scrub the carbon dioxide from the air. They promise that if you place a CO_2 removal order with them, it will be carried out within 6 years. This price is obviously far too high to become viable. A similar service is provided by partners of Carbon Engineering at €336 per ton, one third of the price charged by Climeworks.

As far as costs of Direct Air Capture are concerned, the 2021 report by the Bipartisan Policy Center summarised the current situation as follows: "Recent findings suggest that the economics are within the range of $100–$250 per ton of CO_2 captured based on the systems that are currently under development, with potential for future costs to fall further through learning-by-doing, particularly as technology deployment scales" (Bipartisan Policy Center 2021, p. 9). A study (Keith et al. 2018) published in the journal *Joule*[18] in 2018 cites levelized costs for an industrial, megaton-scale DAC plant at $94–$232 per ton CO_2. This suggests that prices of around $100 per ton may very well be feasible, which is equal to the current (2023) price for a ton of CO_2 in the EU ETS (Fig. 6.1 showing the recent price development).

Much more research is needed, projects have to be started up so that techniques can be developed that work better, are faster and become cheaper. DAC will however always be a cost factor, as cleaning up after having made a mess usually is. Apart from some limited sales of CO_2, no revenue can be expected from DAC. Making society sustainable so that after some time we can do without DAC will cost vast additional sums of money. The way in which the solution to some of the problems is currently approached does not seem to be sensible when looked at more closely. In Chapter 3 and above in this chapter we have mentioned a scheme proposed in the Netherlands for making all houses fossil free (decoupling them from the natural gas grid) by investing the mind-boggling amount of €245 billion. Dutch households burn about 9 billion m³ of natural gas per year for heating and hot water.[19] Since burning one m³ of gas produces 1.8 kg of CO_2, this results in the emission of 16.2 million tons of CO_2 per year for all households combined. Hence the gigantic sum of €245 billion is needed to avoid the annual emission of a mere 16 million tons of CO_2, which comes down to a little over €15,000 per ton. In addition, part of the investment will have to be forked out every 15–20 years, the average lifetime of the solar panels and heat pumps that must do this job.

Let us compare this Dutch plan with scrubbing CO_2 out of the atmosphere as discussed above, at a cost of somewhere between $100 and $250 per ton of CO_2. At

---

[18] It should be noted that all authors of the paper are employees of Carbon Engineering, with an ownership stake in the company. There is nothing wrong with this, but should be kept in mind when reading the paper. David Keith is also a professor at Harvard University and has written extensively about the cost-effectiveness and geopolitical and ethical implications of approaches to climate change and climate mitigation.

[19] 2019 figure, Dutch Central Bureau of Statistics.

€200 per ton it would cost €3.2 billion per year (on average €457 per household) to scrub 16 Mt of CO_2 out of the air. At that price the €245 billion mentioned above as the minimum investment needed to carry out the energy transition from natural gas to carbon-free renewables, if at all possible, would be sufficient to scrub out all CO_2 emissions from the burning of natural gas by Dutch households for 76.5 years. In the meantime, the insulation of houses and the installation of solar panels and heat pumps can be pursued at full speed where this is economically sound; all new houses should be built in a sustainable way as is already done in practice, hydrogen can be added to the natural gas supply, and various other cost-effective measures will undoubtedly be devised in future to bring CO_2 emissions down, while the cost of scrubbing CO_2 out of the air will probably also go down, so that these 76.5 years can be stretched far beyond the lifetime of any house and home owner presently existing in the Netherlands. The conclusion from this must clearly be that DAC is by far the most affordable option currently available to solve the problem of such CO_2 emissions. Installing heat pumps and solar panels and insulating the existing stock of houses is in any case vastly more expensive. The money left over could be used to help fit or retrofit CCS installations on coal-fired power stations in the developing world; €100 billion can do a lot of good in that respect and probably avoid over 200 million tons of CO_2. This is the pitiful state we are in at the moment as regards climate mitigation, with rich Western countries being prepared to spend billions in avoiding the emission of tiny amounts of carbon dioxide while the same money spent elsewhere would guarantee a bountiful harvest.

In its recent report on Direct Air Capture (IEA 2022, p. 28) the IEA estimates the costs of DAC for large-scale applications (1 $MtCO_2$/year) in the range of \$125–335/tCO_2, depending on capture technology (solid- or liquid-based technologies), energy costs (price of heat and electricity), financial assumptions, specific plant configuration, and whether the captured CO_2 is stored or used. With low heat and low electricity prices, projected capture costs can be lowered to just above \$100/tCO_2. And with some form of carbon pricing scheme the levelized capture cost for DAC could fall well below \$100/tCO_2. DAC-based capture could be made *profitable* at a carbon price above around \$160/tCO_2. The profitability argument is irresistible it seems. For some perverse reason it is a must to quote a situation in which DAC could become *profitable*. In any case, it shows, as has also been argued in Chapter 6, that it is imperative for carbon emissions to be priced.

Carbon Engineering claims that with the first plants DAC will cost between \$300 to \$425 per ton, including storage, with a target of \$125 to \$150 a ton for larger-scale facilities. The cost figure of \$150 assumes large-scale deployment (one Mt per year capacity). At this price a home owner burning 3,000 m³ of gas to heat their home and consequently annually emitting 5.4 tons[20] of CO_2 into the air, would have to pay \$810 per year to erase the carbon footprint from heating their home. At such a price

---

[20] Assuming natural gas is all methane, CH_4, 1 kg of natural gas generates 2.75 kg of CO_2. This follows from the ratio of the molecular weights of CH_4 (12 + 4 = 16) and of CO_2 (12 + 2 x 16 = 44). One m³ of methane at atmospheric pressure weighs 0.671 kg. Hence, 1 m³ of natural gas generates about 1.8 kg of CO_2.

it would even hardly be worthwhile to insulate homes to drive the amount of carbon emissions down.[21]

According to the IEA, the industry target for DAC appears to be $100 per ton of CO_2. The US Department of Energy has chosen this target for the Carbon Negative Shot launched in November 2021 and aims to bring the cost below $100 within a decade. In any case, capture costs are expected to decrease substantially in the next five to ten years, when deployment of DAC increases worldwide from the 1,000 ton scale to the million ton scale.

It is not yet clear either at present which technology for DAC is to be preferred. In this respect, a recent analysis (Sabatino et al. 2021) allows for the tentative conclusion that solid sorbent approaches, as the one proposed by Climeworks, are to be preferred over aqueous-scrubbing processes, such as the alkali scrubbing process employed by Carbon Engineering. The alkali-scrubbing and solid sorbent processes were selected for analysis as they are comparatively technically mature among DAC processes, with the conventional and mature amine scrubbing process added as a benchmark by extending its operating range from flue gas to air capture. These three processes were compared on the basis of energy demand and productivity. The results show that absorption-based processes perform generally worse than a solid sorbent process, both economically and as regards productivity.

Transport and storage

Compared to capture the costs of transport and storage are relatively small. In the US, for example, onshore pipeline transport costs are in the range of $2–14 per ton of CO_2, while the cost of onshore storage shows a somewhat wider spread. However, more than half of onshore storage capacity is estimated to be available below $10 per ton of CO_2. In some cases, storage costs can even be negative if the CO_2 is injected into (and permanently stored in) oilfields to enhance production, thus generating revenue from oil sales.[22]

Concluding remarks

From the above it can be deduced that much is still unclear and uncertain about costs for both CCS and DAC. A lot has been said about costs in all kinds of theoretical scenarios but lacking any real practical experience it is almost impossible to say anything definite and worthwhile about costs. The picture is rather confusing and the content of this chapter may have brought you little illumination in this respect. This is for a great part because we have been pussyfooting around a lot and failed to take the bull by the horns by actually building a number of CCS and DAC facilities. The lack of support for carbon capture in the last twenty years or so is the reason

---

[21] This depends of course on the price of natural gas. The war in Ukraine has caused energy prices go through the roof, which would make it attractive to stop burning natural gas not for environmental reasons but for commonplace and still most persuasive pecuniary reasons.

[22] https://www.iea.org/commentaries/is-carbon-capture-too-expensive.

why there is still relatively little experience with capture, transport and storage in a fully integrated CCS system. CCS is still not being used to any relevant extent in large-scale power plants. This is probably going to change soon with the Biden administration having made available billions of dollars for pilot CCS and DAC projects within the framework of the *Infrastructure Investment and Jobs Act* and the *Inflation Reduction Act*. Two programs—Carbon Capture Large-Scale Pilots and Carbon Capture Demonstration Projects—aim to significantly reduce CO_2 emissions from electricity generation and hard-to-abate industrial operations.[23]

When planning the construction of a new power plant, the cost implications of adding a CCS system could influence the type of plant chosen. CCS can be applied to current generation technologies such as pulverized coal or natural gas combined cycle (NGCC).[24] However, the additional costs will be lower when CCS is integrated into emerging technologies such as integrated gasification combined cycle (IGCC) and pre-combustion facilities. While most existing facilities could be retrofitted to accommodate CCS systems, the costs will be significantly higher than for new plants with CCS. It must also be kept in mind that 10–40% more energy is needed for producing the same amount of electricity, so CCS can only be seen as a temporary solution for the transition period to an energy system completely or mainly based on renewable or other carbon-free sources.

It is commonly believed that the cost of building and operating CCS systems will decline over time as a result of learning-by-doing (from technology deployment) and sustained R&D. Historical evidence also suggests that costs for first-of-a-kind capture plants could exceed current estimates before costs subsequently decline. But before any of this can become true we must actually start some projects at a relevant scale.

---

[23] National Energy Technology Laboratory (NETL) Newsletter, Volume 23, No. 3 (2023); https://netl.doe.gov/.

[24] For NGCC and IGCC, see Glossary.

CHAPTER 15

Conclusion

This book is about repair, how we can repair the Earth system from the damage caused by the excesses that have occurred in the past and are still continuing. Repair means restoring as much as possible to an earlier situation. That is why a broken leg is splinted, to allow the bone to heal. That is why a malignant tumour is removed in an operation, to allow the body to recover from the cancer. As far as the Earth system is concerned, the ozone hole has shown that we can indeed damage it and that world-wide efforts can set a healing process in motion. In our case the currently too high concentration of greenhouse gases released into the atmosphere in the past, in particular carbon dioxide, is causing the damage through global warming. This concentration is increasing at a rapid rate and to repair the Earth system we must remove a great part of these gases, while at the same time reducing further release, and more broadly move towards a sustainable society or even better a regenerative society. To achieve this, two types of remedy have been described and recommended in this book: Carbon Capture and Storage (CCS) of emissions from point sources *and* Direct Carbon Removal, and in particular Direct Air Capture (DAC) for scrubbing the surplus CO_2 out of the air. Both remedies aim at the same result, making sure that the carbon-dioxide concentration in the atmosphere goes down, preferably to about 350 ppm. They must be applied in combination. The second remedy makes no sense without the first.

Only very recently the insight has emerged that these two technologies are essential if we want to keep the planet liveable. And still CCS and DAC are often dismissed out of hand. An example is Mark Jacobson, whom we met in Chapter 3 as the great champion of a world powered by 100%-renewable energy. This is a laudable aim but, as apparent from his recent book *No Miracles Needed* (Jacobson 2023), he does not seem to realize the obvious fact that the carbon-dioxide concentration in the air is already far too high and will, even if the world were to switch to renewables in the fastest way imaginable, increase to over 500 ppm in the next three to four decades. The remedy of renewables cannot be administered fast enough to save the patient.

Judging from the statement in his book that "a transition to WWS will (…) slow, then reverse, global warming" (p. 382), Jacobson apparently also holds the erroneous belief that WWS (wind, water, solar) can actually reverse global warming. Well, it cannot as even a full WWS energy system will not reduce the concentration of the carbon dioxide and other greenhouse gases already in the air. That can only be done if they are actively scrubbed out of the air, e.g., by DAC. When a full WWS energy system as envisaged by Jacobson will possibly have been realized in 2050, the CO_2 concentration will be above 500 ppm and remain at that level for several centuries. Global warming will continue until a new equilibrium for the Earth climate system will have established itself at a higher temperature. And even then, it is very probable that due to emissions from natural sources, like melting permafrost and agricultural sources unconnected to fossil-fuel combustion, the greenhouse-gas concentration will continue to rise, making it even more pertinent to develop carbon-capture technologies. When exceeding the 1.5°C mark, a few tipping points may perhaps be triggered and the situation will rapidly get out of control (Armstrong McKay et al. 2022). DAC in particular can come to the rescue. One of the especially worrying greenhouse gases, apart from carbon dioxide, is methane.[1] Methane has a large warming effect but for a relatively brief period, having an estimated mean lifetime of 9.1 years in the atmosphere. The methane cycle is quite complex, but part of it, especially from soils, reacts with oxygen to form water and carbon dioxide, which carbon dioxide DAC can deal with.

Not only individuals like Mark Jacobson who, in a blinkered pursuit of a single approach, refuse to see that the climate problem can only be resolved by mobilizing any and all means available for reducing emissions *and* lowering greenhouse-gas concentrations, are scathing about CCS and DAC, but also many organizations, such as the IEEFA in its *Carbon Capture Crux* report mentioned earlier (IEEFA 2022, p. 12). In this report both BECCS and DAC are dismissed with the remark: "Bioenergy with carbon capture and storage (BECCS) and direct air carbon capture and storage (DACCS) are not well advanced technically and commercially. Theoretically, these technologies could offer environmental and social usefulness by capturing carbon from the atmosphere, thus providing the option of negative emissions, should they prove cost-competitive and commercially robust technologies." This contrasts sharply with what the NGO Carbon180,[2] which sees Direct Carbon Removal as crucial for climate mitigation, has to say about the matter: "DAC has significant potential as a carbon removal solution—estimates suggest that DAC could remove up to 5 gigatons of carbon dioxide per year by 2050. In addition to mitigating climate change, DAC offers other benefits, such as flexibility in facility siting, new market opportunities for repurposed carbon products, and job opportunities in developing, deploying, and operating DAC plants." The difference in perception could hardly be greater.

In closing this book let us dwell a little on the question of how one would sensibly tackle a problem like global warming by harnessing all possible ways of attack. Since it is a global problem, a global approach with coordination on a global scale is needed. Although there is a lot of international getting-together, such coordination has so far clearly been lacking. The Kyoto protocol was a good early

---

[1]   40% of human-caused methane emissions are due to fossil-fuel exploration, production and transportation, 40% to agriculture and 20% to rotting waste sites (*Guardian Weekly*, 10 March 2023). More precise data are provided by the Global Methane Initiative (https://www.globalmethane.org/documents/gmi-mitigation-factsheet.pdf).

[2]   https://carbon180.org/.

attempt but failed as it focused only on bringing down emissions by developed countries. In this respect it succeeded, albeit mainly for reasons that had little to do with the protocol, but its main failure lay in the fact that it did not put a brake on emissions by less-developed nations, in particular China, so that the current situation is worse than it ever was and a huge gap has arisen between action by developed and less-developed countries, exemplified by the fact that in the former coal combustion for power is on the verge of being phased out while in the latter coal is still king with no sign of abdicating soon. The 2015 Paris Agreement failed to improve the situation and further fragmented the problem. It has essentially left it to individual countries to each formulate their own plans and contributions to solving the problem in an uncoordinated way. We are now faced with a hotchpotch of plans and actions whereby each nation or collection of nations, i.e., the EU, focuses on its own emissions without paying regard to the bigger picture. This has resulted in a lot of promises, a lot of action where it hardly matters, and the random setting of net-zero dates, while on a global scale emissions just continue to rise. It typifies the situation that it needed a global pandemic to temporarily slow down emissions. The Covid-19 virus presided with ease over a steep emissions decline, which was reversed with the same ease as soon as it departed the stage.

As we have seen in Chapter 2, the developed countries are no longer the largest contributors to climate change and their obstinate fascination with net zero for their particular small piece of Earth by 2050 or some other date, while losing sight of the bigger global picture, will not contribute much to a solution, while costs are gigantic (see, e.g., the example of the Urgenda scheme in the Netherlands mentioned a few times in earlier chapters which envisages expenditure of hundreds of billions of euro for the avoidance of just 16 million tons of CO_2).

In similar fashion, many people writing about climate change argue for a very drastic and completely unrealistic approach towards a zero-carbon economy. They want to solve the problem by getting rid of the cause: ban the burning of fossil fuels *now* and carbon emissions will automatically reduce to zero. Carbon-free energy sources (in particular wind and solar) are magically conjured up out of thin air to satisfy our energy needs. At first sight, it is indeed perfectly reasonable to replace the burning of fossil fuels by carbon-free (sustainable) sources, as it will do away with the problem once and for all, but on the required timescale it is a very drastic solution that will put society, and especially the energy generating infrastructure, upside-down. Solely focusing on curbing emissions and pushing this, irrespective of the costs, *locally* down to zero is not the most cost-effective, efficient and practical way to achieve the goal of *globally* mitigating the effects of climate change.

We should apply a varied approach to the problems and disasters it causes and will cause, although some of them have not even become reality but have only attached a statistic to them, albeit a rather scary one. For instance, shouldn't we put more effort and money into preventing future emissions in developing countries, where providing clean energy can do a lot of good in other ways as well, instead of just going for the last molecule of CO_2 in our own countries? It could kill several birds with one stone. In this respect it is telling that *Climate Action Tracker* rates the international climate finance contributions to which Western countries have committed themselves under the Paris Agreement as insufficient (for the EU), highly insufficient (for the UK) and even critically insufficient (for the US), while being much more positive about all these countries' domestic efforts. The insufficiency is due to the fact that they just refuse to come up with the money to help developing countries adjust their economies to climate change.

The basic problem is that there is not enough room for developing countries to follow the same path of development as the industrialized countries have done in the past, in spite of the fact that they have every right to try to do so. Human population growth is simply too high and resources are too scarce. As an example, the US has more than 800 vehicles per thousand inhabitants, a total of about 287 million.[3] If the entire world is aiming to reach that level, there will be driving around at some point the stupendous number of 6.5 billion vehicles (compared to the 'mere' 1.4 billion at present), and this only if world population remains stable at 8 billion! Even half the number of cars per thousand inhabitants as in the US would be unsustainable and, what is more, completely unnecessary as well. Unlike Americans most other people on the planet have learned to walk, ride a bicycle, take a train, etc. Still, the developing countries do not give the impression of choosing another path and the developed ones are all too happy to sell them all the junk they want.

Electric vehicles are not a lasting solution either in this respect. It will just mean swapping one precious resource (oil) for other ones (lithium, nickel, chrome) and the ugly picture of traffic jams will remain a common sight, only the smell and the noise will be different and the air perhaps slightly cleaner. Electric vehicles require 207 kgs of rare earth materials, lithium, copper, cobalt, nickel, etc., against just 34 kgs for conventional cars.[4] The price for these metals will undoubtedly rise in the coming decades (the price of lithium already took the lead by reaching an all-time high in November 2022, six times the value of mid 2021). Moreover the environmental damage due to the mining of these metals, in particular lithium, will also increase. An analysis from 2007 already stated that "Analysis of lithium's geological resource base shows that there is insufficient *economically recoverable* lithium available in the Earth's crust to sustain electric vehicle manufacture in the volumes required, based solely on lithium-ion batteries. Depletion rates would exceed current oil depletion rates and switch dependency from one diminishing resource to another" (Tahill 2007).[5] The same applies for other metals, such as nickel and cobalt. It will put irresistible pressure for exploitation of the few untouched ecosystems that still exist. In particular the ocean floor will soon be the next victim of the rape of Nature when deep sea mining, to dredge up the vast quantities of polymetallic nodules found on the sea bottom, will be falsely argued as being indispensable to stop global warming.

When we try to address one problem, we create other ones. What we need is fewer cars and less driving. The threatening destruction from climate change is a great opportunity to change course and turn societies into public-transport heavens with dense networks of rail and bus routes, swapping the old slogan "guns into ploughshares" for "motorways into railway tracks".

---

[3] Figure for 2020 according to the Hedges Company.

[4] https://www.visualcapitalist.com/sp/how-mineral-supply-will-change-ev-forecasts/; see also Knibb Gormezano Limited, *2022–2040 Powertrain Outlook*, October 2022.

[5] A recent paper in the journal *Joule* (Wang et al. 2023) paints a more optimistic picture as regards materials needed for the energy transition, but strangely enough does not mention lithium at all.

Countries that are developing now must forego the pleasures (are they pleasures?) the developed world has been enjoying in the past or is enjoying now, and the developed world must accept a considerable lowering of such pleasures, usually called high living standards. The Australian, American, Canadian, European ways of life, soon to be joined by the Chinese way of life, all energy-intensive modes of life characterised by overconsumption, exemplified by a silly, moronic car culture, are actually one of the main causes of the climate mess we are in. There is no escape from changing this if we want to continue living safely and comfortably on the planet.

We must also let go of the idea that the problem can be solved without costing us money or even that climate change provides vast opportunities for making money, for savings of trillions of dollars, not to speak of the prospect of countless great job opportunities the public is enticed with. In the Guardian Weekly of 22 October 2021 Bill McKibben quotes a report, without giving a proper reference (it concerns Way et al. 2021), of the Institute for New Economic Thinking at the Oxford Martin Institute that says that: "compared to continuing with a fossil-fuel-based system, a rapid green energy transition will likely result in overall net savings of many trillions of dollars." New economic thinking indeed, but McKibben, a distinguished but apparently rather sloppy scholar, forgot to quote the "if" of the story, the miracles that have to happen for this vista to become reality, namely "if solar photovoltaics, wind, batteries and hydrogen electrolysers continue to follow their current exponentially increasing deployment trends for another decade, we achieve a near-net-zero emissions energy system within twenty-five years." Nothing continues exponentially for the time we want it to and certainly not four technologies at the same time; there are no miracles in the real world. It is just a fairy tale told for gullible romantic environmentalists like McKibben. No wonder that the arch-capitalists are not starting to salivate when reading these stories infused with new economic thinking. They know that the last free lunch was served in the Garden of Eden, and even that one was poisoned from the outset and came with hidden costs. It will cost money; vast amounts of money. In that respect John Doerr is more realistic in claiming to have calculated that $1.7 trillion, that is $1,700 billion, is needed *each year*, for twenty years or more, to reach net zero (Doerr 2021, p. 263). In addition, energy consumption in developed countries, and less-developed countries like China, will have to come down substantially if we want to get rid of our dependence on fossil fuels. Efficiency measures can cushion part of the unpleasant effects, as can the rigorous closure of heavily polluting industries that do more harm than good to local communities, even if they entail the loss of a sizable number of jobs and an increase of the price of the consequently scarcer goods they produce.

But that will not be enough. At the current rate of energy use and growth of this use, wind and solar alone will never be able to generate enough energy to meet global energy demand at the timescale involved, i.e., in thirty to fifty years from today, as discussed in more detail in Chapter 3. Our total energy consumption is expected to increase faster than sustainable energies can grow, mainly because the developing world also wants to enjoy a fair share of the diminished pleasures that an overcrowded planet might still have to offer, even if we in the developed world were to learn to do with less.

According to some a revolution is needed. For instance, Nobuo Tanaka, from 2007–11 executive director of the IEA, thinks so and said in 2008 that the world's energy infrastructure needs to change at a cost of $45 trillion[6] by 2050, updated in 2014 to $53 trillion by 2035, an amount equal to about 60% of the entire 2019 global GDP. Although he cannot have had a precise idea of how this should happen and what should replace it, it implies doing away with almost the entire existing energy infrastructure and bankrupting countless, especially developing countries. However, especially in energy supply a revolution does not seem such a good idea. Energy is society's lifeblood, its supply must be reliable and affordable for society to function. Everything depends on it; it is important for everyone. Our energy supply systems as they currently operate are extremely reliable; blackouts happen but are rare. We hardly give it a thought and take it for granted that there is light and heat at the turn of a switch. This is probably the reason why people can be so casual when talking about changing the energy system from fossil fuels to renewables, as if it were no bigger deal than changing your socks or buying a new pair of shoes. Let's put some solar panels on our roofs and a few wind turbines in our (or even better in our neighbour's) backyard and everything will be fine. How difficult can that be? Already now, when energy from wind turbines and solar panels still accounts for only about 10 per cent of electricity supply,[7] the intermittency inherent to these sources puts an enormous strain on the electricity grid. Record amounts of solar electricity are generated on hot summer days when demand for electricity is minimal, driving prices below zero since the electricity cannot be stored anywhere and is virtually worthless. Their large-scale deployment is only possible when there is reliable backup that can store the generated electricity when there is plenty and switched on at short notice when the need arises. That is why energy systems will change only slowly.

It is clear that in order to get rid of fossil-fuel use a drastic approach is needed, but as history has shown, it is never a good idea to try to turn societies upside-down. Societies that changed gradually, sometimes even rather late, turned out in the end to be more successful, although it must be admitted that, as regards the problem we are faced with now, we are already rather late. That does not mean of course that many, even most of the old and new ideas that are put out to combat climate change, are no good, but in many cases people want too much, too soon, by building castles in the air and launching one half-baked radical idea after the other, backed up by cherry-picked facts or figures and shaky economics. They paint a rosy future if a number of ifs is fulfilled, which they very well know cannot be fulfilled in the timespan they have in mind. This is grossly misleading. There is no rosy future.

It is unlikely that living standards, in particular in the West, will not suffer when the amounts mentioned above have to be spent or written off when a huge part of

---

[6] Total global energy infrastructure is valued at $45 to $55 trillion. When fossil fuels are banned all the fossil-fuel power stations, worth $45 trillion, will become stranded assets, a loss that the world cannot sustain. These power stations must be transformed into carbon capturing or even negative-carbon power plants, which costs money of course, $200 billion/year for 15 years, i.e., $3 trillion, according to Chichilnisky and Bal (2019, p. 251), but can be done with existing technology.

[7] https://www.eia.gov/; Feb 2021 8.4% from wind; 2.3% from solar (half of total renewables), with fossil fuels and nuclear accounting for about 80%.

the energy-producing infrastructure, such as fossil-fuel burning power stations, will become stranded assets.[8] An example of such a misguided approach is Chris Goodall's in his latest book (Goodall 2020). He tells us what we need to do *now*, i.e., immediately, in all kinds of areas of life, to actually completely change our lives. He bases himself on, admittedly very valid, environmental arguments, but completely unrealistic to achieve even in the long term (i.e., in 50 years), if at all, and certainly not in two to three decades from today. Laying out a programme that must lead in twenty years to a 100% carbon-free economy is bound to fail. That time will already be needed to convince a sizable part of the population that such a programme may be feasible, making the '*now*' in the title of his book meaningless.

These writers seem to forget that there is a system in place at the moment, that we have a society that works, at any rate to some extent. To call our system a failing system that should be discarded and built up anew from scratch is ludicrous and, if put into practice, will for sure lead to untold misery and hardship. It is true that the system, although not the energy system, often fails, that many wrong decisions are taken, but most of the time it is capable of accomplishing something positive, by trial and error, by luck or otherwise. And it must be emphasized that this is nontrivial, as can be witnessed all over the world in the way the Covid-19 pandemic has been handled.[9] Things are done in a certain way and for a reason, although it is not always easy to see what that reason is. An organization exists that is working, at any rate to some extent, and should be kept intact as much as possible, giving companies, especially energy-generating and energy-distributing companies, and governments the opportunity to gradually work towards change. It is not helpful to engage in bashing fossil-fuel companies for having been "the largest contributors to the climate crisis for decades, all in pursuit of profit and growth."[10] It is not only unhelpful, it is also untrue. The products that these companies generate or distribute may now be perceived as bad or not good, although we still buy them in undiminished quantities, but that has not always been the case and the organizing and managing power that these companies represent is equally important.

Moreover in large parts of the world there still is energy poverty, a far different energy poverty than the one we just experienced in 2022 due to steeply rising gas and electricity prices, again showing how delicate energy systems are. According to Bill Gates (Gates 2021, p. 68), about 860 million people,[11] as many as live in the EU and the US taken together, don't have reliable access to electricity in their homes, if they have a home, that is. Their children must make their homework by candlelight and

---

[8]   A recent report by Carbon Tracker Initiative found that stranded assets for coal utilities alone amount to $220 bn of which $121 bn involves listed equities, 90% of which are on Asian stock exchanges (with more than half in India).

[9]   Naomi Klein's book *This Changes Everything* (Klein 2014) is based on the premise that neoliberal capitalism is the root cause of the climate problem and the main reason for the problem not being properly addressed. She therefore argues that the system must be thoroughly overhauled. We need another "ten days that shook the world". If you know what the first ten shocking days brought the world, you must be rather dumb to ask for a repeat event.

[10]   Greta Thunberg in praising Michael Mann's book *The New Climate War*.

[11]   According to the *Energy for Growth Hub* (https://www.energyforgrowth.org/; memo by John Ayaburi, Morgan Brazilian, Jacob Kincer and Todd Moss) this number is actually 3.5 billion, 45% of the world's population.

cooking is done on open fires. They are entitled to massive amounts of reliable and affordable electricity. And this energy must be clean, and we have a duty to provide it.

The fact that fossil-fuel companies are blamed for the climate mess we are in has resulted in some environmentalists having developed an unreasonable and apparently uncontrollable hatred towards these companies, rejecting carbon-removal technologies like carbon capture. Michael Mann in the hatred-filled introduction to his latest book (Mann 2021) calls coal burning with carbon capture a "false solution."[12] Others incomprehensibly state "carbon capture and storage is not a solution to the climate crisis, it is part of the problem." They see it as "an extraordinarily expensive pipe dream falsely propagated by the fossil fuel industry in an attempt to save itself."[13] It is beyond me why fossil fuel companies shouldn't be allowed to try to save themselves, if they abide by the new rules. Whatever their faults, such companies are in a better position to achieve goals they have committed themselves to after a realistic assessment than any other entities, operating in society or on its fringes, including political organizations and environmental NGOs, especially in the democracies we are so fond of. It may be going a bit far and be rather arrogant, as Shell's former CEO claimed, that we need them on climate change but it can certainly be a heck of a lot easier if they cooperate in a positive way towards a sustainable energy system. A lot of unrealistic wishful thinking and fantasizing can then be cut out from the start. A future in which fossil-fuel companies like Shell turn themselves in a few decades from oil and gas companies into renewable energy companies is not abhorrent to me. It would keep a vast amount of corporate and technical knowledge intact and in use, avoiding the always painful scenes, both from a human and a societal perspective, that accompany industries going bust.[14] As Goodall observes somewhere (p. 44) in his book "Give Shell a $10bn project (…) and you will find no company in the world better at completing it on time and on budget." There are more of such companies with such capability. The way they do business is as valuable as the business itself. But they need time, although things can be speeded up by some prodding, it is better to be cautious. Don't throw the baby out with the bathwater. These companies are not evil *per se*, and I am sure that they are happy to adapt so long as a level playing field is maintained. They in general do not object against paying any tax, including a carbon tax, if such tax is levied across the field. Harping on their huge profits is not helpful either. As the first year (2020) of the Covid-19 pandemic showed, these profits can easily turn into losses of similar magnitude.[15]

---

[12] I totally agree with Mike Hulme's assessment of the book in *Science and Technology* (Spring issue, vol. 37, 2021) as "offering an incoherent and distinctly unhelpful narrative on climate change." Mann acts as the little Putin of climate change, viewing the world through a distinctly Hobbesian lens whereby the fight against climate change is a zero-sum game.

[13] Carbon capture. A solution, or part of the problem? *The Guardian Weekly*, 9 October 2020.

[14] This is contrary to Naomi Klein's view who wants to break up and phase out these centralized monsters, both the privately owned and the big publicly owned oil companies (Klein 2014, p. 130). She is altogether opposed to the private ownership of most things, I believe, but in any case of energy. With others, she is all in favour of small-scale community-owned and democratically run cooperative ventures. She is also in favour of creating huge numbers of jobs; jobs for the sake of jobs, an attitude that has always puzzled me. I wonder whether these jobs come with salaries and, if so, who pays for them.

[15] See https://zerotracker.net/.

When asking for a sweeping and drastic end to the extraction of fossil fuels, it should not be forgotten either that fossil fuels are not just burned for heat or power but also used for other purposes such as the manufacture of plastics and other useful products. Oil is hardly burned at all for heat. Such non-energetic uses of especially oil will remain important. Hence oil production cannot be brought to zero, far from it. Without being combusted, fossil fuels are used directly as construction materials, chemical feedstocks, lubricants, solvents, waxes, and many other products, the 500,000 referred to by Alice Friedemann (Friedemann 2021). More applications will undoubtedly be developed in the future. In 2017, about 13% of total petroleum products consumed were for non-combustion use. Oil reserves at present (early 2023) are estimated at about 1.4 trillion barrels,[16] enough for about 40 years at the rate we are consuming it now.[17] If we stop burning it in cars and such like and just spend it for the other useful products (the 13% mentioned above) we will still only have enough for a little over three centuries, underscoring the fact that oil is just too precious to be used for combustion purposes.

In my view we should be grateful to fossil fuels and cherish the little we still have. They are and will always remain immensely valuable. But they have been used too much, without restraint, as if there is no tomorrow, not because fossil-fuel companies took them out of the earth, but due to man's insatiable demand for energy, for big cars and trips to foreign countries. It is for instance preposterous to blame the fossil-fuel industry for "China's insatiable demand for energy" (Sinha 2020). The Chinese will not pay much attention to such statements and continue to grow at their current pace and burn coal in the more than 1,000 coal power stations they operate, for which they do not need and have never needed Western fossil-fuel companies. China's dependence on coal has already taken a heavy toll on its environment and air quality, and soon the insight will emerge that this cannot continue and has to be remedied, but it will not do this by just shutting down its coal plants. As we have seen in Chapter 2, China is currently by far the world's largest carbon emitter. In May 2021 it pledged to hit peak emissions by 2025, where the earlier date was 2030. It is a first step, but still just a promise and a recent study (Wu and Zhu 2021) has indicated that it will be very hard to keep. It is in any case at odds with the news[18] in a report from the Centre for Research on Energy and Clean Air (CREA) and the Global Energy Monitor (GEM) that in 2022 China issued more permits for new coal plants than at any time since 2015, all of them plants without CCS. Local governments approved 106 GW of new coal-fired power capacity (up to two plants per week) and the construction of new coal power capacity starting in China was six-times larger than the rest of the world combined. These new plants will remain in operation for at least 40 years. Such unabated coal expansion is completely at variance with the need to phase out fossil fuels and/or to reach net zero by 2060, as China has promised.

---

[16] https://www.worldometers.info/oil/, which keeps track of existing reserves from minute to minute.

[17] For a recent balanced and more pessimistic view on oil reserves, see Laherrère et al. 2022.

[18] https://energyandcleanair.org/publication/china-permits-two-new-coal-power-plants-per-week-in-2022/.

For China CCS is indispensable for being able to retain at least part of its huge fleet of coal-fired power stations and efforts to speed up research and development are strongly recommended: "2035 is regarded a suitable time for CCS commercialization in China with a retrofit potential of 143.63–431.04 GW and 31.46 billion USD cost input. Moreover, at the regional level, there is a great CCS retrofit potential of coal-fired power plants in Shaanxi, Hebei, and Inner Mongolia. Policymakers should provide greater support for the second-generation CCS technologies and promote them actively in 2030–2035, especially in Shaanxi, Hebei, and Inner Mongolia, to achieve CCS commercialization and control the CO_2 emissions of coal-fired power plants in China" (Fan et al. 2018). Indeed they should.

We need action, and although it is true that promises mostly precede action, it is also true that promises are seldom followed by action. Global emissions must be brought down rapidly if the Paris climate goals have any chance of being achieved, but at the same time there is an ever more pressing need to scrub part of the carbon we have emitted and are emitting out of the atmosphere, and put it somewhere safe where it can do no further harm. Carbon capture and storage is a must for both!

As said, climate change is a global problem and should be solved by a global approach which in a sensible, acceptable, varied and measured way works towards an achievable goal. When tackling such a huge problem it seems sensible to first investigate what should be the aim. This has essentially been set by the Paris conference, viz. ensuring that global temperature rise due to anthropogenic greenhouse emissions stays below 2°C. It can then be determined how high the CO_2 concentration in the atmosphere is still allowed to rise in order not to exceed this limit.

In 2004, when this concentration was around 375 ppm, Pacala and Socolow (Pacala and Socolow 2004) came up with 500 ppm as the limit for the CO_2 concentration to keep warming below 2°C. They proposed a systematic approach to ensure that emissions would be stabilised at the 2004 level and an atmospheric CO_2 concentration of 500 ± 50 ppm would be reached after 50 years (by 2054). So, their approach does not involve a frantic, uncoordinated race to net zero at gigantic expense in some parts of the world while leaving the rest of the world fend for themselves. They calculated that the mitigation effort needed for this can be carried out with at that time available commercial technologies, of which CCS is just one, but at that stage still without Direct Air Capture or other Direct Carbon Removal technologies. This means that we don't have to wait for new revolutionary technologies as some are fond of exclaiming. Although the authors emphasize the importance of research into new technologies that are needed in the second half of the century, they brought the hopeful message that current technologies are adequate to start solving the climate problem. This is important as there is no single technology that can or must do the job alone. In their framework the period of flat emissions from 2004 to 2054 (at the 2004 level) accomplished by these technologies implies that efforts to reduce emissions only cover the potential emissions increase due to economic growth. For instance, the number of cars on gasoline is allowed to increase if the efficiency of such cars increases accordingly, keeping total emissions at the 2004 level. This will then be followed by a linear decline of about two-thirds in the subsequent 50 years after 2050, helped on its way by revolutionary or otherwise new and promising technologies, like the use of hydrogen, electric vehicles, better wind turbines and storage of electricity, to be developed from 2004 to 2050, and a further slower decline in the next century after 2100. The whole scheme therefore extends over a period of one to one and a half centuries.

The task of tackling climate change and achieving net zero is so big that it will need to be tackled by a wide range of approaches—akin to the many wedges of a pie—each of which may contribute only a small part to the solution. In their paper, which has been much quoted but unfortunately little acted upon, possibly apart from a *Stabilization Wedges Game* to teach students about the scale of the greenhouse-gas problem and about existing technologies that can reduce carbon emissions, Pacala and Socolow identified fifteen areas of activity which in their opinion are capable of contributing to the CO_2 reductions necessary for limiting global warming to less than 2°C by 2054 (or limiting the CO_2 concentration to 500 ± 50 ppm). They selected seven of these areas (energy efficiency and conservation (of buildings and vehicles), fuel shift from coal to natural gas, CCS, nuclear power (their paper is pre-Fukushima), renewable energy and fuels (wind, PV solar, and biomass), management of forests and agricultural soils). They then divided the necessary CO_2 emissions reductions in seven equal "wedges", whereby each wedge represents one of these areas of activity for reducing emissions such that during the 50-year period from 2004 to 2054 emissions will remain stable at roughly 28.6 Gt of CO_2[19] (the 2004 level) per year. They also assume that 'business as usual' (1.5% growth of annual carbon emissions as seen in the 30 years preceding 2004) would correspond to a straight-line rise from 28.6 Gt of CO_2 in 2004 to 51.4 Gt of CO_2 in 2054.[20] Each activity starts at zero in 2004 and makes the same contribution (wedge) to emissions stabilization such that for each wedge total emissions remain at the 2004 level, ending up in 2050 with 3.67 Gt/year of reduced emissions in carbon dioxide for each wedge, and a cumulative total of 92 Gt of CO_2 per wedge ($50 \times 3.67/2$) in reduced emissions from 2004 to 2054, so a grand total of 644 Gt of CO_2. The approach has been schematically visualised in Fig. 15.1.

Their approach is a simple and intuitive way to divide the climate problem into parts and visualise the mitigation efforts that are required. This simplicity and flexibility has resonated with a broad audience, as evidenced by its widespread use in education, as referred to above.

The precise numbers taken as starting point are not that important, nor is the number of seven wedges. It is however clear that the more wedges are necessary or the longer we wait with implementing such an approach, the more difficult the task will be. A later estimate[21] by the same authors in 2011 increased the number of necessary wedges from seven to nine, due to the continuing increase in emissions

---

[19] The authors of the paper work with carbon C, not CO_2. I have converted their numbers into numbers for CO_2.

[20] If you take the *Our World in Data* 2004 emission level of 28.6 Gt of CO_2 and assume a 1.5% linear rise until 2021, the 2021 emission level will come out at 36.8 Gt of CO_2 which is almost spot on with the number of 37.1 Gt actually reported for 2021.

[21] https://thebulletin.org/2011/09/wedges-reaffirmed/.

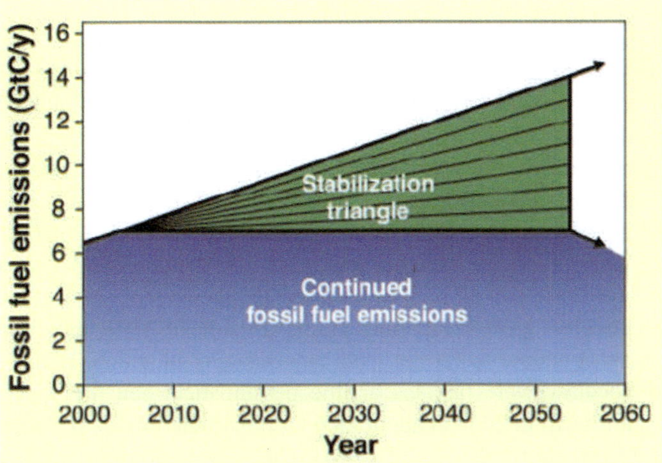

Fig. 15.1. Idealized picture of CO_2 emissions from 2004 to 2054. A stabilization triangle of avoided emissions (green) and allowed emissions (blue) at the 2004 level. The stabilization triangle is divided into seven wedges, each of which reaches 3.67Gt CO_2/year in 2054 (*from* Pacala and Socolow 2004).

since the original paper and since no serious action to mitigate climate change had been taken. If we wait still longer with taking action or if carbon emissions are growing faster than the assumed 'business-as-usual' case, seven or nine wedges will not be enough and additional ones will have to be added.

In a 2021 paper (Johnson et al. 2021) it was shown, in addition to fairly irrelevant criticism of the Pacala-Socolow approach,[22] that the world has not made much progress in stabilizing emissions in the way suggested by Pacala and Socolow and is on track to achieve only 1.5 ± 0.9 wedges of the seven required to achieve the required stabilisation by 2054 and that 14 wedges would be needed to as yet achieve this. Emissions savings account for just 4%–16%. It does not need such a paper or any analysis at all, for that matter, to conclude that Pacala and Socolow's scheme has failed as it has not been pursued, and emissions have not stabilised at the 2004 level. Far from it, mankind now emits over 37 Gt of CO_2 instead of the 28.6 Gt in 2004. Most of the wedges Pacala and Socolow identified have not been utilized at all. This also applies for CCS: "so few CCS facilities have been built that their impact on global carbon emissions has been imperceptible. CCS is not being deployed at scale because the incremental costs of capture and the development of transport and storage infrastructures are not adequately compensated by market or government incentives. While the technological maturity of CO_2 capture options has improved considerably over the past decade, costs have barely fallen due to limited learning in commercial settings and increased resource and energy costs. The perceived risk of long-term CO_2 storage is a further barrier to deployment. (…) Pacala and Socolow showed that the portfolio of technologies required to 'solve the climate problem' has long been ready for deployment—the choice was simply between 'action and delay'

---

[22] It is not Pacala and Socolow that are in the wrong, but the world that has failed to take note of their suggestions.

(…) 15 years on, the world has opted to delay" (Johnson et al. 2021, p. 9; see also Von Hirschhausen et al. 2012; Bui et al. 2018; Bassi et al. 2015).

In 2004 the CO_2 concentration in the atmosphere was about 375 ppm and the authors aimed at a goal of 500 ± 50 ppm by 2050, a value in line with the assumption that this is sufficient to keep global warming below 2°C. In Chapter 2 we have seen that 2.37 ppm is added each year to the CO_2 concentration, at this rate ending up in 2054 at about 495 ppm. Currently, early 2023, the concentration has risen to about 420 ppm, in close agreement with a 2.37 ppm rise per year from 2004. This implies that at a rate of 2.37 ppm per year we will reach 490 ppm by 2050 if we continue to just talk and conclude that all these solutions are fake as they fail to fundamentally challenge the endless growth economy that is at the heart of global climate change, and use this as an excuse to refrain from any meaningful action. It also shows that Pacala and Socolow were perhaps a little pessimistic when they demanded that emissions stabilise from 2004 at about 28.6 Gt of carbon dioxide per year.

We could still start today, or let's say in 2024, just 20 years late. In addition to (retro)fitting power plants with CCS installations, Direct Air Capture and other Direct Carbon Removal approaches, which have gone through serious development since 2004 and, as we have seen, are enthusiastically pursued by a myriad of companies all over the world, could be added as a separate wedge. Since 2004 annual carbon emissions[23] have risen to 37 Gt of CO_2. Stabilizing emissions at the 2022 level until 2054 would increase the atmospheric CO_2 concentration by about 2.4 ppm per year, so 77 ppm in 32 years. Added to the 420 ppm already in the air we would still end up just below 500 ppm by 2054. Assuming, as Pacala and Socolow did, that annual carbon emissions in the business-as-usual scenario grow linearly by 1.5%,[24] the contribution of the combined wedges to emissions avoidance will have to grow from zero in 2024 to 20 Gt of CO_2 in 2054. This will be a more modest achievement than the original Pacala and Socolow scheme, but we will only have 30 years (instead of 50) for the wedges to work towards this goal. It also implies accepting a higher level of residual carbon emissions, namely the 2022 level of 37 Gt instead of the 2004 level of 28.6 Gt, but still such that the CO_2 concentration in the air (500 ppm) is compatible with a temperature rise of 2°C. This must be feasible, perhaps with more than seven wedges, but it certainly can be done. A global organization will have to be set up (the International Energy Agency or another already existing organization could perhaps fulfil this role), provided with funds from the developed countries in particular and/or from a (low) tax on emissions starting for instance at \$1 per ton of CO_2 emitted, to start the necessary activities in the countries where it is most effective. This organization will for instance be charged with the task to select suitable coal and/or gas plants anywhere in the world, e.g., India, China, Indonesia, to be fitted

---

[23] The rise in CO_2 emissions was especially pronounced from 2000 (25.2 Gt) to 2011 (34.4 Gt) after which it started to slow down with just a rise of 2Gt in about ten years (2018 (36.7 Gt), 2019 (36.7 Gt) and 2020 (34.8 Gt, impact of the Covid-19 pandemic)). If the downturn due to the pandemic is not wiped out in one or two years and we start the Pacala and Socolow approach now, we can still easily end up below 500 ppm by 2050.

[24] In such a business-as-usual approach carbon-dioxide emissions in 2054 would reach 57.8 Gt of CO_2 per year, so to keep emissions stable at the 2022 level about 20 Gt must be avoided by 2054.

with CCS facilities to achieve the goal of that particular wedge, but will also have to supervise the implementation of the other wedges chosen to achieve the goals set. After having stabilised CO_2 emissions by 2050, at the admittedly rather high 2022 level of 37 Gt, the effort can then be speeded up by deploying new technologies and achieve a further two-thirds reduction in accordance with Pacala and Socolow's original scheme. It is a completely different approach from each country working towards its own carbon-reduction goals or discarding the fossil-fuel infrastructure as soon as possible without further ado. It gradually incorporates sustainable energy sources into the energy infrastructure, softens the disadvantages of the continued use of fossil fuels and transits in a measured and equitable way towards a reliable energy infrastructure that no longer burdens the planet with greenhouse-gas emissions.

So, let's get started. There is nothing that should hold us back and there is no time to lose. It should be realized that we have no other option. The alternatives that some put forward are just theoretical, in particular immediately stopping to burn fossil fuels and living off the sun and wind. It does not even come close to a solution of the problem, as it will not reduce the amount of carbon dioxide already in the atmosphere and will make most lives miserable. So far the problem has not even been brought to a standstill but actually moves faster than the solution. If we wait longer, repair will become ever more difficult and in the end very likely impossible. This book has attempted to show that nothing, apart from money and political will, stands in the way of applying these two types of remedy. In 2009 poor countries were promised $100 billion per year from 2020 to 2025 to help them cope with the negative consequences of climate change. Rich countries have not kept that promise but could still do so and create a fund filled with these promised billions for the IEA or another dedicated organization to help poor countries cutting emissions in the way as set out above and by Pacala and Socolow. We know how to fit CCS installations to power stations and other industrial point-source emitters and also know how to filter carbon dioxide out of the air. The technology is all there. It can be improved for sure, but that will happen. We will learn as we go along as we have done many times before with all kinds of technology. Wind and solar were decried as too expensive just a few decades ago and currently they generate energy at bottom prices. In the process of deploying these carbon-reducing methods we will rapidly learn how to solve any remaining problems.

More and more people are waking up to the fact that solely reducing emissions will not be enough, but that legacy emissions, i.e., the CO_2 emitted in the last century and still residing in the air, will also have to be tackled. For this purpose, more than 80 businesses and organizations involved or interested in the deployment of carbon capture, transport, use, removal and storage have been brought together in the Carbon Capture Coalition[25] by the Great Plains Institute,[26] a non-partisan organization with the laudable objective to make the way we produce, distribute, and consume energy both environmentally and economically sustainable. The Carbon Capture Coalition promotes carbon-capture technologies and works to achieve economy wide

---

[25] https://carboncapturecoalition.org/.

[26] https://www.betterenergy.org/.

deployment of carbon capture, removal, transport, utilization, and storage, with the aim of reducing carbon emissions to meet mid-century climate goals, foster domestic energy and industrial production.

As mentioned before, the Biden Administration has launched an extensive programme for DAC and in its Regional Direct Air Capture (DAC) Hubs Program it envisages carbon capture to take place in the US by setting up four large-scale, regional DAC hubs that each comprise a network of carbon capture projects.[27]

The IEA report on Direct Air Capture (IEA 2022) states that its Net Zero Scenario requires the immediate and accelerated scale-up of DAC, calling for an average of 32 large-scale plants (1 MtCO2/year each) to be built each year between now and 2050. That sounds a lot, in total some 900 plants, but would only scrub out about 1 Gt per year and cumulatively some 15 Gt by 2050, less than half a year's worth of global emissions, but already quite close to the 20 Gt demanded for it as a wedge in the revised Pacala and Socolow scheme described above. On itself it will not significantly affect the CO_2 concentration in the air, as this will still increase until 2050 and remain too high for a couple of centuries, unless accompanied by a huge reduction in emissions by CCS, renewables, energy efficiency, etc.

As far as carbon capture and storage in general is concerned, the following can in summary be stated:[28]

1. The technology has been in use for more than 50 years, and around 300 million tons of CO_2 have already been successfully captured and injected underground;

2. It is one of the few proven technologies that can deliver deep emissions reductions in industrial sectors;

3. The capture, transport, and storage of CO_2 is well regulated and empirically proven to be safe;

4. Its cost is quickly declining as the breadth of deployment increases and additional policy and financial incentives are made available;

5. It is a necessary tool for reducing the emissions of fossil fuels already in use and putting the world on a path to net-zero;

6. The world has more than enough capacity for storing the captured CO_2.

Let me finish this book with a quote from James Lovelock's book *The Revenge of Gaia*, by far the wisest of all the books about our carbon predicaments I have read and one everybody should read: "Despite all the warnings, we carry on destroying and seem to worry only about the nearly trivial, even imaginary, risk of cancer from mobile telephones, power lines, pesticide residues in food, or sunlight, topping them all is a fear of anything to do with nuclear energy. We are indeed straining at a gnat but swallowing a camel with ease" (Lovelock 2007, p. 156–157).

---

[27] NETL Carbon Storage Newsletter 22(6): 2022.

[28] CCS Mythbusters: dispelling myths around Carbon Capture and Storage (CCS), Global CCS Institute; https://www.globalccsinstitute.com/wp-content/uploads/2022/06/MythBusters-Flyer_FINAL-5.pdf.

Glossary

Amines compounds that contain a basic nitrogen atom and a pair of valence electrons that are not shared with another atom.

Amine scrubbing Amine gas treating, also known as amine scrubbing, gas sweetening or acid gas removal, refers to a group of processes that use aqueous solutions of various alkylamines (commonly referred to as amines) to remove hydrogen sulphide (H_2S) and carbon dioxide (CO_2) from gases. With current technology it is the most effective method of CO_2 capture from the flue gas of a pulverized coal plant and for other applications that require CO_2 removal.

CAES Compressed Air Energy Storage is a way to store energy by using compressed air which is pumped into a large tank or, if available, an underground salt cavern. When electricity is needed, the high pressure air is slowly released and used to drive a turbine; a bit like blowing up a balloon and then using the air that comes rushing out to generate electricity.

This type of storage has not been tested at meaningful scale yet, but the expectation is that if you have access to a suitable salt cavern, it will be cheap, as it is simple and, apart from the salt cavern, uses off-the-shelf components (e.g., compressors, pumps and turbines). Safety is a concern as storing anything at extremely high pressure is always a risk. One further unresolved challenge in large-scale design is the management of thermal energy since the compression of air leads to an unwanted temperature increase that not only reduces operational efficiency but can also lead to damage. The lifetime of a CAES project could be as long as 50 years—allowing for a very low levelized cost of storage by spreading capital costs over a long period.

Carbon credit A carbon credit is any tradable certificate or permit representing the right to emit one ton of carbon dioxide or the equivalent amount of a different greenhouse gas.

Carbon-dioxide removal (CDR) anthropogenic activities removing CO_2 from the atmosphere and durably storing it in geological, terrestrial, or ocean reservoirs, or in products. It includes existing and potential anthropogenic enhancement of biological or geochemical sinks and direct air capture and storage, but excludes natural CO_2 uptake not directly caused by human activities.

CO_2e or CO_2-equivalent CO_2-equivalent or CO_2e is a unit of measurement that is used to standardise the climate effects of various greenhouse gases. Apart from water vapour, the main greenhouse gases are carbon dioxide (CO_2), methane (CH_4), nitrous oxide (N_2O), hydrochlorofluorocarbons (HCFCs), hydrofluorocarbons (HFCs) and ozone (O_3) (the six gases of the Kyoto greenhouse-gas basket). To make the effects of the various greenhouse gases comparable, the IPCC has defined the so-called "Global Warming Potential" for each gas. It expresses the warming effect of a certain amount of a greenhouse gas over a set period of time (usually 100 years) in comparison to CO_2. For example, over a 100-year period the effect of methane on the climate is 28 times more severe than CO_2, but it doesn't stay in the atmosphere as long. The environmental impact of nitrous oxide also exceeds that of CO_2 by almost 300 times. In this way, greenhouse gases can be calculated as CO_2-equivalents. CO_2-equivalents are abbreviated as CO_2e.

CSP Concentrated solar power (CSP, also known as concentrating solar power, concentrated solar thermal) systems use mirrors or lenses to concentrate sunlight into a receiver. Electricity is generated by converting the concentrated light to heat, which drives a heat engine (usually a steam turbine) connected to an electric power generator or powers a thermochemical reaction.

Direct Air Capture (DAC) (Chemical) process by which a pure carbon dioxide (CO_2) stream is produced by capturing CO_2 from ambient air.

Discount rate The interest rate used to determine the present value of future cash flows. The discount is the difference between the original amount owed in the present and the amount that has to be paid in the future to settle the debt. It defines the relative value of present costs and future damages. The lower the discount rate the lower the social cost.

Electrolytic hydrogen This refers to hydrogen obtained by the splitting of water by electrolysis. This produces hydrogen and oxygen, the components of water. The hydrogen can be stored for later use and the oxygen vented to the atmosphere without negative impact. To achieve electrolysis we need electricity. The idea is to use surplus wind and solar power for this, generated when there is not enough demand. The produced hydrogen is 'green' and can later be recombined with oxygen to produce energy (and water).

Energy poverty In the UK a household is said to be fuel poor if it needs to spend more than 10% of its income on fuel to maintain an adequate level of warmth; the French "Grenelle II" Act defines energy poverty as a situation in which a person has difficulty obtaining the necessary energy in their home to meet their basic needs because of inadequate resources or living conditions. In Europe even middle-income families are now (2022 due to the Ukraine war) paying up to 20–30% of their budget on energy.

Externality An externality is a cost or benefit for a third party that did not agree to it. The concept was first developed by the British economist Arthur Pigou (1877–1959) in the 1920s, hence a tax on a negative externality, like air pollution, is also called a Pigouvian tax.

Hydrogen comes in various colours nowadays, of which the most common are grey, blue and green. *Grey* hydrogen is the most common, and cheapest, form of hydrogen.

As it is created from natural gas its production process generates greenhouse-gas emissions. The technologies used don't capture the carbon emissions, which are instead released into the atmosphere.

Blue hydrogen is also extracted from natural gas but the carbon emissions are captured and stored, which reduces emissions to the atmosphere. Blue hydrogen is sometimes called 'low-carbon hydrogen' as the production process doesn't avoid the creation of greenhouse gases, just stores them away.

Green hydrogen doesn't generate any carbon emissions in its entire life cycle as it uses renewable energies in the production process, making it a true source of clean energy. It is made by electrolyzing (splitting) water using clean electricity created from surplus renewable energy from wind and solar power. The main challenge is reducing the production costs.

Beyond these three biggest types of hydrogen, there's a whole array of other colour options.

Black and *brown* hydrogen is created using coal in the extraction process. This process, called gasification, is on the opposite end of the spectrum from green hydrogen's electrolysis. It is used in many industries converting carbon-rich materials into hydrogen and carbon dioxide. The emissions are released into the air.

Pink hydrogen is extracted through electrolysis powered by nuclear energy. It is also referred to as purple or red hydrogen.

Turquoise hydrogen is very new and made using a process called 'methane pyrolysis', which produces hydrogen and solid carbon by using heat to break down a material's chemical makeup. No carbon is released into the air, instead it's stored in the solid carbon created. If proven to be effective, turquoise may join blue as a 'low-carbon hydrogen' if the carbon can be permanently stored in an environmentally safe way.

Yellow hydrogen is made through electrolysis specifically using solar power, similar to the process used to create green hydrogen, but with a sunnier name.

Finally, *white* hydrogen is found naturally in underground deposits of geological hydrogen. It is extracted through fracking, the process of drilling into the earth and directing a high-pressure mix of water, sand, and chemicals at the rock to release the gas inside. At the moment there are no plans to use this type of hydrogen as an energy source.

Integrated Gasification Combined Cycle (IGCC) Power plant using gas produced from high-sulphur coal, heavy petroleum residues or biomass. IGCC is an advanced power generation technology which reduces emissions of NOx, SO_2, and particulate matter and improves fuel efficiency of coal. It is a combination of two technologies: coal gasification, which uses coal to create a clean-burning gas (syngas), and combined-cycle, which is the most efficient method of producing electricity from gas commercially available today (a gas turbine generator generates electricity and the waste heat is used to make steam to generate additional electricity via a steam turbine), like in NGCC plants.

Intergovernmental Panel on Climate Change (IPCC) The panel was established in 1988 by the World Meteorological Organization (WMO) and the United Nations Environment Programme (UNEP) and was later endorsed by the United Nations General Assembly. It is dedicated to providing the world with objective, scientific information relevant to understanding the scientific basis of the risk of human-induced climate change, its natural, political, and economic impacts and risks, and possible response options. Membership is open to all members of the WMO and UN, and it currently has 195 members (two more than the UN itself, implying that all countries of the world are members, from the very small to the very large), so it is a truly global organization. Thousands of scientists and other experts contribute to the writing and reviewing of its reports, which are then scrutinized by governments. Since 1988 the IPCC has issued six so-called assessment reports, the last one having appeared in 2022 (the first part already in August 2021), which are comprehensive and balanced assessments of the state of agreed knowledge on topics related to climate change. All its reports include a 'summary for policymakers' which is subject to line-by-line approval by delegates from all participating governments. In addition, special reports, like the 2005 report on CCS or the 2018 report on Global Warming of 1.5°C, are published. In this way the IPCC provides an internationally accepted authority on climate change, with its reports having the agreement of leading climate scientists and consensus from participating governments. It also implies that pronouncements in IPCC reports are far from revolutionary, but on the contrary rather conservative, many think too conservative.

Levelized Cost of Electricity (LCOE) represents the average revenue per unit of electricity generated that would be required to recover the costs of building and operating a generating plant during an assumed financial life and duty cycle. It is a measure of the average net present cost of electricity generation for a generating plant over its lifetime. For technologies with no fuel costs and relatively small variable operation and management costs, such as solar and wind technologies, LCOE changes nearly in proportion to the estimated capital cost of the technology. For technologies with significant fuel cost, both fuel cost and capital cost estimates significantly affect LCOE.

Natural gas is a mixture of four naturally occurring gases, all of which have different molecular structures. It consists for 70–90% of methane (CH_4) and some ethane (C_2H_6), butane (C_4H_{10}) and propane (C_3H_8). These compounds nicely illustrate the versatility of carbon. So, natural gas essentially is just another name for methane but sounds much nicer. After all, how can anything that is natural be bad? And if it is, there is not much we can do about it since it is 'natural'. "Don't call it methane, call it 'natural' gas." In this connection it must be pointed out that in fracking operations, both for oil and gas, a lot of methane leaks into the environment, so much that, according to some, such gas is in the end as polluting as coal. During its life in the atmosphere, albeit rather short, methane acts as a very strong greenhouse gas (80 times stronger than CO_2 calculated over a 20 year period). The fracking boom could actually be the explanation for the puzzling rise in global methane levels.

Natural gas combined cycle (NGCC) advanced gas-fired power plant technology, which improves the fuel efficiency of natural gas. Most new gas power plants in North America and Europe are of this type. A gas turbine generator generates electricity and the waste heat is used to make steam to generate additional electricity via a steam turbine. See also IGCC.

Photovoltaics (PV) the conversion of light into electricity using semiconducting materials that exhibit the photovoltaic effect, i.e., the generation of voltage and electric current by the absorption of light which excites an electron or other charge carrier into an excited state. This technique is applied in solar cells in solar panels, which are therefore also called photovoltaic of PV modules. An alternative technology is concentrated solar power (CSP; see above) whereby solar power is generated by using mirrors or lenses to concentrate a large area of sunlight onto a receiver.

Power-to-Liquid Power to Liquid technology is a form of Power-to-X (also known as PtX or P2X) which is a collective term for conversion technologies that turn electricity into carbon-neutral synthetic fuels, such as hydrogen, synthetic natural gas, liquid fuels, or chemicals. They can be used in sectors that are hard to decarbonize or, unlike electric power, be stored for later use.

Pre-industrial This term is used very often in the IPCC reports in combination with level(s), time(s), era, baseline or temperatures, but a definition is never given. A few times the IPCC speaks of the pre-industrial period 1850–1900, although of course everything prior to 1850 is even more pre-industrial, and it can be argued that shortly after 1850 is no longer pre-industrial. According to Merriam-Webster the first use of pre-industrial in the sense of "occurring, existing, or originating before the development or adoption of industry" dates from 1883, while the advent of the Industrial Revolution is usually placed from 1750–1850. Others have argued that the period 1850–1900 includes some large volcanic eruptions and that greenhouse-gas concentrations had already started to rise before that. From 1850–1900 carbon-dioxide emissions rose by a factor of 10 from 200 to 2,000 Mt, while in 1800 it was 28 Mt. According to the data provided by the website *Our World in Data*, carbon-dioxide emissions have been on the rise from 1800, but really took off around 1840–50. In view of this it has been suggested that the earlier period of 1720–1800 is a better choice as a baseline. In 1750 all CO_2 emissions originated from the UK which in the first decades of the 19th century still accounted for 80–90% of emissions, going down to a little over 60% in 1850. This of course reflects the fact that industrialization, or at any rate the burning of coal, was already in full swing at that time in Britain. Whatever is chosen does not really matter that much, as long as it is clear, but the fact that we can have this discussion remains baffling and it is decidedly sloppy that such an important baseline is not defined more precisely.

Pumped hydro Energy storage system whereby water is pumped from a lower elevation reservoir to a higher elevation; not a particularly very high-tech solution. Low-cost surplus off-peak electric power is typically used to run the pumps. During periods of high electricity demand, the stored water is released through turbines to produce electric power. Although the losses of the pumping process make the plant a net consumer of energy overall, the system increases revenue by selling more electricity during periods of peak demand, when electricity prices are highest. If the upper lake collects significant rainfall or is fed by a river, the plant may be a net energy producer in the manner of a traditional hydroelectric plant.

Regenerative braking A regenerative braking system is an energy recovery mechanism that slows down a moving vehicle or object by converting its kinetic energy into a form that can be either used immediately or stored until needed. When taking the foot off the accelerator pedal, rotational energy from the rotating wheels is fed into a generator (the motor running in reverse) to generate electricity, which can be stored in the vehicle's batteries. It slows the car without the need of brake pads usually employed to slow down a car (and hence it also reduces the release of fine particulate matter into the atmosphere). In electric vehicles it is employed to charge the battery. Every time you brake, some electricity goes back into the battery. Braking actually helps you get more miles from the car, enabling you to drive further. In the most common form of regenerative braking an electric motor functions as an electric generator and uses the vehicle's momentum to recover energy that would otherwise be lost to the brake discs as heat. The system has been in use on trains already for decades.

Supercritical fluid A supercritical fluid is a substance at temperature (T) and pressure (P) above its critical point, i.e., the point in the PT diagram at which a liquid and its vapour can coexist. At higher temperatures, the gas cannot be liquefied by pressure alone. The *critical temperature* of a substance is the temperature above which vapour of the substance cannot be liquefied, no matter how much pressure is applied, or differently stated a substance cannot exist as a liquid above the critical temperature. The *critical pressure* of a substance is the pressure required to liquefy a gas at its critical temperature. The critical point of CO_2 is at 31°C and 73.8 bar, at which its density is 500 times that of the gaseous state.

Sustainability and regeneration Sustainability means to maintain the current state of the environment, whereas regenerative programs aim to restore the environment to a former state. "Sustainable development is the use of resources to improve society's wellbeing in a way that does not destroy or undermine the support systems needed for future growth. Regenerative development is the use of resources to improve society's wellbeing in a way that builds the capacity of the support systems needed for future growth. To give one example: 'sustainable agriculture' refers to a process of producing food that does not degrade the ecosystems on which agriculture depends. It seeks to farm in ways that keep soil erosion at 'replacement' levels. In this way, future generations will be able to farm the same land. This is a huge improvement over traditional, soil-erosion intensive farming, but does not go far enough. To farm regeneratively is to produce food while simultaneously leaving the plot of land better off—to farm in ways that not only leave roughly the same amount of soil after harvest, but actually increase the quantity and quality of soil after harvest. Regeneration builds capacity; sustainability, at best, maintains it" (Gabel 2015).

A degenerative society, as we are having at present and is bound to collapse at some time, maintains its life support systems through interactions with its environment that tend to draw more energy and resources, over time, from the environment than it can produce in that period of time, while a **sustainable society** maintains its life support systems in such a way that it draws from the environment over time only as much as the environment can produce in that period of time, and no more, and a **regenerative society** does even better by maintaining its life support systems in such a way that its actions of drawing from the environment actually serve to help create more production, more health, more resilience, and more longevity in its ecosystems than otherwise would have been created without the participation of that society as a feature of the ecosystem. In a regenerative society therefore mankind's participation in it actively regenerates the ecosystems which it is part of and depends on (Meyers 2022). Looking around you it will soon be clear that we are very far away indeed from such a society, no matter how desirable it may be. Actually, mankind never in all its history has behaved otherwise than as a destroyer of Nature. But, if the regenerative idea attracts you, please visit the website of Koru Collaborative to learn more.

Ultramafic and mafic rock The term mafic is a portmanteau of 'magnesium' and 'ferric' and refers to rock rich in magnesium and iron. Igneous rock (derived from the Latin word *ignis* meaning fire) is formed through the cooling and solidification of magma or lava. They are classified by their silicon dioxide (SiO_2) content, with ultramafic rock having < 45% SiO_2 and mafic rock 45–52%. An example is olivine which is a magnesium iron silicate and the most abundant type of rock in the crust of the Earth. It weathers quickly on the surface. For this reason, olivine has been proposed as a good candidate for enhanced weathering to sequester carbon dioxide from the Earth's oceans and atmosphere.

References and Literature

Allam, R., S. Martin, B. Forrest, J. Fetvedt, X. Lu, D. Freed et al. 2017. Demonstration of the Allam Cycle: An update on the development status of a high efficiency supercritical carbon dioxide power process employing full carbon capture. Energy Procedia 114: 5948–5966.

Alvizo, O., L.J. Nguyen, C.K. Saville, J.A. Bresson, S.L. Lakhapatri, E.O.P. Solis et al. 2014. Directed evolution of an ultra-stable carbonic anhydrase for highly efficient carbon capture from flue gas. Proceedings of the National Academy of Sciences of the United States of America 111: 16436–41.

American Physical Society. 2011. Direct Air Capture of CO2 with Chemicals. June 2011.

Anthony, E.J. 2008. Solid looping cycles: A new technology for coal conversion. Industrial & Engineering Chemistry Research 47: 1747–1754.

Armstrong, C. 2018. Climate change and justice. *In*: Thompson, W.R. [ed.]. Oxford Research Encyclopedias: Politics. Oxford University Press.

Armstrong, J.M. 2021. The Future of Energy – The 2021 Guide to the Energy Transition. Energy Technology Publishing.

Armstrong McKay, D.I., A. Staal, J.F. Abrams, R. Winkelmann, B. Sakschewski, S. Loriani et al. 2022. Exceeding 1.5°C global warming could trigger multiple climate tipping points. Science 377: eabn7950.

Ayres, E. 1999. God's Last Offer. Four Walls Eight Windows.

Bachman, E., A. Tavasoli, T.A. Hatton, C.T. Maravelias, E. Haites, P. Styring et al. 2022. Rail-based direct air carbon capture. Joule 6: 1368–1389.

Bassi, S., R. Boyd, S. Buckle, P. Fennell, N. Mac Dowell, Z. Makuch et al. 2015. Bridging the gap: improving the economic and policy framework for carbon capture and storage in the European Union. Grantham Institute.

Bayer, P. and M. Aklin. 2020. The European Union Emissions Trading System reduced CO_2 emissions despite low prices. Proceedings of the National Academy of Sciences 117: 8804–8812.

Beerling, D.J., E.P. Kantzas, M.R. Lomas, P. Wade, R.M. Eufrasio, P. Renforth et al. 2020. Potential for large-scale CO_2 removal via enhanced rock weathering with croplands. Nature 583: 242.

Beller, J.N. and M. Beller. 2019. Artificial photosynthesis: An introduction, Faraday Discussions 215: 9.

Benson, S.M. and D.R. Cole. 2008. CO_2 sequestration in deep sedimentary formations. Elements 4: 325–331.

Berners-Lee, M. 2019. There is No Planet B. Cambridge University Press.

Berners-Lee, M. 2022. The Carbon Footprint of Everything. Greystone Books.

Bipartisan Policy Center. 2021. The Commercial Case for Direct Air Capture 2021. Washington DC.

Bobeck, J., J. Peace and F.M. Ahmad. 2019. Carbon Utilization—A Vital and Effective Pathway for Decarbonization. Center for Climate and Energy Solutions.

Boehm, S., L. Jeffery, K. Levin, J. Hecke, C. Schumer, C. Fyson et al. 2022. State of Climate Action 2022. Climate Action Tracker.

Book, A. 2022. 7 green technologies we need to save the world. Medium Daily Digest 10.10.2022.

Bozal-Ginesta, C. and J.R. Durrant. 2019. Artificial photosynthesis—concluding remarks. Faraday Discussions 215: 439.

Brandl, P., M. Bui, J.P. Hallett and N. Mac Dowell. 2021. Beyond 90% capture: Possible, but at what cost? International Journal of Greenhouse Gas Control 105: 103239.

Breakthrough Energy. 2021. US Climate Policy Playbook.

Broecker, W. and R. Kunzig. 2008. Fixing Climate: The story of climate science—and how to stop global warming. Profile Books, London.

Brohé, A. 2017. The Handbook of Carbon Accounting. Routledge, London.

Brown, T.W., T. Bischof-Niemz, K. Blok, C. Breyer, H. Lund and B.V. Mathiesen. 2018. Response to 'Burden of proof: A comprehensive review of the feasibility of 100% renewable-electricity systems'. Renewable and Sustainable Energy Reviews 92: 834–847.

Bruvoll, A. and B.M. Larsen. 2004. Greenhouse-gas emissions in Norway: Do carbon taxes work? Energy Policy 32: 493–505.

Buck, H.J. 2021. Ending Fossil Fuels – Why Net Zero Is Not Enough. Verso, London.

Bui, M., C.S. Adjiman, A. Bardow, E.J. Anthony, A. Boston, S. Brown et al. 2018. Carbon capture and storage (CCS): the way forward. Energy & Environmental Science 11: 1062–1176.

Callahan, Ch.W. and J.S. Mankin. 2022. National attribution of historical climate damages. Climatic Change 172: 40.

Cárdenas, B., L. Swinfen-Styles, J. Rouse, A. Hoskin, W. Xu and S.D. Garvey. 2021a. Energy storage capacity vs. renewable penetration: A study for the UK. Renewable Energy 171: 849–867.

Cárdenas, B., L. Swinfen-Styles, J. Rouse and S.D. Garvey. 2021b. Short-, Medium-, and Long-Duration Energy Storage in a 100% Renewable Electricity Grid: A UK Case Study. Energies 14: 8524.

Cheng, L., G. Forster, Z. Hausfather, K.E. Trenberth and J. Abraham. 2022. Improved quantification of the rate of ocean warming. Journal of Climate 35: 4827–4840.

Chenier, P.J. 1986. Survey of Industrial Chemistry. Wiley, New York.

Chichilnisky, G. and P. Bal. 2019. Reversing Climate Change. World Scientific.

Choi, W.-J., H.-J. Park, Y. Cai and S.X. Chang. 2021. Environmental risks in atmospheric CO_2 removal using enhanced rock weathering are overlooked. Environmental Science & Technology 55: 9627–9629.

Ciais, P., C. Sabine, G. Bala and W. Peters. 2013. Carbon and other biogeochemical cycles *In*: Stocker, T.F., D. Qin, G.K. Plattner, M. Tignor, S.K. Allen, J. Boschung et al. [eds.]. Climate Change 2013: The Physical Science Basis. Contribution of Working Group I to the Fifth Assessment Report of the Intergovernmental Panel on Climate Change. Cambridge University Press.

Clack, C.T.M., S.A. Qvist, J. Apt, M. Bazilian, A.R. Brandt, K. Caldeira et al. 2017. Evaluation of a proposal for reliable low-cost grid power with 100% wind, water, and solar. Proceedings of the National Academy of Sciences 114: 6722–6727.

Climate Action Tracker. 2022. State of Climate Action 2022.

Connolly, D., H. Lund, B.V. Mathiesen and M. Leahy. 2011. The first step towards a 100% renewable energy-system for Ireland. Applied Energy 88: 502–507.

Cook, P.J. 2012. Clean Energy, Climate and Carbon. CSIRO Publishing, Collingwood, Australia.

Cox, E.M, N. Pidgeon, E. Spence and G. Thomas. 2018. Blurred lines: the ethics and policy of greenhouse gas removal at scale. Frontiers of Environmental Science 6: 18.

Cravens, G. 2007. Power To Save The World—The Truth About Nuclear Energy. Vintage Books, New York.

Crippa, M., D. Guizzardi, M. Banja, E. Solazzo, M. Muntean, E. Schaaf et al. 2022. CO_2 emissions of all world countries – JRC/IEA/PBL 2022 Report. Publications Office of the European Union, Luxembourg.

Davidson, R.M. 2011. Pre-combustion capture of CO_2 in IGCC plants. IEA Clean Coal Centre.

Del Grosso, S.J. and M.A. Cavigelli. 2012. Climate stabilization wedges revisited: can agricultural production and greenhouse-gas reduction goals be accomplished? Frontiers in Ecology and the Environment 10: 571–578.

De Luna, P., C. Hahn, D. Higgins, S.A. Jaffer, T.F. Jaramillo and E.H. Sargent. 2019. What would it take for renewably powered electrosynthesis to displace petrochemical processes? Science 364(3506): 1–9.

Dickel, R. 2022. Achieving net zero plus reliable energy supply in Germany by 2045: the essential role of CO2 sequestration, OIES Paper: ET13, June 2022.

Doerr J. 2021. Speed & Scale—A Global Action Plan for Solving Our Climate Crisis Now. Penguin Business.

Dooley, K., H. Keith, A. Larson, G. Catacora-Vargas, W. Carton, K.L. Christiansen et al. 2022. The Land Gap Report 2022; available at: https://www.landgap.org/.

Duan, H., S. Zhou, K. Jiang, C. Bertram, M. Harmsen, E. Kriegler et al. 2021. Assessing China's efforts to pursue the 1.5 degrees C warming limit. Science 372: 378–385.

Ellis, K., N. Cantore, J. Keane, L. Peskett, D. Brown and D.W. te Velde. 2010. Growth in a carbon constrained global economy. Overseas Development Institute.

Emun, F., M. Gadalla, T. Majozi and D. Boer. 2010. Integrated gasification combined cycle (IGCC) process simulation and optimization. Computers and Chemical Engineering 34: 331–338.

English, J.M. and K.L. English. 2022. An overview of carbon capture and storage and its potential role in the energy transition. First Break 40: 35–40.

Epstein, A. 2014. The Moral Case for Fossil Fuels. Portfolio/Penguin, New York.

European Environment Agency. 2011. Technical Report no. 14. Air pollution impacts from carbon capture and storage (CCS).

Fan, J.-L., M. Xu, F. Li, L. Yang and X. Zhang. 2018. Carbon capture and storage (CCS) retrofit potential of coal-fired power plants in China. Applied Energy 229: 326–335.

Fasihi, M., O. Efimova and C. Breyer. 2019. Techno-economic assessment of CO_2 direct air capture plants. Journal of Cleaner Production 224: 957–980.

Foster, J.B. and B. Clark. 2015. Crossing the River of Fire, Monthly Review 66: 1–17.

Fräss-Ehrfeld, C. 2009. Renewable Energy Sources. A chance to combat climate change. Kluwer Law International.

Friedlingstein, P., M. O'Sullivan, M.W. Jones, R.M. Andrew, L. Gregor, J. Hauck et al. 2022. The Global Carbon Budget 2021. Earth System Science Data 14: 1917–2005.

Friedemann, A.J. 2021. Life After Fossil Fuels. Springer.

Fthenakis, V., J.E. Mason and K. Zweibel. 2009. The technical, geographical, and economic feasibility of solar energy to supply the energy needs of the US. Energy Policy 37: 387–399.

Fuss, S., W.F. Lamb, M.W. Callaghan, J. Hilaire, F. Creutzig, T. Amann et al. 2018. Negative emissions—Part 2: Costs, Potentials and Side Effects. Environmental Research Letters 13: 063002.

Fuss, S. and F. Johnsson. 2021. The BECCS Implementation Gap—A Swedish case study. Frontiers in Energy Research 8: 553400.

Gabel, M. 2015. Regenerative Development: Going Beyond Sustainability. Kosmos XV 1: 38.

Gale, J. and P. Freund. 2001. Coal-bed methane enhancement with CO_2 Sequestration Worldwide Potential. Environmental Geosciences 8: 210–217.

Gasser, T., C. Guivarch, K. Tachiiri, C.D. Jones and P. Ciais. 2015. Negative emissions physically needed to keep global warming below 2°C. Nature Communications 6: 7958.

Gates, B. 2021. How to Avoid a Climate Disaster. Allen Lane.

GCI. 2016. Global Roadmap for Implementing CO2 Utilization. The Global CO_2 Initiative.

Gilbertson, T. and O. Reyes. 2009. Carbon Trading – How it Works and Why it Fails, Critical Currents. Dag Hammarskjöld Foundation Occasional Paper Series.

Gilbraith, N., P. Jaramillo, F. Tong and F. Faria. 2013. Comments on Jacobson et al.'s proposal for a wind, water, and solar energy future for New York State. Energy Policy 60: 68–69.

Glaser, B., M. Parr, C. Braun and G. Kopolo. 2009. Biochar is carbon negative, Nature Geoscience 2: 2.

Global CCS Institute. 2016. Special Report: Introduction to Industrial Carbon Capture and Storage.

Global CCS Institute. 2017. Global Cost of Carbon Capture and Storage.

Global CCS Institute. 2021a. Global Status of CCS 2021.

Global CCS Institute. 2021b. Technology Readiness and Costs of CCS.

Global CCS Institute. 2022. Global Status of CCS 2022.

Goodall, C. 2020. What We Need To DO Now - For a Zero Carbon Future. Profile Books, London.

Grant, D., D. Zelinka and S. Mitova. 2021. Reducing CO_2 emissions by targeting the world's hyper-polluting power plants. Environmental Research Letters 16: 094022.

Griffith, S. 2021. Electrify: An Optimist's Playbook for Our Clean Energy Future. MIT Press.

Grubert, E. 2020. Fossil electricity retirement deadlines for a just transition. Science 370: 1171–1173.

Gür, T.M. 2022. Carbon dioxide emissions, capture, storage and utilization: Review of materials, processes and technologies. Progress in Energy and Combustion Science 89: 100965.

Hansen, J., M. Sato, P. Kharecha, D. Beerling, R. Berner, V. Masson-Delmotte et al. 2008. Target Atmospheric CO_2: Where should humanity aim? The Open Atmospheric Science Journal 2: 217–231.

Hansen, J. 2009. Storms of My Grandchildren. Bloomsbury, London.

Hansen, K., C. Breyer and H. Lund. 2019. Status and perspectives on 100% renewable energy systems. Energy 175: 471–480.

Haoyang, C. 2018. Algae-based carbon sequestration. Earth and Environmental Science 120: 012011.

Hargraves, R. 2021. Electrifying Our World: for climate, for people, with fission. Independently Published.

Harvey, H., R. Orvis and J. Rissman. 2018. Designing Climate Solutions : A Policy Guide for Low-Carbon Energy Account. Island Press, Washington DC.

Hawken, P. [ed.]. 2017. Drawdown: The Most Comprehensive Plan Ever Proposed to Reverse Global Warming. Penguin Books.

Hayhoe, K. 2021. Saving us—A Climate Scientist's Case for Hope and Healing in a Divided World. One Signal Publishers, New York.

Heard, B.P., B.W. Brook, T.M.L. Wigley and C.J.A. Bradshaw. 2017. Burden of proof: A comprehensive review of the feasibility of 100% renewable-electricity systems. Renewable and Sustainable Energy Reviews 76: 1122–1133.

Hebert, M. 2015. New technologies for EOR offer multifaceted solutions to energy, environmental, and economic challenges. Oil & Gas Financial Journal (13 January).

Hermans, J. 2011. Energy—Survival Guide. Leiden University Press, Leiden.

Herzog, H.J. 2018. Carbon Capture. MIT Press.

Heyes, A. and B. Urban. 2019. The economic evaluation of the benefits and costs of carbon capture and storage. International Journal of Risk Assessment and Management 22: 324–341.

Holt, N. and G. Booras. 2009. Updated cost and performance estimates for advanced coal technologies including CO2 capture. Electric Power Research Institute.

IEA. 2016. Ready for CCS retrofit—The potential for equipping China's existing coal fleet with carbon capture and storage.

IEA. 2018. World Energy Outlook 2018.

IEA. 2019. Putting CO_2 to Use.

IEA. 2021a. Net zero by 2050—A Roadmap for the Global Energy Sector.

IEA. 2021b. Global Energy Review 2021.

IEA. 2021c. Energy Technology Perspectives 2020 – CCUS in Clean Energy Transitions.

IEA. 2021d. Direct air capture Tracking Report.

IEA. 2022. Direct Air Capture—A key technology for net zero.

IEEFA. 2022. The Carbon Capture Crux. September.

IPCC. 2005. IPCC Special Report on Carbon Dioxide Capture and Storage. Prepared by Working Group III of the Intergovernmental Panel on Climate Change. Cambridge University Press.

IPCC. 2018. Global Warming of 1.5°C. Cambridge University Press.

IPCC. 2021. Climate Change 2021: The Physical Science Basis. Contribution of Working Group I to the Sixth Assessment Report of the Intergovernmental Panel on Climate Change. Cambridge University Press.

IRENA. 2021. Renewable capacity statistics 2021. International Renewable Energy Agency (IRENA), Abu Dhabi.

Irlam, L. 2017. Global Costs of Carbon Capture and Storage, Update. Global CCS Institute.

Irvine, M. 2011. Nuclear Power—A Very Short Introduction. Oxford University Press.

Jaccard, M. 2005. Sustainable Fossil Fuels. Cambridge University Press.

Jaccard, M. 2020. The Citizen's Guide to Climate Success: Overcoming Myths that Hinder Progress. Cambridge University Press.

Jackson, S. and E. Brodal. 2018. A comparison of the energy consumption for CO_2 compression process alternatives, IOP Conf. Series: Earth and Environmental Science 167: 012031.

Jacobson, M.Z. and M.A. Delucchi. 2011. Providing all global energy with wind, water, and solar power, Part I and Part II, Energy Policy 39: 1154–1169; 1170–1190.

Jacobson, M.Z., M.A. Delucchi, M.A. Cameron and B.A. Frew. 2015. Low-cost solution to the grid reliability problem with 100% penetration of intermittent wind, water, and solar for all purposes. Proceedings of the National Academy of Sciences 112: 15060–15065.

Jacobson, M.Z., M.A. Delucchi, Z.A.F. Bauer, S.C. Goodman, W.E. Chapman, M.A. Cameron et al. 2017a. 100% Clean and renewable wind, water, and sunlight all-sector energy roadmaps for 139 countries of the World. Joule 1: 108–121.

Jacobson M.Z., M.A. Delucchi, M.A. Cameron and B.A. Frew. 2017b. The United States can keep the grid stable at low cost with 100% clean, renewable energy in all sectors despite inaccurate claims. Proceedings of the National Academy of Sciences 114: E5021–E5023.

Jacobson, M.Z., M.A. Delucchi, M.A. Cameron, S.J. Coughlin, C.A. Hay, I.P. Manogaran et al. 2019. Impacts of green new deal energy plans on grid stability, costs, jobs, health, and climate in 143 countries, One Earth 1: 449–463.

Jacobson, M.Z., A-K. von Krauland, S.J. Coughlin, E. Dukas, A.J.H. Nelson, F.C. Palmer et al. 2022. Low-cost solutions to global warming, air pollution, and energy insecurity for 145 countries. Energy & Environmental Science 15: 3343–3359.

Jacobson, M.Z. 2023. No Miracles Needed – How Today's Technology Can Save Our Climate and Clean Our Air. Cambridge University Press.

Jeffery, S., F.G.A. Verheijen, C. Kammann and D. Abalos. 2016. Biochar effects on methane emissions from soils: A meta-analysis. Soil Biology & Biochemistry 101: 251–258.

Jiang, Y., P.M. Mathias, C.J. Freeman, J.A. Swisher, R.F. Zheng, G.A. Whyatt et al. 2021. Techno-economic comparison of various process configurations for post-combustion carbon capture using a single-component water-lean solvent. International Journal of Greenhouse Gas Control 106: 103279.

Johnson, N., R. Gross and I. Staffell. 2021. Stabilisation wedges: measuring progress towards transforming the global energy and land use systems. Environmental Research Letters 16: 064011.

Joppa. L., A. Luers, E. Willmott, S.J. Friedmann, S.P. Hamburg and R. Broze. 2021. Microsoft's million-ton CO2-removal purchase—lessons for net zero. Nature 597: 629–632.

Keim, S. 2009. *In*: Leadership for Climate Change Times Three, notes for a paper presented to the Resources Management Law Association Conference.

Keith, D.W., G. Holmes, D. St. Angelo and K. Heidel. 2018. A Process for Capturing CO_2 from the Atmosphere. Joule 2: 1573–1594.

Kelemen, P., S.M. Benson, H. Pilorgé, P. Psarras and J. Wilcox. 2019. An overview of the status and challenges of CO_2 storage in minerals and geological formations. Frontiers in Climate 1(9): 1–20.

Kelly, S. 2018. How America's clean coal dream unravelled, Guardian 2 March.

Kheirinik, M, S. Ahmed and N. Rahmanian. 2021. Comparative techno-economic analysis of carbon capture processes: pre-combustion, post-combustion, and oxy-fuel combustion operations. Sustainability 13: 13567.

Kikstra, J.S., P. Waidelich, J. Rising, D. Yumashev, C. Hope and C.M. Brierley. 2021. The social cost of carbon dioxide under climate-economy feedbacks and temperature variability. Environmental Research Letters 16: 094037.

Kim, S., M.P. Nitzsche, S.B. Rufer, J.R. Lake, K.K. Varanasi and T.A. Hatto. 2023. Asymmetric chloride-mediated electrochemical process for CO_2 removal from oceanwater. Energy & Environmental Science 2023.

Kivi, I.R., R.Y. Makhnenko, C.M. Oldenburg, J. Rutqvist and V. Vilarrasa. 2022. Multi-layered systems for permanent geologic storage of CO_2 at the Gigatonne Scale. Geophysical Research Letters 49: e2022GL100443.

Klein, N. 2011. Climate rage. pp. 239–248. *In*: McKibben, B. [ed.]. The Global Warming Reader. Penguin Books.

Klein, N. 2014. This Changes Everything—Capitalism vs The Climate. Simon & Schuster, New York.

Kolbert, E. 2021. Under a White Sky—The Nature of Our Future. Crown, New York.

Kolstad, C., K. Urama, J. Broome, A. Bruvoll, M. Cariño-Olvera, D. Fullerton et al. 2014. Social, economic and ethical concepts and methods. *In*: Climate Change 2014: Mitigation of Climate Change. Contribution of Working Group III to the Fifth Assessment Report of the Intergovernmental Panel on Climate Change. Cambridge University Press.

Koonin, S.E. 2021. Unsettled—What Climate Science Tells Us, What It Doesn't, And Why It Matters. BenBella Books, Dallas.

Kothandaraman, J., J.S. Lopez, Y. Jiang, E.D. Walter, S.D. Burton, R.A. Dagle et al. 2022. Integrated capture and conversion of CO_2 to methanol in a post-combustion capture solvent: heterogeneous catalysts for selective C-N bond cleavage. Advanced Energy Materials 12: 2202369.

Kramer, D. 2022a. A windfall for US carbon capture and storage. Physics Today 75(1): 22–24.

Kramer, D. 2022b. Carbon dioxide removal is suddenly obtaining credibility and support, Physics Today 75(6): 26–29.

Krauss, L.M. 2021. The Physics of Climate Change. Head of Zeus, London.

Küngas, R. 2020. Review—Electrochemical CO_2 reduction for CO production: comparison of low- and high-temperature electrolysis technologies. Journal of The Electrochemical Society 167: 044508.

Lackner, K.S., C.H. Wendt, D.P. Butt, E.L. Joyce and D.H. Sharps. 1995. Carbon dioxide disposal in carbonate minerals. Energy 20: 1153–1170.

Lackner, K.S., H.-J. Ziock and P. Grimes. 1999. Carbon dioxide extraction from air: is it an option? Proceedings of the 24th International Conference on Coal Utilization & Fuel Systems. Clearwater, Florida.

Lackner, K.S., P. Grimes and H.-J. Ziock. 1999. Carbon Dioxide Extraction from Air? Los Alamos National Laboratory, LAUR-99-5113, Los Alamos, NM.

Lackner, K.S., S. Brennan, J.M. Matter and B. van der Zwaan. 2012. The urgency of the development of CO2 capture from ambient air. Proceedings of the National Academy of Sciences 109: 13156–13162.

Laherrère, J., C.A.S. Hall and R. Bentley. 2022. How much oil remains for the world to produce? Comparing assessment methods, and separating fact from fiction. Current Research in Environmental Sustainability 4: 100174.

Langholtz, M., I. Busch, A. Kasturi, M.R. Hilliard, J. McFarlane, C. Tsouris et al. 2020. The economic accessibility of CO_2 sequestration through bioenergy with carbon capture and storage (BECCS) in the US. Land 9: 299.

Leahy, S., H. Clark and A. Reisinger. 2020. Challenges and prospects for agricultural greenhouse gas mitigation pathways consistent with the paris agreement. Frontiers in Sustainable Food Systems 4: 1–8.

Leonzio, G., P.S. Fennell and N. Shah. 2022. Analysis of technologies for carbon dioxide capture from the air. Applied Sciences 12: 8321.

Le Quéré, C., J.I. Korsbakken, C. Wilson, J. Tosun, R. Andrew, R.J. Andres et al. 2019. Drivers of declining CO_2 emissions in 18 developed economies. Nature Climate Change 9: 213–217.

Letcher, T.M. [ed.]. 2019. Managing Global Warming—An Interface of Technology and Human Issues. Academic Press.

Li, J. and Y. Wan. 2019. Present state of Chinese magnetic fusion development and future plans. Journal of Fusion Energy 38: 113–124.

Li, K., W. Leigh, P. Feron, H, Yu and M. Tade. 2016. Systematic study of aqueous monoethanolamine (MEA)-based CO_2 capture process: Techno-economic assessment of the MEA process and its improvements. Applied Energy 165: 648–659.

Liu, S.-Y., B. Ren, H.-Y. Li, Y.-Z. Yang, Z.-Q. Wang, B. Wang et al. 2022. CO_2 storage with enhanced gas recovery (CSEGR): A review of experimental and numerical studies. Petroleum Science 19: 594–607.

Liu, X., G. Seberry, S. Kook, Q.N. Chan and E.R. Hawkes. 2022. Direct injection of hydrogen main fuel and diesel pilot fuel in a retrofitted single-cylinder compression ignition engine. International Journal of Hydrogen Energy 47: 35864–35876.

Lomborg, B. 1998. The Skeptical Environmentalist: Measuring the Real State of the World. Cambridge University Press.

Lomborg, B. 2008. Cool It—The Skeptical Environmentalist's Guide to Global Warming. Vintage Books, New York.

Lomborg, B. 2020. False Alarm—How Climate Change Panic Costs Us Trillions, Hurts The Poor, And Fails To Fix The Climate. Basic Books, New York.

Lovelock, J. 2007. The Revenge of Gaia. Penguin Books.

Lovelock, J. and C. Rapley. 2007. Ocean pipes could help the Earth to cure itself. Nature 449 (27.09.2007) 403.

Luderer, G., R.C. Pietzcker, C. Bertram, E. Kriegler, M. Meinshausen and O. Edenhofer. 2013. Economic mitigation challenges: how further delay closes the door for achieving climate targets. Environmental Research Letters 8: 034033.

Lund, H. and B.V. Mathiesen. 2009. Energy system analysis of 100% renewable energy systems—the case of Denmark in years 2030 and 2050. Energy 34: 524–31.

Lynas, M. 2020. Our Final Warning—Six Degrees of Climate Emergency. 4th Estate, London.

Lyngfelt, A., A. Brink, O. Langorgenc, T. Mattissona, M. Rydena and C. Linderholm. 2019. 11,000 h of chemical-looping combustion operation—Where are we and where do we want to go? International Journal of Greenhouse Gas Control 88: 38–56.

Mac Dowell, N., D.M. Reiner and R.S. Haszeldine. 2022. Comparing approaches of carbon dioxide removal. Joule 6: 2233–2239.

MacKay, D.J.C. 2009. Sustainable Energy—Without the Hot Air. UIT Cambridge.

Mann, M.E., R.S. Bradley and M.K. Hughes. 1998. Global-scale temperature patterns and climate forcing over the past six centuries. Nature 392: 779–787.

Mann, M.E. 2021. The New Climate War: The Fight to Take Back Our Planet. Scribe Publications, London.

Marcucci, A., S. Kypreos and E. Panos. 2017. The road to achieving the long-term Paris targets: energy transition and the role of Direct Air Capture. Climatic Change 144: 181–193.

Martin, R., M. Muûls and U.J. Wagner. 2016. The impact of the European Union Emissions Trading Scheme on regulated firms: what is the evidence after ten years? Review of Environmental Economics and Policy 10: 129–148.

Maslin, M. 2021. How to Save Our Planet. Penguin Life.

McKibben, B. 2003. The End of Nature. Bloomsbury, London.

McKibben, B. 2011. Eaarth—Making a Life on a Tough Planet. St. Martin's Griffin, New York.

McKibben, B. [ed.]. 2012. The Global Warming Reader. Penguin Books.

McQueen, N., K.V. Gomes, C. McCormick, K. Blumanthal, M. Pisciotta and J. Wilcox. 2021. A review of direct air capture (DAC): scaling up commercial technologies and innovating for the future. Progress in Energy 3: 032001.

McQueen, N., M. Ghoussoub, J. Mills and M. Scholten. 2022. A scalable direct air capture process based on accelerated weathering of calcium hydroxide. Heirloom Carbon Technologies.

Meyers, G. 2022. What is the Difference Between "Regenerative" and "Sustainable"? Medium 22 March 2022.

Michaels, D. 2008. Doubt is Their Product: How Industry's Assault on Science Threatens Your Health. Oxford University Press.

Miller, J. 2021. The Amazon is reaching its carbon tipping point. Physics Today 74: 9 12–14.

Moore, F.C. and D.B. Diaz. 2015. Temperature impacts on economic growth warrant stringent mitigation policy. Nature Climate Change 5: 127.

Moosdorf, N., P. Renforth and J. Hartmann. 2014. Carbon dioxide efficiency of terrestrial enhanced weathering. Environmental Science & Technology 48: 4809–4816.

Mosleh, M.H., M. Sedighi, M. Babaei and M. Turner. 2019. Geological sequestration of carbon dioxide. *In*: Letcher, T.M. [ed.]. Managing Global Warming. Academic Press, London.

Nasheed, M. 2012. Speech at Klimaforum. *In*: Bill McKibben [ed.]. The Global Warming Reader. Penguin Books.

National Academies of Sciences, Engineering, and Medicine (NAS). 2019. Negative Emissions Technologies and Reliable Sequestration: A Research Agenda. Washington, DC: The National Academies Press.

NETL. 2007. Cost and Performance Baseline for Fossil Energy Plants. DOE/NETL-2007/1281.

NETL. 2010. Cost and Performance Baseline for Fossil Energy Plants. DOE/NETL-2010/1397.

NETL. 2019. Cost and Performance Baseline for Fossil Energy Plants Volume 1: Bituminous Coal and Natural Gas to Electricity. NETL-PUB-22638.

Nicholas, K. 2021. Under the Sky We Make. Putnam, London.

Norton, M., A. Baldi, V. Buda, B. Carli, P. Cudlin, M.B. Jones et al. 2019. Serious mismatches continue between science and policy in forest bioenergy. GCB Bioenergy 11: 1256–1263.

Núñez-López, V. and E. Moskal. 2019. Potential of CO_2-EOR for near-term decarbonization. Frontiers in Climate 1: 5.

Onyebuchi, V.E., A. Kolios, D.P. Hanak, C. Biliyok and V. Manovic. 2018. A systematic review of key challenges of CO_2 transport via pipelines, Renewable and Sustainable Energy Reviews 81: 2563–2583.

Oreskes, N. and E.M. Conway. 2020. Merchants of Doubt: How a Handful of Scientists Obscured the Truth on Issues from Tobacco Smoke to Global Warming. Bloomsbury Press.

Ozkan, M., S.P. Nayak, A.D. Ruiz and W. Jiang. 2022. Current status and pillars of direct air capture technologies. iScience 25: 103990.

Ozkan, M. and R. Custelcean. 2022. The status and prospects of materials for carbon capture technologies. MRS Bulletin 47: 390.

Pacala, S. and R. Socolow. 2004. Stabilization wedges: solving the climate problem for the next 50 years with current technologies. Science 305: 986–972.

Parsapur, R.K., S. Chatterjee and K.-W. Huang. 2020. The insignificant role of dry reforming of methane in CO_2 emission relief. ACS Energy Lett. 5: 2881–2885.

Partanen, R. and J.M. Korhonen. 2020. The Dark Horse: Nuclear Power and Climate Change. Independently published.

Peña, J.A., E. Lorente and J. Herguido. 2010. "Steam-Iron" process for hydrogen production: recent advances. *In*: Stolten, D. and T. Grube [eds.]. 18th World Hydrogen Energy Conference 2010. WHEC.

Perissi, I., A. Lavacchi and U. Bardi. 2021. The role of energy return on energy invested (EROEI) in complex adaptive systems. Energies 14: 8411.

Powers, M. 2020. Ethical challenges posed by climate change. *In*: Miller, D.E. and B. Eggleston [eds.] Moral Theory and Climate Change. Routledge.

Promes, E.J.O., T. Woudstra, L. Schoenmakers, V. Oldenbroek, A. Thallam Thattai and P.V. Aravind. 2015. Thermodynamic evaluation and experimental validation of 253 MW Integrated Coal Gasification Combined Cycle power plant in Buggenum, Netherlands. Applied Energy 155: 181–194.

Realmonte, G., L. Drouet, A. Gambhir, J. Glynn, A. Hawkes, A.C. Köberle et al. 2019. An inter-model assessment of the role of Direct Air Capture in deep mitigation pathways. Nature Communications 10: 3277.

Reece, S.Y., J.A. Hamel, K. Sung, T.D. Jarvi, A.J. Esswein, J.J. Pijpers et al. 2011. Wireless solar water splitting using silicon-based semiconductors and earth-abundant catalysts. Science 334: 645–648.

Rees, M. 2003. Our Final Hour - A Scientist's Warning: How Terror, Error, and Environmental Disaster Threaten Humankind's Future in This Century—On Earth and Beyond. Basic Books, New York.

REN21. 2021. Renewables 2021 Global Status Report (Paris: REN21 Secretariat).

Renfrew, S.E., D.E. Starr and P. Strasser. 2020. Electrochemical approaches toward CO_2 capture and concentration. ACS Catalysis 10: 13058–13074.

Ringrose, P. 2020. How to Store CO_2 Underground: Insights from early-mover CCS Projects. Springer.

Ritchie, S. 2020. Science Fictions—Exposing Fraud, Bias, Negligence and Hype in Science. The Bodley Head, London.

Rochelle, G.T. 2009. Amine scrubbing for CO_2 capture. Science 325: 1652–1654.

Rogelj, J., G. Luderer, R.C. Pietzcker, E. Kriegler, M. Schaeffer, V. Krey et al. 2015. Energy system transformations for limiting end-of-century warming to below 1.5°C. Nature Climate Change 5: 519–528.

Rogelj, J., D. Shindell, K. Jiang, S. Fifita, P. Forster, V. Ginzburg et al. 2018. Mitigation pathways compatible with 1.5°C in the context of sustainable development. *In*: Masson-Delmotte, V., P. Zhai, H.-O. Pörtner, D. Roberts, J. Skea, P.R. Shukla et al. [eds.]. Global Warming of 1.5°C. An IPCC Special Report on the impacts of global warming of 1.5°C above pre-industrial levels and related global greenhouse gas emission pathways, in the context of strengthening the global response to the threat of climate change, sustainable development, and efforts to eradicate poverty.

Romm, J. 2018. Climate Change—What Everyone Needs to Know. Oxford University Press.

Roussanaly, S., A.L. Brunsvold, E.S. Hognesa, J.P. Jakobsen and X. Zhang. 2013. Integrated techno-economic and environmental assessment of an amine-based capture. Energy Procedia 37: 2453–2461.

Rubin, E.S. and H. Zhai. 2012. The cost of carbon capture and storage for natural gas combined cycle power plants. Environmental Science & Technology 46: 3076–3084.

Rubin, E.S., J.E. Davison and H.J. Herzog. 2015. The cost of CO_2 capture and storage. International Journal of Greenhouse Gas Control 40: 378–400.

Sabatino, F., A. Grimm, F. Gallucci, M. van Sint Annaland, G.J. Kramer and M. Gazzani. 2021. A comparative energy and costs assessment and optimization for direct air capture technologies. Joule 5: 2047–2076.

Sandalow, D., R. Aines, J. Friedmann, P. Kelemen, C. McCormick, I. Power et al. 2021. Carbon Mineralization Roadmap. ICEF Innovation Roadmap Project.

Sanz-Pérez, E.S., C.R. Murdock, S.A. Didas and C.W. Jones. 2016. Direct capture of CO_2 from ambient air. Chemical Reviews 116: 11840–11876.

Schmelz, W.J., G. Hochman and K.G. Miller. 2020. Total cost of carbon capture and storage implemented at a regional scale: north-eastern and midwestern United States, Interface Focus 10: 20190065.

Scholten, D. [ed.]. 2018. The Geopolitics of Renewables. Springer.

Sekera, J. and A. Lichtenberger. 2020. Assessing carbon capture: public policy, science, and societal need. Biophysical Economics and Sustainability 5: 14.

Shapiro, J. 2002. Mao's War Against Nature—Politics and the Environment in Revolutionary China. Cambridge University Press.

Sherwood, S.C., M.J. Webb, J.D. Annan, K.C. Armour, P.M. Forster, J.C. Hargreaves et al. 2020. An assessment of Earth's climate sensitivity using multiple lines of evidence. Reviews of Geophysics 58: e2019RG000678.

Shi, X., H. Xiao, H. Azarabadi, J. Song, X. Wu, X. Chen et al. 2020. Sorbents for the direct capture of CO_2 from ambient air. Angewandte Chemie International Edition 59: 6984–7006.

Shi, X., H. Xiao, K. Kanamori, A. Yonezu, K.S. Lackner and X. Chen. 2020. Moisture-Driven CO_2 Sorbents. Joule 4: 1823–1837.

Shrader-Frechette, K. 2011. What Will Work – Fighting Climate Change with Renewable Energy. Oxford University Press.

Singer, S. [ed.] 2011. The energy report: 100% renewable energy by 2050. WWF, Gland, Switzerland.

Singh, D., E. Croiset, P.L. Douglas and M.A Douglas. 2003. Techno-economic study of CO_2 capture from an existing coal-fired power plant: MEA scrubbing vs. O_2/CO_2 recycle combustion. Energy Conversion and Management 44: 3073–3091.

Singh, J. and D.W. Dhar. 2019. Overview of carbon capture technology: microalgal biorefinery concept and state-of-the-art. Frontiers in Marine Science 6: 29.

Sinha, A. 2020. China's insatiable demand for energy. December. United News of India.

Smil, V. 2017. Energy – A Beginner's Guide. One World.

Smil, V. 2018. Energy and Civilization – A History. MIT Press.

Smil, V. 2022. How the World Really Works. Penguin Random House, London.

Smit, B., J.A. Reimer, C.M. Oldenburg and I.C. Bourg. 2014. Introduction To Carbon Capture and Sequestration. Imperial College Press, London.

Smith, P. 2016. Soil carbon sequestration and biochar as negative emission technologies. Global Change Biology 22: 1315–24.

Somerville, R. 2022. Weaning a house and the world from fossil fuels: lessons learned. Bulletin of the Atomic Scientists. August 24.

Sørensen, B. 1975. Energy and Resources: A plan is outlined according to which solar and wind energy would supply Denmark's needs by the year 2050. Science 189(4199): 255–260.

Steiner, M. 2021. Capturing carbon one cooling tower at the time. Who What Why 28 Sep. 2021.

Strefler, J., T. Amann, N. Bauer, E. Kriegler and J. Hartmann. 2018. Potential and costs of carbon dioxide removal by enhanced weathering of rocks. Environmental Research Letters 13: 034010.

Sullivan, I., A. Goryachev, I.A. Digdaya, X. Li, H.A. Atwater, D.A. Vermaas et al. 2021. Coupling electrochemical CO_2 conversion with CO_2 capture. Nature Catalysis 4: 952–958.

Tahill, W. 2007. The Trouble with Lithium: Implications of Future PHEV Production for Lithium Demand. Meridian International Research.

Talekar, S., B.H. Jo, J.S. Dordick and J. Kim. 2022. Carbonic anhydrase for CO_2 capture, conversion and utilization. Current Opinion in Biotechnology 74: 230–240.

Temple, J. 2020. How green sand could capture billions of tons of carbon dioxide. MIT Technology Review (22 June 2020); https://www.technologyreview.com/2020/06/22/1004218/how-green-sand-could-capture-billions-of-tons-of-carbon-dioxide/.

Teske, S., S. Sawyer and O. Schäfer. 2015. Energy [r]evolution. A sustainable world energy outlook 2015, 5th ed., Greenpeace: Global Wind Energy Council & European Renewable Energy Council.

Tiefenthaler, J., L. Braune, C. Bauer, R. Sacchi and M. Mazzotti. 2021. Technological demonstration and life cycle assessment of a negative emission value chain in the Swiss concrete sector. Frontiers in Climate 3: 729259.

Turner, G.M., B. Elliston and M. Diesendorf. 2013. Impacts on the biophysical economy and environment of a transition to 100% renewable electricity in Australia. Energy Policy 54: 288–299.

Verheggen, B. 2020. Wat iedereen zou moeten weten over klimaatverandering. Prometheus, Amsterdam.

Vince, G. 2012. Sucking CO2 from the skies with artificial trees. BBC Future, 4 October.

Von Hirschhausen, C., J. Herold and P. Oei. 2012. How a 'low carbon' innovation can fail—tales from a 'lost decade' for carbon capture, transport, and sequestration (CCTS). Economics of Energy & Environmental Policy 1: 115–124.

Wallace-Wells, D. 2019. The Uninhabitable Earth. Penguin Books.

Wang, S., Z. Hausfather, S. Davis, J. Lloyd, E.B. Olson, L. Liebermann et al. 2023. Future demand for electricity generation materials under different climate mitigation scenarios. Joule 7: 309–332.

Way, R., M. Ives, P. Mealy and J.D. Farmer. 2021. Empirically grounded technology forecasts and the energy transition. INET Oxford Working Paper No. 2021-01.

Weart, S.R. 2003. The Discovery of Global Warming. Harvard University Press.

Wilcox, J. 2012. Carbon Capture. Springer.

World Bank. 2021. State and Trends of Carbon Pricing 2021. Washington, DC.

Wu, L. and Q. Zhu. 2021. Impacts of the carbon emission trading system on China's carbon emission peak: a new data-driven approach. Natural Hazards 107: 2487–2515.

Xia, C., B. Ye, J. Jiang and Y. Shu. 2020. Prospect of near-zero-emission IGCC power plants to decarbonize coal-fired power generation in China: Implications from the GreenGen project. Journal of Cleaner Production 271: 122615.

Xu, C., Q. Dai, L. Gaines, M. Hu, A. Tukker and B. Steubing. 2020. Future material demand for automotive lithium-based batteries. Communications Materials 1: 99; https://doi.org/10.1038/s43246-020-00095-x.

Yergin, D. 2020. The New Map – Energy, Climate and The Clash of Nations. Penguin Books.

Zeman, F.S. and K.S. Lackner. 2004. Capturing carbon dioxide directly from the atmosphere. World Resource Review 16: 157–172.

Zhang, Z.X., G.X. Wang, P. Massarotto and V. Rudolph. 2006. Optimization of pipeline transport for CO_2 sequestration. Energy Conversion and Management 47: 702–715.

Zhu, X., Q. Imtiaz, F. Donat, C.R. Müller and F. Li. 2020. Chemical looping beyond combustion—a perspective. Energy & Environmental Science 13: 772–804.

Zickfeld, K., D. Azevedo, S. Mathesius and H.D. Matthews. 2021. Asymmetry in the climate–carbon cycle response to positive and negative CO_2 emissions. Nature Climate Change 11: 613–617.

Websites

I list here some of the vast number of websites on climate science, climate change, global warming, carbon capture and suchlike that I have consulted in the course of writing this book, many of which are referred to in the text.

Air Pollution and Climate Secretariat https://www.airclim.org/acidnews/ccs-sidelined-public-opposition
Alice Friedemann https://energyskeptic.com/
Berkeley Earth https://berkeleyearth.org/
BeZero Carbon https://bezerocarbon.com/
Carbon180 https://carbon180.org/ , formerly the Center for Carbon Removal
Carbon Brief https://www.carbonbrief.org/
Carbon Capture Coalition https://carboncapturecoalition.org/
Carbon Capture Journal https://www.carboncapturejournal.com/
Carbon Collect Limited https://mechanicaltrees.com/mechanicaltrees/
Carbon Cure https://www.carboncure.com/
Carbon Dioxide Information Analysis Center https://cdiac.ess-dive.lbl.gov/
now Environmental Systems Science Data Infrastructure for a Virtual Ecosystem https://ess-dive.lbl.gov/
Carbon Neutrality Coalition https://carbon-neutrality.global/
Carbon Plan https://carbonplan.org/
Carbon Tracker Initiative https://carbontracker.org/
CCS+ Initiative https://www.ccsplus.org/
CCS on GreenFacts https://www.greenfacts.org/en/co2-capture-storage/index.htm
Cell Press https://www.cell.com/
Center for Environmental Law https://www.ciel.org/about-us/
Centre for Research on Energy and Clean Air https://energyandcleanair.org/
Clean Energy Wire https://www.cleanenergywire.org/factsheets/quest-climate-neutrality-puts-ccs-back-table-germany
Climate Action Tracker https://climateactiontracker.org/
Climate Change Deniers http://www.co2science.org/index.php
Climate Change Litigation Databases http://climatecasechart.com/
Climate Etc. https://judithcurry.com/
Climate Feedback https://climatefeedback.org/
Climate.gov https://www.climate.gov/news-features/understanding-climate/climate-change-ocean-heat-content
Climate Overshoot Commission https://www.overshootcommission.org/
Copenhagen Consensus https://www.copenhagenconsensus.com/
COP26 https://ukcop26.org/?mc_cid=2c68591ce2&mc_eid=f46d2cf59e
Drawdown Project https://drawdown.org/
Ellen MacArthur Foundation https://ellenmacarthurfoundation.org/
Ember https://ember-climate.org/
Energie Beheer Nederland https://www.ebn.nl/
Energy & Climate Intelligence Unit https://eciu.net/
Energy for Growth Hub https://www.energyforgrowth.org/
Energy Innovation https://energyinnovation.org/

Energy Modeling Forum https://emf.stanford.edu/
Energy Policy Solutions https://energypolicy.solutions/
Energy Transition Commission https://www.energy-transitions.org/
EPA https://www.epa.gov/
Future Earth https://futureearth.org/
Global Carbon Project https://www.globalcarbonproject.org/
Global Commission on Adaptation https://gca.org/about-us/the-global-commission-on-adaptation/
Global CCS Institute https://www.globalccsinstitute.com/
Global Energy Monitor https://globalenergymonitor.org/
Global Thermostat https://globalthermostat.com/
Greenhouse Development Rights http://gdrights.org/
Greenhouse Gas Protocol https://ghgprotocol.org/
Innovation for Cool Earth Forum https://www.icef.go.jp/
Inside Climate News https://insideclimatenews.org/ (non-profit, nonpartisan news organization on climate change, energy and the environment)
Institute for Carbon Removal Law and Policy https://www.american.edu/sis/centers/carbon-removal/
Interest Group Environmental Chemistry https://www.envchemgroup.com/
Intergovernmental Panel on Climate Change https://www.ipcc.ch/
International CCS Knowledge Center https://ccsknowledge.com/
International Energy Agency https://www.iea.org/
James Hansen website http://www.columbia.edu/~jeh1/
Joint Center for Artificial Photosynthesis https://solarfuelshub.org/
Lucid Catalyst https://www.lucidcatalyst.com/
NASA https://climate.nasa.gov/; https://earthobservatory.nasa.gov/world-of-change/global-temperatures
National Academies of Science etc. https://www.nap.edu/content/about-the-national-academies-press
National Energy Technology Laboratory https://www.netl.doe.gov/coal/carbon-storage/publications; https://www.netl.doe.gov/carbon-management/carbon-storage
National Energy Technology Laboratory https://netl.doe.gov/ with
   https://netl.doe.gov/coal/carbon-capture on Carbon Capture
   https://netl.doe.gov/coal/carbon-utilization on Carbon Utilization
   https://netl.doe.gov/coal/carbon-storage on Carbon Storage
Nature Conservancy https://www.nature.org/en-us/about-us/who-we-are/
Net Zero Tracker https://zerotracker.net/
NOAA www.noaa.gov/climate
Ocean Grazer https://oceangrazer.com/
Open Air Collective https://openaircollective.cc/
Oxford Institute for Energy Studies https://www.oxfordenergy.org/publication-topic/papers/
Our World in Data https://ourworldindata.org/co2-emissions
Peak Oil https://www.peakoil.net/
Real Climate https://www.realclimate.org/
Remaining Carbon Budget https://www.mcc-berlin.net/en/research/co2-budget.html
Rocky Mountain Institute https://rmi.org/
Shell https://blogs.shell.com/category/ccs/
Science Based Targets https://sciencebasedtargets.org/
Skeptical Science https://skepticalscience.com/
Stop Fossil Fuels https://350.org/
The Solutions Project https://thesolutionsproject.org/
Understanding Money https://howmuch.net/
U.S. Geological Survey https://www.usgs.gov/
US Energy Information Administrations https://www.eia.gov/, especially https://www.eia.gov/todayinenergy/
Verra https://verra.org/
World Economic Forum https://www.weforum.org/
World Nuclear Organisation https://world-nuclear.org/
World Climate Research Programme https://www.wcrp-climate.org/

Index